The Tangled Field

The Tangled Field

Barbara McClintock's Search for the Patterns of Genetic Control

Nathaniel C. Comfort

Harvard University Press
Cambridge, Massachusetts, and London, England | 2001

Copyright © 2001 by Nathaniel C. Comfort
All rights reserved
Printed in the United States of America

Library of Congress Cataloging-in-Publication Data

Comfort, Nathaniel C.
    The tangled field : Barbara McClintock's search for the patterns
of genetic control / Nathaniel C. Comfort.
        p.   cm.
Includes bibliographical references and index.
ISBN 0-674-00456-6 (cloth : alk. paper)
    1. McClintock, Barbara, 1902–1992.   2. Women geneticists—
United States—Biography.   I. Title.

QH429.2.M38 C66 2001
576.5′092—dc21
[B]   00-069712

*To Valborg, Lanny, Louise,
Honore, Susan, Carol, Ruth,
Jane, and now Gwendolyn*

Contents

Photographs follow page 68.

Acknowledgments

I have been fortunate in having guidance, support, and assistance in all phases of this project, from kernels to commas. This book grew out of my dissertation, so chronologically first is Ruth Schwartz Cowan, my Ph.D. advisor, who overcame initial skepticism to support my project strongly. Ruth guided me to the feminist and history-of-science literature, questioned me on genetics, and gave me valuable criticism as I was writing my dissertation. Many friends and colleagues at Cold Spring Harbor encouraged me, talked with me, and made resources available to me: Clare Bunce, Susan Gensel Cooper, Margaret Henderson, Ludmila Pollack, David Stewart, Bruce Stillman, Jan Witkowski. While James D. Watson did not advise me on this project, in a deeper way he helped give me the freedom to do it. Special thanks must go to Robert Martienssen of Cold Spring Harbor Laboratory. Rob spent many hours with me, explaining McClintock's papers and corn, reading early drafts, and offering encouragement. Ronald Phillips generously invited me to his laboratory at the University of Minnesota, where I spent a week learning cytology under his and his students' tutelage.

I thank David Botstein, Susan Gensel Cooper, Harriet Creighton, James F. Crow, Nina Fedoroff, Bentley Glass, Melvin M. Green, Ira Herskowitz, Rollin Hotchkiss, Jerry Kermicle, Joshua Lederberg, Oliver E. Nelson, M. Gerald Neuffer, Robert A. Nilan, Robert Pollack, Drew Schwartz, James Shapiro, A. Lawrence Taylor, Waclaw Szybalski, James D. Watson, Diter von Wettstein, Evelyn Witkin, Judson John van Wyk, and Norton Zinder for consenting to be interviewed. Many of these individuals showed me great kindness, often spending many hours with me, sometimes inviting me into their home, and granting me follow-up interviews and helping me make sure I got the story right. Botstein, Herskowitz, Lederberg, Shapiro, Taylor, and Witkin also read sections of the manuscript for technical accuracy.

Others who read portions of the manuscript include Richard Burian, Angela Creager, Scott Gilbert, Horace Freeland Judson, Jane Maienschein, Betty Smocovitis, and two anonymous reviewers who offered valuable comments and criticisms. If any errors remain, it is my responsibility and none of theirs. I also thank Robert Kohler for his valuable comments at an early stage and for his continued support of my project. Ann Downer-Hazell at Harvard University Press shepherded the manuscript through production, and, at the end, Elizabeth Hurwit's manuscript editing buffed it to a high gloss.

The members and staff of the Center for History of Recent Science have spent many hours with this book, chapter by chapter. Anne Fitzpatrick, Carl-Henry Geschwind, David Grier, Mark Lesney, Patrick McCray, Tomoko Steen, and Steve Weiss have all made substantial improvements of fact, structure, argument, style, and punctuation. Mary Jenifer has assisted in many ways, direct and indirect. Horace Judson's extraordinary editing skills have helped excise clichés and sentimentality and have tuned my ear for language; his conversation has given me more confidence to think—and write—in pictures.

Carol Greider has been my most diligent reader—she has read and listened to every chapter many times, almost always cheerfully, and has offered hundreds of suggestions on science and style. In addition, she has supported me in countless ways: picking up the kids, making dinner, listening to my thinking out loud—all while running a research laboratory. This book could not have been done, and life simultaneously lived, without her.

The Tangled Field

MYTH

Among scientists are collectors, classifiers, and compulsive tidiers-up;
many are detectives by temperament and many are explorers; some
are artists and others are artisans. There are poet-scientists and
philosopher-scientists and even a few mystics.

—SIR PETER MEDAWAR

Here is the Barbara McClintock most people think they know. In the late
1920s and early 1930s, a brilliant young geneticist made a string of important discoveries concerning the chromosomes of maize, or Indian corn.
She was one of a small group of stellar maize geneticists at Cornell University, but when the other members of the group, men, got jobs, she could
not. When she did finally get a job, at the University of Missouri in 1936,
she quit after five years because she had no prospect of tenure and was
about to be fired.

Unable to find another job at a university, she landed at Cold Spring
Harbor, Long Island, New York, where the Carnegie Institution of Washington maintained a small private research laboratory. In the late 1940s,
isolated from her colleagues, working alone without even a graduate student to help her, she discovered something revolutionary. Nearly fifty
years of genetics had established that genes were independently acting
units, fixed in position on the chromosomes like pearls on a necklace.
McClintock found that one of the pearls near the clasp jumped spontaneously to a new site toward the center. She called this transposition; the
jumping genes were transposable elements. She presented her findings
in 1951, at the Cold Spring Harbor Symposium, a prestigious scientific
meeting held each year at her home institution. No one believed her; a
few scientists were outright hostile. A second presentation five years later
brought if anything a worse reaction. She ceased publishing and retreated
into her laboratory, pursuing her meticulous experiments in isolation for
decades.

McClintock could challenge the canonical view of the gene—the story
goes on—because she was not bound by dogma as other geneticists were.
She attended to her corn plants with sensitivity, even empathy. Free from

1

the ossified theory that constrained other scientists' vision, she could see what others could not: genes were dynamic, interactive, flexible.

In the 1970s, a new generation of molecular biologists found transposition in the geneticist's stalwart creatures, bacteria and fruitflies. Now, when transposition was shown, by men, in bacteria, people listened. Like Gregor Mendel, the father of genetics, whose ideas had been ignored for the last third of the nineteenth century, McClintock was rediscovered. Her premature ideas had at last found their milieu. She began to win prizes, and in 1983 an eighty-one-year-old McClintock traveled to Stockholm to shake the hand of the king of Sweden and receive a Nobel Prize in Physiology or Medicine. She died less than nine years later, in 1992.

The McClintock of this story would seem no match for the powerful machine of twentieth-century genetics, yet by quiet persistence and personal integrity, not to mention sheer longevity, she outlasted the narrow-mindedness and bigotry that blinded her colleagues. She seems in every way the opposite of the archetypal molecular biologist: senior, humble, intuitive, and female, working in the fields on a large, slow-growing, complex organism—in contrast to the young, rational, arrogant, male biologists working on bacteria and viruses. Her rejection and ultimate legitimation demonstrate the strictures placed on scientific practice by the dominant reductionist, manipulative style of science.

This story is a myth. By this, I do not mean a Greek tale populated with fabulous and capricious anthropomorphic gods. Joseph Campbell distinguished such "traditional myths" from "creative myths." Creative myths often spring from an identifiable individual whose persona may be explicitly attached to the myth. This individual communicates personal experience through symbols, metaphors, and stories. For those touched by this experience, the symbols take on the power and status of myth. By this route, according to Campbell, creative mythology enforces a moral order, and shapes an individual to her historical time and place.[1]

For Roland Barthes, myth had political connotations. He used an example of an image of a black soldier saluting the French tricolor. The components of the image—man, uniform, flag—cease to have historical significance. A new, symbolic meaning emerges from their juxtaposition. This image creates a myth that incorporates the power of the military, the benevolence of colonialism, the honor of the nation-state. Barthes emphasized that myths are not deceptive: they do not claim to represent truth or history. Rather, their purpose is to distort history, to create a new and pristine "nature."[2]

In biography, creative myths operate on an inner and an outer level. François Jacob, a Nobel laureate in genetics and among the most reflective and graceful memoirists science has produced, refers to the inner level as "the statue within." He recognizes within himself a "secret image that from the deepest part of me guides my tastes, desires, decisions." This image, for Jacob, has solidity, it is "a statue sculpted since childhood, that gives my life a continuity and is the most intimate part of me, the hardest kernel of my character."[3] The strength of Jacob's reflection lies at the point at which his metaphor breaks down. He saw the developmental character of the statue. Its initial themes are established in childhood, but throughout one's life they are polished and shaped, and new ones are laid down on top.

For Jacob, the statue within is the deepest, truest part of himself. The memoirist, however, cannot distinguish between those elements of the statue that are real and those he has constructed to mask some part of himself of which he may be ashamed or frightened. That is the biographer's task. Biographer Leon Edel incorporates both the true and the fabricated aspects of one's personal story in his concept of "private myth." The private myth is a personal vision that guides the way we act and present ourselves to others. The private myth, Edel writes, is "the figure in the carpet," a pattern woven into our life's tapestry. The biographer's duty is first to describe this figure.[4]

But the biographer, Edel continues, must take a step further: he must turn back the carpet and expose the figure beneath, on the reverse of the tapestry. In Edel's image, the figure under the carpet gives form to that *in* the carpet precisely because it is the reverse of the image most people see. He takes the example of Ernest Hemingway, whose chest-beating bravado, Edel suggests, emerged from a deep-seated insecurity about his manhood.[5] Though such an analysis can quickly degenerate into amateurish psychohistory, Edel is correct to charge biographers with looking beyond the stories individuals tell about themselves, to question the motivations for private myths.

The private myth may be contrasted with a "public myth." Gerald Geison, in his study of Louis Pasteur, has distinguished between public and private science—that which is published and that which is only in the notebooks or the mind of the scientist.[6] Similarly, there may be public and private myths. When it becomes public, the private myth mutates or evolves into a story that others use in ordering their lives. It may share elements with the private myth, but, as in the parlor game of "telephone," as the myth passes from mouth to mouth, elements are dropped, new associations are made, and meaning shifts. When private myths become public, they may take on the character of Campbell's creative myths or Barthes's political myths. The myth of McClintock with which I began is her public myth.

Barbara McClintock is mythical in all of these senses. Barthes would recognize that she has become a composite of symbols, each abstracted from their history and gaining new meaning in juxtaposition to one another. A recent publication by the American Association of University Women, which honored McClintock with their Achievement Award in 1947, presents vignettes of historical women for role-playing by elementary-school children. McClintock is caricatured as a little old lady with cropped iron-gray hair, dressed in simple slacks, a tailored white shirt, plain sweater, and sturdy shoes, and bearing a microscope, an ear of Indian corn, and a clipboard.[7] Her slacks and white shirt symbolize her humility and spartan lifestyle, her clipboard her scientific rigor, and her ear of corn her connection to nature. Campbell would see that McClintock's story imposes a moral order on twentieth-century science. It has become a cautionary tale of the exclusion of women from science's vaunted democratic and self-correcting nature, and a model of what an ideal science might look like. And, as I shall explain, this public myth stems from an uncritical acceptance of McClintock's own private myth, the story she told about herself.

When I refer to the standard McClintock story as a myth, do not mistake me to be saying the story is wholly false. Like any creative myth, this myth contains both fact and fiction. Individual assertions of fact may be false, but the interpretive content of a myth cannot be—though, Campbell points out, not everyone may acknowledge it as true. This book is an effort to peel back the mask, to place the symbols of the McClintock myth back into historical context, and to open new possibilities for interpreting the life of an extraordinary and flawed scientist.

McClintock herself is the source of the public myth, although, like a river, it swelled, meandered, and gathered debris downstream. When interest in her work was renewed in the late 1970s, historians and journalists began to request interviews. Like anyone asked to tell and retell a story, she settled into a pat version, easy to understand and ornamented with anecdotes and humor. In it, she was always the maverick, flouting convention, misunderstood. For many of her discoveries—not just transposition—she said she had been thought crazy, laughed at, or ignored by colleagues. She told such stories even about research questions that were widely understood as central problems in genetics.

Evelyn Fox Keller listened attentively to McClintock's autobiography. Trained in bacterial and viral genetics and in physics, Keller was steeped in science but became interested in the social problems faced by women in the laboratory.[8] Her interviews with McClintock in 1978 and 1979 formed the basis of a profile for a popular science magazine.[9] She expanded that article into a brief biography, *A Feeling for the Organism,* which appeared in 1983.[10] Keller took McClintock at face value. She had little choice. None of her other interview subjects contradicted the story: not George Beadle, Marcus Rhoades, or Harriet Creighton, McClintock's graduate school friends; not Helen Crouse, a student of McClintock's from Missouri; not Evelyn Witkin, a Cold Spring Harbor colleague. McClintock's scientific papers are notoriously difficult to read. And Keller did not have access to McClintock's correspondence and research notes, which tell a much more complex story than the one McClintock gave to reporters and interviewers.

McClintock and Keller fueled each other's fame. Within months of the appearance of *A Feeling for the Organism,* McClintock won a Nobel Prize. Keller's book, elegantly written and with a timely message, was widely read. It was by then beyond dispute that women struggling for acceptance in male-dominated fields suffer professional setbacks, financial hardship, and humiliation. McClintock's story seemed to epitomize women's experience in science. Indeed, Keller's book led many women scientists to the realization that they were not alone. McClintock became for them a symbol of the discrimination they had experienced and an emblem of hope for legitimation.

Keller's book became a founding text of the feminist critique of science. The feminist critique grew out of the period of great feminist activism and consciousness-raising of the 1970s and 1980s. Women's anger at discriminatory treatment, simmering through the 1960s and 1970s, boiled over into demands for equal treatment and equal pay. In science departments around the country, women began to hammer at the glass ceiling keeping them out of the upper ranks of salary, prestige, and respect.[11]

The feminist critique soon went beyond the scandal of sex discrimination in science to develop gender as an analytical tool for dissecting scientific ideas and institutions. Recognizing that women had been systematically excluded from modern science, feminists began to ask whether the very nature of science was masculine or androcentric. Through the 1980s, exposure of the biases in science gave way to the notion not just of a different science, but of a feminine science. What would science look like, some feminists asked, if it were done by women as well as—or instead of—men?

Would it be less reductionist, less rigidly rational, more empathic, more intuitive?[12]

Scientists themselves also began to apply a feminist perspective to the study of nature.[13] Women biologists exposed gendered assumptions in the portrayal of the relative roles of sperm and eggs and of females' roles in animal societies.[14] Women anthropologists asked new kinds of questions of primate societies, came to new views on behaviors such as parental care, and overturned the dogma that animal societies are exclusively male-dominated and trade only in the currencies of violence and aggression.[15] Psychologists and others developed an argument for gender differences in thought patterns. Reading such studies, many feminists accepted the traditional division of the sexes, in which males are seen as detached, controlling, and manipulative, and females as nurturing, attentive, and empathic. But they rejected the idea that such qualities made females weaker than or subordinate to males; instead, they celebrated these allegedly feminine traits as a source of power.[16] Carol Gilligan's *In a Different Voice,* published in 1982, and the multiauthor *Women's Ways of Knowing,* 1986, advanced the notion of scientific sexual essentialism: the argument that women inherently think and do science differently from men.[17]

Keller's biography became the primary case study of this view. McClintock, she wrote, provided a glimpse of what science might look like if the matter of gender dropped away. She, as others were beginning to do, saw the history of modern science as the expression of a masculine intellectual style dating back to Francis Bacon.[18] This style was dominating, controlling, reductionist, rational, and linear. Keller argued that McClintock represented an alternative to this style: holistic, intuitive, interactionist, even mystical. McClintock, Keller said, did not force her material into logical schemata; she let nature speak to her, tell her what to see. She was concerned with individuals, in a way that seemed to fuse the rational and the emotional sides of the intellect. This was what McClintock meant, Keller explained, when she spoke of having a "feeling for the organism."[19]

Keller had been careful to state that the alternative to masculine science was not "feminine" science but nongendered science. But in the context of the feminist critique, her shades of gray were sharpened into stark black and white. Though in her preface Keller cautioned against reading her book "as a tale of dedication rewarded after years of neglect—of prejudice or indifference eventually routed by courage and truth,"[20] most read *A Feeling for the Organism* in precisely that way. In the pages of scholarly journals, Keller bemoaned the fact that her book had become a "manifesto of feminine science."[21] She debated her feminist colleagues, who insisted that the only logical alternative to masculine science was feminine science. Keller wondered why some people could only count to two. They responded in effect that the only alternative to right was left.[22]

Meanwhile, the public McClintock myth degenerated into sentimental fancy. McClintock's style of science was portrayed as nonmasculine, holistic, intuitive, noninterventionist, nonreductionist.[23] Joan Dash called her "a mystic by nature, who tended to think intuitively," in a style "utterly different from the logical thinking to which experimental scientists are accustomed." According to Mary Frank Fox, her science was different from "inherently masculine" normal science in its views of "nature, domination, and control." Linda Jean Shepherd wrote that "rather than separate herself emotionally from her objects of study, she became intimately involved with her corn plants." She worked "virtually in isolation for over thirty years," Shepherd continued, and Deborah Felder said McClintock "never gave lectures to build her career or consulted with her colleagues." Through such portraits, which often directly contradicted the facts, McClintock's private myth of being a loner whom everyone thought was crazy mutated into the public myth of McClintock as feminist idol.[24]

Scientists also read Keller's book, and they developed their own version of the McClintock myth. Among scientists, the mysticism attributed to McClintock in the feminist version of the myth transmuted into brilliance and a cryptic style of communication, more Wittgenstein than Gurdjieff. Allan Spradling, a geneticist, wrote that McClintock's "powers derived from reasoning so well developed that the therefores and QEDs of her papers and conversations left others straining to catch up." In *Molecular Biology of the Gene,* a widely used multiauthor textbook based on James D. Watson's 1965 text of the same name, McClintock's work was not ignored or dismissed, but rather its truth "could not be ascertained." Rebutting the feminists' charge that they had ignored McClintock's revolutionary but straightforward discovery, the scientists said they merely had been baffled by a brilliant but knotty concept, puzzlingly presented.[25]

McClintock has become an "inevitable example" of a woman who achieved scientific eminence.[26] Among those harshly critical of mainstream science, she symbolizes both the risks women run and the prices they pay for challenging masculine norms. To them, her late awards are simultaneously a legitimation of her and a critique of her detractors.

More recently, the tone has softened, but McClintock continues to be stereotyped, as by Helen Longino, as a scientist with a "loving identification with various aspects of the plants she studied."[27] It is now axiomatic, an assumption seemingly no longer requiring proof, that McClintock faced obstacles throughout her career and that resistance to her findings was rooted in sex discrimination.[28] Her portrait adorns the dust jacket of the second volume of Margaret Rossiter's comprehensive history of women scientists in America.[29] That this admirable and even-handed book bears McClintock's image on its cover shows how McClintock has become, like Barthes's black soldier, a symbol: in her case of hardship, persistence, and legitimation. Few historical studies since Keller have examined McClintock as an individual; fewer still have questioned the standard interpretation of her science and life.[30]

The myth became publicized further. Journalists romanticized McClintock as a sort of prophet in the wilderness, portraying her as "long-neglected" (Victor Cohn, *Washington Post*), a "brilliant loner" (John Noble Wilford, *New York Times*). Cold Spring Harbor became a "secular monastery of science" (David Zinman, *Newsday*).[31] Children's stories have distilled her into a cartoon character, an exemplar of determination, perseverance, and individualism.[32]

In many ways, the myth has done much good. Science benefits from female role models; scholarly studies of science are strengthened by attention to alternative ways science might be practiced. The story has inspired girls and women to enter (or remain in) science—a population as deserving of role models as it has been deprived of them.

Yet mythology is not history. In the course of my research, many people have told me that McClintock detested the status she attained as a feminist icon.[33] Peeling back the mask reveals what is left out of the myth: the intimidating intellect, biting humor, and fierce independence of McClintock the woman; her high visibility and continuing presence in the genetics community; and the complexity and historical situation of her science. In short, by examining her not as a representative of a class but as a unique individual, we get to know a new Barbara McClintock, one with a distinctive scientific style and a misunderstood contribution to the history of biological thought.

This book dismantles the McClintock myth in seven steps. First, McClintock was neither "marginal" nor socially "deviant" among geneticists.[34] Her early publications established her reputation as a cytogeneticist of the first rank. Professional achievements, such as her election to the vice presidency (1939) and presidency (1945) of the Genetics Society of

America and her election to the National Academy of Sciences (1944)—only the third woman to be so honored—reflected the immense respect she earned even while still a young scientist.

Second, transposition was neither ignored nor disbelieved. Following McClintock's discovery in 1948, it was confirmed independently by 1950 (published in 1952) and again in 1953, and by the mid-1950s it was accepted as generally true for maize.

Third, McClintock called her movable elements *controlling* elements, not *transposable* elements. Her primary research interest, from 1944 until the end of her life, was the genetic control of development. Transposition was merely part of the mechanism by which she believed this control to be exerted. It was of secondary importance to her.

Fourth, McClintock's allegedly holistic, intuitive scientific style was in fact highly rational and based on immense experience and reading. All good scientists develop a feel for their experimental material. McClintock was distinctive in the speed and facility with which she solved problems and in her emphasis on the synthesis that follows reduction in any complete solution to a scientific problem.

Fifth, Cold Spring Harbor was not a backwater but possibly the only institution at which McClintock could have been happy and productive. The Carnegie Institution of Washington, her employer, gave her freedom she could not have had at a university. She was hardly isolated at Cold Spring Harbor; indeed, she sometimes complained of the incessant distractions of gregarious colleagues who interrupted her work during the summer meetings season.

Sixth, the "rediscovery" of McClintock's work was not a simple, forehead-slapping recognition that she had been right all along. It was a complex social and intellectual process, by which her theory of genetic control of development, never accepted, was at last formally rejected, while transposition, never the main aspect of her science, was found to be of great biological significance. The scientific community has been consistent over the decades: McClintock was right about transposition but wrong about development. Molecular evidence largely bears out this judgment (though with some revealing exceptions).

Finally, McClintock recognized that this shift had occurred, and so for her the Nobel Prize was not a long-awaited legitimation but rather a bittersweet epitaph to her theory.

This new story will not be welcomed by those who have a stake in the myth. Already, in talks and lectures I have given, some have heard in it an attack on women scientists, a denial of the discrimination against women in science, or a defense of masculinist science. It is none of these. It is, rather, an effort to remove the stigma of victimization from, to restore the dignity of agency to, a vigorous, intelligent, and complex historical actor. Nothing in McClintock's story denies the prejudice, discrimination, and marginalization experienced by other women, either in or out of science. But she is a poor symbol for those hardships. In particular, her time at Cold Spring Harbor illustrates the hopeful fact that in science's meritocracy occasionally a woman who had no betters intellectually was treated as a peer.

Nor does her story support the idea of a nonmasculine science. The many women scientists I know reject the argument, made ostensibly on their behalf, that women do the intellectual work of science differently from men. Rationality and intuition, experiment and observation, manipulation and attentiveness are components of all good science. The chapters that follow will show that McClintock's scientific methods and intellectual style were no different from those of any other leading cytogeneticist—gender is simply not a relevant category here. What sets her apart is her unique mixture of elements and the extraordinary ways in which she manifested them.

Recently, some feminist authors have come to similar conclusions. Londa Schiebinger, in her survey of the influences of feminism on science, sees the abstract feminist critique giving way to a "more positive task" of investigating the useful changes brought about in science by feminism, both directly and indirectly. Schiebinger notes that sexual essentialism, or difference feminism, has led to what she calls blind alleys that harm both women and science. She also recognizes that McClintock has become "an icon for a supposed 'feminine' or, at times, even a 'feminist' science."[35] Noretta Koertge observes, as did Stephen Jay Gould in 1984 and Horace Freeland Judson in 1980, that developing a feeling for the organism is an experience shared by many working scientists. The vitality of science depends on a dialectic between logic and intuition, abstract reasoning and concrete evidence, and other stylistic pairs claimed by feminists to be gendered. In comparing such pairs, Koertge notes, "it is just silly to try to label one side as more feminist, or more progressive, or whatever, merely on the basis of their methods and approaches."[36]

This book operates on three levels. On the biographical level, Barbara McClintock was a complex, fascinating character. She was both warm and emotionally distant, immensely successful professionally and yet a failure in what mattered most to her—persuading her colleagues that she had found the source of genetic control. By any measure, she was of genius caliber, and her startlingly original approach to almost any problem struck everyone who knew her. Indeed, this became part of her private myth. Later in her career, her confidence at being able to solve problems grew so great that she tended to forgo the rigorous experimental tests on which she had built her reputation. The second part of the story is mainly about that reputation, how it was negotiated and how it was transformed into myth. I show that McClintock herself had a role in these negotiations; her fate was not decided by some faceless group of scientific power brokers.

Historically, this is an effort to correct and deepen our understanding of the history of genetics. Up to now, McClintock has been portrayed almost exclusively from the vantage of early molecular biology. I place McClintock within the history of cytology, genetics, and—for the first time—embryology, which she claimed as an influence and which indeed is the source of the scientific problem she believed she solved with controlling elements. By the 1950s, the string-of-pearls vision of the gene came to be seen as simplistic by classical geneticists, who had begun to focus on problems of gene action. A few even rejected the classical gene outright. The string-of-pearls gene had been taken up, however, by a cadre of young bacterial and viral geneticists, who were concentrating instead on the material nature of the gene. By 1951, the very nature of the gene was hotly in dispute. That year, the Cold Spring Harbor Symposium, by then the premier professional meeting of geneticists, provided a debating platform for the old radicals, classical geneticists, arguing against the gene, and the young republicans of bacterial and viral genetics, arguing for it. McClintock's now-famous paper at the symposium and the reaction to it cannot be properly understood without positioning her in this debate, firmly on the side of the old radicals.

Further, McClintock's contributions to science during her Cold Spring Harbor years have been misunderstood: they are in some ways less yet in other ways greater than previous accounts have portrayed them. Once we grasp that the Nobel was a rejection, not a legitimation, of McClintock's controlling elements (in addition to being a tribute to her as a scientist), it becomes possible to understand her enigmatic late work. She put forward a remarkable vision of organic change, much of it untestable as she phrased it, but nevertheless one that could lead to a new synthesis of biology's most profound problems: heredity, development, and evolution.

On the analytical level, I provide the first detailed reconstruction of McClintock's experiments on controlling elements, as derived from her field notes, lab notebooks, and other research materials. This has been impossible up until now, because McClintock's published works do not contain the detail needed to identify which plants she used, what genes they carried, and what progeny they produced. These details, as well as thousands of pages of hand-written thoughts, letters, memoranda, and draft reports are now available in the McClintock collection at the American Philosophical Society Library, in Philadelphia. This book is the fruit of hundreds of hours spent with that material. In addition, I have combed through some ten other archival collections, some public, some opened to me by individual scientists.

The importance of the science, though, goes beyond the experiments themselves. Understanding the science is crucial to the dismantling of the McClintock myth. In recent years, historians of science—including those allied with the feminist critique—have shown that science cannot be understood without placing it in social context.[37] This book demonstrates that the social context cannot be understood without grounding it in the science.

This book is also an essay in the history of recent science, defined roughly as the history of those one can still talk to.[38] Historians of recent science have access to a rich and highly problematic source of data unavailable to historians of more distant periods: interviews. Mnemosyne is a notoriously fickle archivist: she stores selectively, catalogs capriciously, retrieves inconsistently. She shelves fact and fiction side by side, and the unwary patron is often handed both in a stack of unbound and sometimes shuffled sheaves. Historians of dead people often scoff at our precious tapes and transcripts, and well they might: their actors may have written letters every day and kept journals and detailed, organized lab notebooks. Increasingly, the task of the historian of recent science is becoming more like that of the medievalist: We get all the documents we can, then scrabble for any shard of pottery or piece of flint that can fill a gap in our fragmentary tale. With no offense intended to scientists, their brains are our midden: we troll through the accumulated junk to find evidence of daily life recorded nowhere else. Often the process is immensely pleasurable. One of the joys of doing recent history is the interview that clicks, when both parties come away exhilarated with having learned something surprising.

This book makes extensive and cautious use of interviews. From archives and through the generous cooperation of colleagues, I have assembled a set of ten interviews with McClintock, given to various people between 1970 and 1980. Among them are the interviews she gave to Evelyn Fox Keller. Rarely does the historian have the opportunity to compare and contrast so many versions of an individual's personal narrative. They are an invaluable tool for reconstructing McClintock's private myth and exposing its origins. A major subtext of Chapter 2, then, is an argument-by-demonstration of the critical use of interviews. I tell McClintock's story as she herself told it. The basic outlines of the narrative have been confirmed, but the details and embellishments are presented as artifacts. This approach allows me to show first, how McClintock created her private myth and, second, how the very transcripts that gave us the feeling-for-the-organism story can yield quite a different narrative.

With this skeptical tone set early on, the rest of the book uses the interviews with McClintock, as well as many I conducted with her friends and colleagues, in a variety of ways: as indicators of how scientists reacted to McClintock's work; as pointers to and elaborations of issues verifiable with documents; and as guides to her character and personality such as could not be obtained any other way.

This book is narrated chronologically but organized thematically. Each of the next five chapters develops a theme in McClintock's life and science. As I trace McClintock's development, new themes layer on the old. Each one covers a period in her life; the themes have a development that parallels the life. The reader is cautioned not to overinterpret them or their assignment to a given period. I do not suggest that a given theme was absent either before or after the period for which it serves as an emblem. Rather, the themes are associated with phases in McClintock's life in which they seem especially important, or because they illuminate a crucial aspect of her science.

McClintock said that to do good science one had to listen to one's material, let it guide one's interpretation. Good history must work the same way. I have drawn the themes of this book from McClintock's life and in many cases from her own language; I have organized this book to reflect the structure of McClintock's thought as I understand it. The private myth can be a guide to that structure.

Chapters 2 and 3 examine McClintock's first forty years with a low-power lens. Chapter 2 uses McClintock's early years, and particularly her own reconstruction of those years, to illustrate what I see as the foundation theme of her life and work: freedom. The desire for freedom was a dominant—perhaps the dominant—motivator in her life and career and explains many of her disparate and sometimes contradictory traits. It was also, I believe, what she felt when she was thinking. Chapter 3 provides an introduction to the history of genetics and cytology up to the time McClintock entered the field, and then integrates her into it. Viewing McClintock as it does from the perspective of molecular biology, the public McClintock myth has a presentist quality that distorts her context. Her intellectual heritage was that older tradition of microscopes and cells, of "colored bodies" (chromosomes) and genetic factors and elements. Further, integration describes her intellectual style. She had a synthetic turn of mind, an intellectual style of assembling complex, multidimensional puzzles in her head with extraordinary speed. Some colleagues called this ability wizardly, others called it mystical. McClintock called it integration.

In Chapters 4–6, I switch to a high-power lens and examine the development of her theory of controlling elements during the middle and late 1940s. Each is organized around a theme of her science. The first, pattern, is the basis of all of McClintock's discoveries and the starting point for controlling elements. Special patterns in her plants suggested to her how the genes might be controlled to produce patterns and indeed all traits of the plant. Control, then, is the theme that follows pattern. For McClintock, pattern always indicated an underlying system of control. Later, she developed a dialectical view of pattern and control, as two biological forces arranged in alternating logical layers: control underlay pattern, but controlling systems were themselves produced by deeper patterns in gene-gene and gene-environment interactions.

As she pursued these ideas, she sought to incorporate increasing numbers of variables into her model. She amazed and sometimes confused colleagues with her seeming ability to account for the influences of many different genes on one gene's action. Indeed, some doubted she was actually doing this and suspected her of being misled by confounding or spurious variables. In short, McClintock developed a thematic style of complexity, the theme of Chapter 6. Complexity distinguished her style from the simple-systems approach adopted by many of the bacterial and viral geneticists who were then coming to dominate genetics.

The central argument of these three chapters is that developmental control was McClintock's primary interest. She was searching for the source of developmental control when she discovered transposition. She interpreted transposition in light of developmental control. She was interested in transposition only insofar as it explained developmental control.

Having reconstructed McClintock's initial controlling-element experiments, I then switch back to lower magnification to observe the results. The remaining chapters interpret the reception and renaissance of controlling elements in light of the five themes developed in the preceding chapters. I show that the reception of controlling elements was not simply one of rejection; that her reputation remained high, even among those who did not accept that controlling elements regulated development; that she had opportunities and offers during the 1950s and 1960s that few biologists—perhaps no other biologist—could have received, and turned them down to stay at Cold Spring Harbor.

I examine McClintock's work on the races of maize of the Americas, a twenty-year project during which she developed her interest in evolution. I show how, in her later work on controlling elements, she all but abandoned transposition, although control remained central to all her experiments and writings. Although McClintock's files in the Nobel archives will be closed until 2033, I have used other sources to provide a first account of her four nominations and a glimpse into some of the deliberations over them. These reveal how scientists constructed their own McClintock myth—a benevolent myth, intended to restore credit where it seemed lacking.

Having thus dismantled the McClintock myth and analyzed her science, in the last chapter I offer a synthesis, in which I weave McClintock into a new narrative of twentieth-century biology. Whereas a major task of the rest of the book is to demonstrate that McClintock was not marginal in the way she has been held to be, the goal of Chapter 10 is to establish how radical she in fact was, in the broader context of the questions of growth and form.

By this time, with McClintock's theory having been described as rejected by most scientists and her Nobel portrayed as honoring work that had little connection with her experiments, the reader may wonder whether she was really right or wrong, whether she deserved the accolades she received. An appendix attempts to resolve these questions by briefly surveying recent work on the molecular basis of transposable elements.

This book, then, is about several kinds of controlling elements. At its core is a history of the pieces of chromosomes McClintock named controlling elements. This history is a transect through twentieth-century biology. Tracing it lashes together disciplines whose connections are poorly understood and reveals clefts within disciplines often portrayed as unitary in outlook and approach. This is also the story of the elements that controlled McClintock, the themes that gave structure to her life and science.

And it is the story of the control she exerted over her own life. Again, it must be emphasized that *control* is McClintock's word, her theme.

In this story, it is often left ambiguous who or what is doing the controlling. Although McClintock was often more in control than she has been given credit for, there were times when she convinced herself that she was more in control than she was. The ambiguity of control is both the brilliance and the flaw in McClintock's science, and the root of the complexity of her position in the scientific community. Elements, too, are ambiguous. They may be genetic, chemical, meteorological, social. At points, all of these senses are implied or addressed here. What we want to know, then, is this: When did McClintock control the elements, and when did the elements control her?

2

FREEDOM

The sense of freedom was more important than any aspiration.
—BARBARA MCCLINTOCK

We begin near the end. Colleagues and friends who knew Barbara McClintock during the 1970s and 1980s remember a woman who dressed androgynously, in simple collared shirts, loose-fitting slacks, and work shoes. She was small and wiry, a fraction under five feet two, about one hundred pounds. Her hair was short and neat and gray. Though not poor, she lived with great economy, which reflected both frugality and thought. She wore a good lab coat threadbare, repaired it with iron-on tape, and wore out the patches before replacing it. She owned seven clothes irons, decades old, six still in their original boxes.[1] She spoke in a distinctive accent of obscure provenance, upper-class, almost haughty, and her speech was idiosyncratic, elliptical, epigrammatic. Her sense of humor was well-developed, bawdy, self-effacing. She had an almost morbid dislike of writing in books, although she underlined scientific papers with a straightedge and in four colors.

An air of supreme confidence combined with a sharply focussed pursuit of truth—others' feelings be damned—contributed to her reputation for ferocity. Yet those who won her over found a warm, attentive friend. She was especially fond of younger scientists of a gentle nature. Though she enjoyed company, she lived and worked alone. She advised many but never had full-time students or technicians. Depression may have visited her, at intervals of two or several years. She kept a cot in her office. She slept little.

In her later years, McClintock was a living scientific legend. In the 1970s, transpositions and rearrangements of genetic material became one of biology's hot topics, an aspect of chromosome behavior in many organisms that had profound implications for disciplines ranging from cancer biology to immunology. McClintock was recognized as the forebear of this fruitful line of research. Although her heyday had run from the late 1920s to the mid-1950s, she became one of those rare scientists whose work underwent a renaissance, its implications being recognized fully only decades after the fact.

There were other examples. In 1944, Oswald Avery, a serologist at the Rockefeller Institute in New York, and his colleagues Colin MacLeod and Maclyn McCarty found that strains of "rough" Pneumococcus bacteria were transformed into smooth strains by the transfer of thymonucleic acid, today known as deoxyribonucleic acid, DNA. The "transforming principle," the hereditary material, they concluded, was DNA. Though read and cited, their paper did not comparably transform biologists' thinking about the nature of the genetic material. That happened eight years later, when viral geneticist Alfred Hershey and Martha Chase, his technician, reached the same conclusion using the tools, techniques, and language of viral genetics, which was concerned—as serology was not—with the nature of the gene. Today, their paper, known as Avery, MacLeod, and McCarty 1944 is widely acknowledged as the first demonstration that DNA is the material of the gene.[2]

Peyton Rous, a pathologist at the Rockefeller Institute, underwent an even more dramatic renaissance. In 1911, Rous had discovered a virus in a Plymouth Rock hen that produced tumors. Leading biologists rejected Rous's findings. Rous's biographical sketch in the *Biographical Memoirs of the Royal Society* states that for years "pathologists refused to consider that the current discovery was relevant to the problem of cancer." Ralph W. Moss wrote that Rous was "one of those scorned by the establishment" and that "his findings were ignored." Nathaniel Berlin, writing from within the cancer research community, wrote that "relatively little attention was paid to [Rous's] findings." For years, other cancer researchers "did not even try to repeat his experiments," wrote Charles Oberling, another cancer researcher.[3] Fifty-five years after his discovery of tumor viruses, Rous shared the 1966 Nobel Prize in Physiology or Medicine.[4]

Like Rous, McClintock was seen as a scientist whose discovery had been premature, in biologist Gunther Stent's term.[5] Like Rous—but unlike the most famous example in genetics, Gregor Mendel, the father of the field—McClintock outlived the eclipse of her discovery.

In her late seventies, during the ten years prior to her Nobel Prize, McClintock gave numerous interviews. She sometimes told the interviewer what he wanted to hear, but more often she built for herself a version of her past with which she could be comfortable. The interviewer who obtained the most detailed account of her childhood and early adulthood is Evelyn Fox Keller. Other interviewers obtained in-depth accounts of other parts of her life. Close reading of these interviews is a useful way to begin exploring McClintock's life, mind, and science. These expose her private myth, a creation story, part fact part fiction, from which she formed her identity.[6]

The foundation of this myth is freedom. Freedom underlies her intense need for solitude. It is the basis of her approach to science and to her career. Later themes of her life and science—integration, pattern, control, complexity—are elaborations and implications of her need for freedom. In short, freedom is the key to her personal and intellectual life, and to the relation between them.

As McClintock told her own tale, she had from infancy an intense sense of independence. This seems to have been fostered by her parents, who had, in Keller's apt phrase, a "profound respect for self-determination."[7] McClintock cast this independence as a need for freedom: freedom of action and, especially, freedom of thought. She could not bear to be constrained by conventions, rules, or organizations. In interviews late in life she refused to acknowledge influences, role models, or heroes. She presented herself as a maverick, misunderstood, ahead of her time, lucky.

"Don't Touch Me"

She was born Eleanor McClintock on 16 June 1902, in Hartford, Connecticut. Soon after, the family moved to Flatbush, Brooklyn, where she spent most of her childhood. Her mother, Sara Handy, was a Boston blueblood who traced her lineage to the Mayflower. Her father, Thomas Henry McClintock, was the son of British immigrants. Although he was a physician, to the Handys he seems to have been NOCD ("not our class, dearie"). Eleanor was the third in a rapid succession of daughters, following Marjorie (October 1898) and Mignon (November 1900). A year and a half after her birth, in December 1903, appeared a son, named Malcolm Rider.

The last two children were not called by their given names. Malcolm was called Tom, and Eleanor, Barbara. "I showed some kinds of qualities (that I do not know about, nor did I ask my mother what they were) that made them believe that the name Eleanor, which they considered to be a very feminine, very delicate name, was not the name that I should have. And so they changed it to Barbara, which they thought was a much stronger name."[8] McClintock used the name change as an example of having not been understood or quite approved of—in a way that was not entirely negative. The name Barbara, she said, "was not very common, and this change was objected to very much by the relatives." For the first half of her life, she lived with an unofficial identity. Not until 1943 did she legally change her name, and even then she ascribed the impetus to her father.[9] "My father became disturbed and fearful that I would not be able to get a passport, that I would not be able to prove myself, so he legally had the name changed to Barbara."[10] There must be at least a little fiction in this story, since in 1933, still legally "Eleanor," she went to Germany and would have required a passport. Nevertheless, from this story McClintock extracted a sense that her self had never matched her formal identity.

In seeking the roots of her independence, she drew on another family story. "My mother used to put a pillow on the floor and give me one toy and just leave me there. She said I didn't cry, didn't call for anything."[11] Many babies can be thus entertained, but McClintock told the story to demonstrate her lack of need of attention. As she grew, this indifference became an aversion. According to McClintock family history, Barbara's parents had wanted a boy.[12] In Tom they got their son, but Barbara used the story to defeminize herself and set herself apart from her family. "Somewhere along the line (probably this early stage) I sensed my mother's dissatisfaction with me as a person, because I was a girl or other-wise, I really don't know, and I began to not wish to be too close to her."[13] By the time she was three, she said, she refused to let her mother touch her. "I remember we had long windows in our house, and curtains in front of the windows, and I was told by my mother that I would run behind the curtains and say to my mother, 'Don't touch me, don't touch me.'"[14] It may be impossible to reconstruct Mrs. McClintock's reasons for telling such stories, but they shaped Barbara's construction of her life's arc.

McClintock's sense of alienation from her mother sometimes conflicted with the accepted version of the family history. According to her oldest sister, their mother was severely stressed around the time of Barbara's birth, partly because of financial difficulties as Dr. McClintock struggled to es-tablish a practice. Young Tom's arrival overtaxed Mrs. McClintock and strained relations with her eighteen-month-old daughter. Barbara, how-ever, thought her mother just did not like babies. Her mother, she claimed, considered motherhood "just a period to go through. Though she had four children she did not like babies, though people are supposed to like them." In Marjorie's version, Mrs. McClintock's stress was the rea-son Barbara was sent off to Massachusetts at about the age of three to live with an aunt and uncle. But for Barbara, the cause lay in her be-draped protestations. Her alienation from her mother "was so serious that my father decided that I should go and live for a while with his sister," who was married to a wholesale fish dealer and lived in Campello, Massachusetts. She lived with them "on and off over several years during those preschool years," she said. She had some "very unusual experiences" with her uncle and aunt. "I don't regret that part at all."[15]

Whether or not McClintock's mother liked babies, she was perceptive with children. Both she and her husband respected Barbara's need for freedom, no matter how idiosyncratic her desires, even in the face of neighborhood disapproval. McClintock did not enjoy "girl" things. She liked sports. Her mother defended her right to play with boys, even having bloomers made for her so she could play unhindered by dresses.[16] "My parents supported everything I wanted to do, even if it went against the mores of the women on the block. They wouldn't let anybody interfere."[17] Once, when a neighbor told McClintock she ought to be more ladylike and not play such rough games, her mother telephoned the woman and gave her a scolding. When McClintock showed an interest in ice-skating, her parents bought her skates and allowed her to play hooky and go skating. McClintock believed that her parents respected her judgment. When she had an odious teacher and came to dread going to school, her parents sanctioned her truancy.[18]

Socially, she was neither girl nor boy; she did not feel like the one and was not accepted as the other. She was a group of one. "You had to be alone," she said. "You couldn't be in a society you didn't belong to." Nor did she feel part of her family, though she liked them well enough. "I didn't belong to that family, but I'm glad I was in it. I was an odd member."[19]

Alone, McClintock was free. She loved to "just sit alone. I do not know what I would be doing when I was sitting alone, but I know that it disturbed my mother, and she would sometimes ask me to do something else, because she did not know what was going on in my mind while I was sitting there alone."[20] Was the child emotionally disturbed? First curtain-hiding, now this. "Although I knew that it was perfectly safe, she felt there was something wrong. But there was really nothing wrong; my sitting there was related to things that I was thinking about, and had nothing to do with any psychological difficulty."[21] In the summers, the family would go out to the end of Long Beach. She would take the dog and stroll the strand. "I used to love to be alone, just walking along the beach." Alone on the beach, she said, she developed a technique for detaching herself, body from sand, mind from body, which she later likened to a technique developed by Hindu mystics. "You stood quite straight with your back completely straight, and you practically floated. Each step was rhythmically floating, without any sense of fatigue, and with a great sense of euphoria."[22]

At Erasmus Hall high school in Brooklyn, McClintock revealed a passion for learning. Perhaps the solitude of reading and problem-solving appealed to her. But she also must have discovered that her mind was a fast and powerful tool; thinking was fun. "I loved information," she said. "I loved to know things." In her recollection, she solved problems the way she had done everything: unconventionally. "I would solve some of the problems in ways that weren't the answers the instructor expected." She would then ask to re-solve the problem, to see if she could find the standard answer. Problem-solving was "just pure joy."[23]

McClintock thought her blossoming intellectual side worried her mother. The well-bred Mrs. McClintock had indulged her daughter's love of sports, but life for Barbara would soon get serious and require more conventional behavior. McClintock did not explain this shift in her mother's attitude—this was the woman who let her daughter wear bloomers?—but she used it to define herself. She said her mother feared Barbara would become a "strange person, a person that didn't belong to society,"[24] which is indeed how the adult McClintock looked upon her own life. Her native inclinations, her parents' attitudes, her social milieu all contributed to her deep-seated need to follow her own course, alone. She could not abide the constraints imposed by groups. She did not remember feeling great pain as a child, except when she had been required to conform. "That kind of pain I felt very seriously, and that happened mainly in relationship to standard procedures, in relationship to having to conform in a group way. I think that always caused me trouble."[25]

A person is freest when sitting alone, thinking. No group rules or pressures constrain her. Social conventions fall away; she is no longer "she," or young, or odd. Her body imposes no limitations of strength or speed. Even time ceases to exist. McClintock discovered she could think anything she wanted, solve any problem, fast. As she discovered the pleasures of her mind, social relations came to seem not anchors but tethers. Though she was never the hermit she has sometimes been portrayed as being, she did choose her relationships carefully, so as to maintain control and preserve her freedom.

Free Agent

McClintock graduated from high school in 1919. She wanted to go to college, but apparently her mother was opposed. "Mother was quite sure I was going to be a professor, and they were all very much concerned by it." It is unclear whether her mother denigrated the life of the mind on principle, or whether she had emotional, historical reasons for her opposition. McClintock remembered her mother as an obstacle to college and an opponent of college education for women, and recalled that there was a woman professor in the family of whom her mother did not approve.[26] Dr. McClintock was overseas, serving as an army surgeon. When he returned from Europe that September, he supported her wish for higher education.

It was her mother, though, who McClintock remembered as breaking the good news. "So on a Friday morning my mother woke me up early and said they had been talking about it, and that they want me to go to college."[27] She had her heart set on the state school at Cornell, one of the great bargains in higher education. Cornell was and is two colleges: a private arts and sciences college and a public agricultural school. Students in either college can enroll in any course. As a New York resident, McClintock could matriculate at the agricultural college—known locally as the Ag School—for free and receive an Ivy League education. Her mother called up to Ithaca to ask when school started. Registration begins Monday, they replied. The M's register Tuesday.

McClintock liked to portray her life as having been visited by mysterious little miracles that opened doors for her. She never questioned them. Good luck imposes no debts. As she described the event to Franklin Portugal in 1980, getting into college was one such miracle. "I went up on the Lehigh Valley train. I got a place to stay and the next morning I was on the line with all the M's at 8:00. And then—the registrar was the one that really saw everybody. And I came up to him and he asked me for my papers. I said I didn't have any. And he said, 'Well how do you expect to get in here without any papers?' And just at that time my name was called out in the room, and he heard it and I heard it. So he stopped and went back to see what this meant. And he came back and he handed me my papers. He says, 'You can register.' Now, I never found out what happened, but I thought that was quite a coincidence. I had a series of interesting coincidences. That began it."

Pressed on whether her mother might have had something to do with it, McClintock admitted, "I don't know. She must have." She easily imagined a logical scenario in which her mother had sent a telegram that was only opened and sent up to the registrar when the M's were to come through.

In McClintock's narrative, exceptions were made for her. After she had received her doctorate, she said, "Some girl was trying to get in the Ag. school, and they wouldn't let her in. So I went over and saw the dean and said, 'Why aren't you letting her in? I came through on Ag. Why aren't you letting her in?' He said to me—now this is strange too. He said, 'You were an exception; we made an exception for you.' Why did they make an exception when I first entered?"[28] She did not pursue the answer, preferring simply to accept such gifts and race toward the next hurdle.

More than men, women tend to see their successes as the outcome of luck.[29] But whereas for many women this tendency is detrimental to success, McClintock used this sense of her own luck to advantage. She imagined herself as being privileged yet without debts to others. She was free because she was special, and vice versa.

However she got in, to McClintock "college was just a dream." New social and intellectual opportunities presented themselves on all sides. Socially, she continued to flout convention. She took up smoking—risqué behavior for a woman at the dawn of the flapper era. She bobbed her hair. In her stories, she was the first woman on campus with short hair. It "made quite a noise on campus the next day." She could have been (and quite likely also was) talking about her science when she said it "just happened to be a little ahead of something that was coming anyway!"[30]

McClintock liked to stand out. She had what David Botstein, a bacterial geneticist who befriended her in the 1970s, characterized as panache. She socialized with new and interesting groups of people who were outside the social mainstream. For example, "At that time there was a great separation between the Gentiles and Jews," she said.[31] McClintock enjoyed crossing that gap. She even learned to read Yiddish, doubtless as much for the language puzzle as for the social benefit. The Jewish girls were different from the Ivy League socialites clamoring for acceptance.

The in crowd wanted McClintock, though. Confident, pretty, unusual, and brilliant, she must have appealed to them. Although she had never aspired to high social status, she could not reject that world without trying it first. So McClintock experimented with popularity. The girl who had loved nothing more than sitting alone was elected president of the women's freshman class. She had an active social life. She was rushed by a sorority, which was flattering, but she did not take long to recognize the exclusiveness, the snobbishness of sorority life. "So I thought about it for a while, and broke my pledge, remaining independent the rest of the time. I just couldn't stand that kind of discrimination."[32] Never again would she feel comfortable joining groups, though sometimes her professional life demanded it. "I have never been able to function well in an organized group where the organization is handed down to you. Under any circumstances I can't function in a group like that." Though she could hardly proscribe such associations for others, "I can't function in a personally successful way. Sooner or later I upset the apple cart, somewhere along the line, and somebody else's apple cart, not always mine."[33] She always preferred loose, temporary affiliations that she formed herself—like the sports teams she played on as a child, or like music groups.

Instead of the sorority girls, she found diversion with the "townies." Geography gives Cornell an unusually wide town-gown divide. The university is perched on West Hill, "high above Cayuga's waters." Glacial gorges separate the campus from surrounding neighborhoods. To get to town, one descends one or another steep avenue. University folk are literally above the nonacademic Ithacans, and few of the former condescend to do more than shop or drink downtown. Such a barrier was irresistible to McClintock. She joined a jazz combo that played in cafés and bars downtown. She played banjo. In a casual conversation I had with her in 1991, she recalled how one icy winter evening she slid on her bottom down Stewart Street cradling her banjo to get to a gig.

Scientists with an affinity for music are legion. Stereotypically, however, they are drawn to classical music, especially the rigorous and mathematical Bach. McClintock loved the improvisation and interplay of jazz. The best jazz players break the rules imaginatively. Some of the best players in jazz history have been self-taught, playing by ear without ever learning the formal rules of music. Idiosyncrasy is prized among jazz musicians. Jazz fit McClintock's developing intellectual style and her private myth as an idiosyncratic rule-breaker.

As a freshman, McClintock took music 1 but failed the course. She took another music course in the fall of her senior year, which, she said, "stood me in good stead when I was in my senior year in college and used to play in a little jazz combo." Learning a bit of music theory helped her improvisations. But in class, once again she portrayed herself as the maverick. "The professor in harmony used to play my compositions that I had to write. We all had to write compositions. I don't know why—and he would ask me, 'How did you ever think of that?' Well, I didn't tell him the reason I thought of that was that I had no other way of thinking. I hadn't any experience, nothing to lead me into this or the other thing."[34] She received a C+.

Overall, McClintock excelled in college when she was motivated and prepared. She greedily signed up for large overloads of interesting courses, dropping those she didn't enjoy. Until 1933, Cornell used a "Z" to denote withdrawal from a class with official cancellation.[35] "By the time I was a junior, I discovered I had quite a lot of 'Zs.'"[36] Later in life, McClintock bore these Zs proudly as a mark of her independence, a sign of her naughty rejection of formal constraints.

Other than the F and the C+ in music, a D in political science, and Cs in bacteriology and biology 1, McClintock earned As and Bs. She dabbled widely in science and took meteorology in each of her last three years, receiving an A each time. Overall, she earned just under a B average.[37]

McClintock's personal myth of freedom was part of the foundation of her sense of herself as a scientist. She had, she said, a desire to be anonymous, "to be completely free of what I called 'the body.' The body was something you dragged around. I always wished that I could be an objective observer; instead of going to a party and being there, the body could be forgotten and I could be just an objective observer, and not be what is known as 'me' to other people."[38] Though McClintock said this in 1978, in earlier correspondence she often referred to her physical parts as "the" rather than "my," suggesting that this concept of "the body" was an old one for her.

The results of getting free of the body could be comic. Albert Einstein once fell into a Princeton storm sewer; Barbara McClintock forgot her own name. In her first semester at Cornell, she took geology 1, which she remembered as having taken much later. "I remember when I was, I think, a junior in college, or a sophomore, I was taking geology, and I just loved geology. Well, everybody had to take the final; there were no exemptions. Usually I could get an exemption from a final if they had them, but this was 'no exemption.'" This exam caused her no worry. "I just *knew* the course; I knew more than the course." Bluebooks were handed out. The front page had a blank for the student's name. "Well I couldn't bother with putting my name down; I wanted to see those questions. So I looked at the questions and oh, I was delighted, and I started writing right away. I just enjoyed it immensely. I knew everything was fine, and I got to turn it over and write my name down, and I couldn't remember it. I couldn't remember to save me, and I waited there. I was much too embarrassed to ask anybody what my name was, because I knew they would think I was a screwball. I got more and more nervous, and so finally (it took about twenty minutes) my name came to me."

Of course, if her address had been required on the bluebook, she might have blanked on that; but to McClintock, the event symbolized her immersion in the material of nature and the thrill of learning. To her, this forgetfulness was "related to the fact that the body was a nuisance."[39] In her private myth, she was too interested in the outside world and in her own thoughts to attend to what she was called.

McClintock segued into graduate school in 1923. Already treated as a graduate student by the botany and plant breeding faculty, she found that her diploma hardly altered her effective status. For her research, she had settled on the cytology and genetics of maize, or Indian corn. With few diversions, she studied maize genetics and chromosomes for the next sixty-nine years. She went on to postdoctoral research, academic jobs. To appearances, she was a driven, goal-oriented woman. Not in her own mind. She "was a free agent," she told Horace Freeland Judson in 1973. "I was not strangled by the *need* to compete, nor by a family."[40] In her recollection, it dawned on her only after the fact that she had chosen a career. "I never thought about it being science," she said. "I remember waking up in my mid-thirties, when I had gone some period of time, and saying 'Oh, my goodness, this is what they call a career for women.' At no time did I ever feel that I was required to continue something, or that I was dedicated to some particular endeavor."[41]

She looked at life through a microscope, not a telescope. Her greatest concern was to keep her options open so that she could follow what interested her at the moment. "I think the sense of freedom was more important than any aspiration. The sense of being able to do what you want to do. It doesn't make any difference whether it's genetics or something else, but the sense of freedom to be able to pursue an extraordinary type of investigation was the most important aspect. No other aspect could compete with it."[42]

The Bearable Lightness of Being

We would be wrong to assume McClintock had an intellectual love affair with genetics or with maize. Maize offered her a set of interesting puzzles;

it became special to her only in the sense that she grew familiar with it. When Evelyn Fox Keller asked her about her feeling for maize in 1979, she said she had no emotional feeling for the plants; she took care of them in order to extract information from them. True, she tended her plants carefully. "I don't want to have any wind blow it over, or anything like that happen to it, so in that sense I'm caring for it, but I'm caring for it so it will be certain to give me the answer that I know it can give me."

Did she have an intimacy with her plants? "I really wouldn't put it that way, no," she said. "I think you'd have the same feeling towards some piece of apparatus if you had used it enough that it was giving you something and you knew that it had to be taken care of well." For example, "I like this microscope here very much; I could substitute with my other ones, but this one here I want." Calling it a fondness for her microscope did not quite fit. "It looks like I have an affection for it, but it isn't that."[43]

For people she may have had affection, but she formed no binding ties. "I had not adjusted myself, from the time I was young, to being closely associated with anybody, even members of my family."[44] The "Don't touch me" stories of her early childhood are consistent with this assessment. Looking back on her life, she could not recognize "that strong necessity for a personal attachment to anybody." Marriage, then, was beside the point. Evelyn Witkin, a bacterial geneticist at Cold Spring Harbor from 1945 to 1955, became a close friend. In an interview in 1996, Witkin said that McClintock "found it very hard to imagine how anybody could share a life with someone, for instance in a marriage. This was something that she could not conceive of for herself. She actually told me that it was very difficult for her to imagine how anybody could do it."[45] In 1978, McClintock said the same to Keller: "I never could understand marriage. Let me put it this way: I really do not even now understand it, because I didn't go through that experience of requiring it."[46]

Some speculated that she was a lesbian.[47] There is no evidence for this; she appreciated the physical attractiveness of both men and women but seems not to have been sexually involved with either. She lived amid a circle of admirers, at the core of which was a cluster of men and women who were devoted to her. "When I felt a very strong emotional attachment, it was an emotional attachment, nothing else," meaning, apparently, that it was not an intellectual or physical attachment. "One was a person who loved to dance and played the piano beautifully and just liked esthetic things. I liked him immensely. Another one was a pianist." But for her, emotional attachment was fleeting. "They wouldn't have lasted. I know

that most any man I met, nothing could have lasted."[48] Susan Gensel
Cooper, first the librarian and then also director of public affairs at Cold
Spring Harbor, knew her during the last twenty years of McClintock's life.
She recalled that McClintock enjoyed the aesthetics of tushes. "But they
were not gender-specific," Cooper said. "It wasn't just guys' tushes and it
wasn't just girls' tushes, it was just tushes."[49]

Yet she was a good friend. "She was always sending me a book, or writing little notes," said Witkin in 1996. "Most of the people who knew her
had this kind of experience with her. She really was very attentive to
friends and very warm."[50] She had friends with whom she had dinner,
friends with whom she took long walks. They were all smart, of course—
she did not suffer fools—and most were gentle. Many subordinated themselves to her, seeing that McClintock had something to teach. Cooper said
that if McClintock "really thought someone was bright, she loved to teach
them, she loved to work with them. But she was pretty intense about people listening to her." McClintock soon found in Cooper a good pupil and
confidant. The primary rule for winning McClintock's respect was not to
doubt what she told you: "First of all, I didn't question her."[51] Robert
Pollack, a biologist at Cold Spring Harbor in the early 1970s, described
his conversations with McClintock thus: "You basically said, 'Barbara,
what does this mean?' And then listened a lot and tried to understand
it."[52]

Pollack said he "traded with Barbara in the barter of knowledge"—really, knowledge for work. "I could ask her anything. Anything. And she
could ask me to *do* anything. I installed her TV for her. I set up things in
her office. I moved things around for her. I was trusted to see her as a person and she was trusted to not tell me, 'What a stupid question.' I was her
student. I allowed myself to be her student, she allowed me to be her student. And it was great."[53]

"She really was into her own kind of control," Cooper said in 1997.
Cooper, too, traded with McClintock, exchanging knowledge for emotional support and also help with the quotidian details of life as an aging
woman. McClintock especially enjoyed the company of junior scientists.
They were bright and energetic and never challenged her authority. They
gave her human contact without limiting her freedom. Intellectually,
McClintock held the power in her relationships. In exchange, she allowed
herself to be served, permitting just enough melting of her independence
to make a bond of friendship. But if McClintock ever began to feel dependent on her, Cooper said, "she somehow annihilated it as soon as she
could."[54]

Peeling Back the Mask

The value of any interpretation of a private myth lies in the power of that interpretation to make sense of the events and actions that compose a life. In a recent essay, M. Susan Lindee describes a famous episode from Marie Curie's early years in Paris. In her autobiography, Curie said she dined on mere radishes and her apartment was so cold she had to break the ice in her basin to wash in the morning. Recognized as a potent image, it was picked up by Curie's many biographers. The iced-over basin, Lindee writes, revealed what Curie wanted to show: the monasticism of science, Curie's rejection of her own body, and her own personal awakening.[55]

McClintock, too, told stories that revealed what she wanted to show. Reading and listening to multiple interviews with her, taken by interviewers with different backgrounds and agendas over several years, provides a rare opportunity to view a scientist's private myth with high resolution. What she wanted to convey was freedom, the lack of dependence on anyone else's ideas, favors, efforts, or attentions. Why?

Like independence, freedom implies autonomy, self-determination. In this sense, McClintock's need for freedom may seem to have been merely that which any among us would desire: the lack of constraints, the need to give no person power over us. No accomplished person submits willingly to arbitrary authority. Further, when McClintock gave these interviews, she was being asked to explain what made her different from other scientists, how she came to her startlingly original conclusions.

Yet to McClintock freedom seems to have meant more than mere independence; it was the absence of commitment to anyone or anything, the latitude to move anywhere, mentally or physically, at any time. Family, friends, job, and property are fixed marks on which one is constantly sighting, triangulating one's psychic and social position. McClintock used as few such marks as possible, as though navigating by a single star proved her skill as a pilot. Beyond this, personal ties are themselves sources of meaning. In one dimension, freedom's opposite is constraint, but in another it is intimacy.

"Intimate" has become a key adjective in the McClintock myth, though it has gone through a considerable shift in meaning. What for Keller was McClintock's intimate *knowledge* of her plants became for later writers, such as Linda Jean Shepherd, her intimate *involvement* with them, and even, as for Mary Rosner, an intimate *science*.[56] The failure to acknowledge McClintock's need for freedom has misled authors into attempting to cast McClintock as an exemplar of a feminine, loving intellectual style.

Of course, it might be that what appears to be an incapacity for intimacy was merely a decision that she had to be independent in order to pursue science. Although some women scientists have carved out time for both career and family, many have felt forced to choose between them.[57]

Another aspect of McClintock's private myth, however, was that she never "chose" science: she simply did what she found interesting. When she told Keller in 1978 that she had never understood or gone "through the experience of requiring" marriage, Keller responded that surely she at least had an image of herself as being or becoming a biologist. Not even that, McClintock replied. She followed with her remarks, quoted above, about never having thought about a career and waking up and realizing this was what they meant by a "career for women." She continued, "There was no thought of a career, just enjoying myself immensely because what I was doing was so interesting."[58]

Perhaps, then, an incapacity for intimacy is the figure under McClintock's carpet. Much in her life could be explained if her story of hiding in the drapes as a girl were linked, as she implied, to the events of her career as a scientist. As she developed that career, she constantly had trouble staying in one place or tying herself down. Transforming an inability to commit into a private myth of freedom is an act of power and self-control. The theme of freedom casts McClintock as precisely the opposite of her stereotype: not loving, empathic, or intimate, but strangely cool, detached, and intensely independent.

3

INTEGRATION

It's when you look down and do something with what you see.
—BARBARA MCCLINTOCK

The year before Barbara McClintock was conceived, a new science was born. The foundation—general principles, some of the terminology—had been laid decades earlier by the Moravian monk Gregor Mendel. But Mendelism was the product of a younger generation of scientists who discovered Mendel's rules and papers at the turn of the twentieth century. They applied these concepts not merely in the creation of a new field of study but also in the division of what had once been unified.

Peter Bowler, a historian of Darwinism and Mendelism, has shown that in the nineteenth century, evolution and growth or development were tightly linked; indeed, "evolution" originally referred to the development of the embryo.[1] Mendelism, the formal, mathematical study of heredity, was the decisive event in severing evolution from growth.[2] Before the third decade of the twentieth century, the great problems of growth and form—development, heredity, and evolution—were riven. Change within an individual, within a lineage, and within a species was addressed by distinct sets of techniques and theories. Mapping the efforts—both successful and not—to reintegrate the problems of growth and form provide a chart for a large tract of twentieth-century biology.

Tracing briefly the disintegration of the problems of growth and form and the beginning of their reintegration sets up the scientific problems McClintock addressed and the language and concepts she used. It provides the historical context for her early work, which itself must be understood in order to grasp both the continuities and the discontinuities in her controlling-element studies. And when, in Chapters 8–10, I come to her last, most speculative ideas, the framework of the re-integration of the problems of growth and form will elucidate how McClintock attempted her own integration, quite distinct from that of any other thinker with whom she was in contact.

Further, "integration" was a word with deep significance for McClintock. She used it to characterize her famous ability to solve scientific prob-

32

lems; it therefore describes her intellectual faculty better than the "intuition" of the McClintock myth. Picking up the theme of integration provides clues to McClintock's vision of mind and nature. It is a window on McClintock's intellectual development, both scientifically and personally.

Pater Gregor

In the standard origin story of genetics, Gregor Mendel was a genius ahead of his time who discovered the laws of heredity, only to be ignored for thirty-five years until his work was rediscovered. Mendel had trained in natural philosophy before entering the Augustinian order at Brünn (Brno), in what is now the Czech Republic. As students of genetics know, in the 1850s, Mendel began using garden peas *(Pisum sativum)* to explore the patterns of inheritance. He looked at seven traits: seed shape, albumin color, seed coat color, pod shape, pod color, flower distribution, stem length. Each trait occurred in only two forms, and each strain bred true for one form or the other of a given trait. When individuals of each form were crossed, in each case one form dominated over the other. A round-seeded plant crossed to a wrinkled-seeded plant gave round-seeded progeny; yellow albumin was dominant over green. The other form, Mendel said, did not disappear; rather, it *receded,* only to return in the next generation. When Mendel crossed the hybrids to one another, he found that for each trait, the dominating and recessive forms occurred in a fixed ratio: three fourths dominant to one fourth recessive. When he crossed these first-generation hybrids to one another, one third of the progeny bred true for the dominating trait, while the remaining two thirds were hybrids, in the same 3:1 ratio.[3]

Mendel worked out the mathematics of these experiments, and it boiled down to a few simple axioms and two principles. The plant behaves as though it has two elements for each trait, one from each parent. The dominant form of the trait appears if the plant bears at least one dominant element; the recessive trait is expressed only if the plant has two recessive elements for that trait. Given these assumptions and observations, the observed ratios could only come about if two things were true. First, the elements must assort randomly; a dominant must pair up with another dominant as often as with a recessive. Second, elements for different traits must be independent of one another. Whether the seeds are green or yellow cannot depend on whether the seed pod was arched or constricted.

Mendel wrote up his results and read them at two consecutive meetings of the Brünn Natural History Society in early 1865. Early accounts of Mendel's life and work cast the audience's reaction as stony silence: "The minutes of the meeting inform us that there were neither questions nor discussion. The audience dispersed and ceased to think about the matter."[4] More recently, historians have found other accounts of Mendel's presentation that portray a more favorable reception, referring to the "lively participation" and "enthusiastic response of the audience."[5] Whatever the immediate reaction, Mendel's papers had little impact on the field. Although they were cited several times, they did not stimulate sufficient interest to induce others to confirm or extend his findings.[6]

The biological revelation of the day was Darwin's theory of transmutation of species, what we know as evolution. It was not then obvious, as it seems in retrospect, that a materialist vision of nature, developed so far by Darwin, lacked precisely what Mendel provided: a mechanism for producing heritable variation, the raw material for natural selection. Mendel's work reposed in relative obscurity for the rest of the century.

Mendelism and Cytology

In 1900, two plant hybridizers, Carl Correns, a German, and Hugo de Vries, a Netherlander, demonstrated the same principles of segregation and independent assortment. A third botanist, Erich von Tschermak, also published an account, although his interpretation lacked sufficient depth to consider him on a par with de Vries and Correns.[7]

There followed a dramatic period of naming. Correns's and de Vries's papers were seized upon by the Englishman William Bateson, who instantly became one of Mendelism's most articulate advocates. Bateson contributed the term "allelomorph" (roughly, "parallel form"), shortened to "allele" in the 1930s, to describe one of the alternate forms of a Mendelian element. He also distinguished the two basic genetic conditions: homozygous ("zygous" from the Greek for "yoke," symbolizing the unification of egg and sperm), in which the organism has either two recessive or two dominant alleles for a given gene; and heterozygous, in which it has one recessive and one dominant allele.[8]

The ideas of Mendelism came so fast and were so exciting that names for the objects of this new field and for the field itself were late arrivals. Bateson first suggested "genetics" as the name of the new field in 1905.[9] And it was not until 1909 that Wilhelm Johannsen at last gave us the term "gene." Johannsen also contributed the powerful concepts of "pheno-type" and "genotype." The former refers to an organism's appearance; the latter to its genetic constitution. The distinction relaxes the linkage of genes and traits. More than one genotype can underlie a given phenotype; round peas may be heterozygous (round/wrinkled, with round domi-nant) or homozygous (round/round). Conversely, a given genotype can result in more than one phenotype.[10]

Mendelism was an immediate and direct threat to Darwinism. Under Mendelism, hereditary elements are envisioned as particles, "unit charac-ters" in Bateson's term, which produce traits by combination. Darwinism was based on the observation that phenotypic differences in nature are usually graded, not stepped: height, coat color, seed weight, even behavior are much more often a continuum than a pair of either-or alternatives. In-heritance was assumed, therefore, to be blending. The same year Mendel read his paper to the Brünn Natural History Society, Darwin proposed a theory of "pangenes" that blended in the reproductive cells to pro-duce a spectrum of traits.[11] Bateson and other Mendelians also battled biometricians, whose bell curves and statistical analyses of traits required blending, not particulate, heredity.[12] In 1901, de Vries published the first volume of his *Mutationstheorie,* which argued, based on observations of wild evening primrose *(Oenothera),* that new species arose suddenly, out of sweeping changes affecting the entire organism. He called these muta-tions. Ironically, given his decidedly nongradual view of evolution, de Vries was one of the few pro-Darwinian Mendelians of this period.[13]

Long before heredity was reintegrated with evolution or development, it was integrated with cytology, the microscopic study of the cell. In doing so, however, it wrenched part of cytology away from embryology. In 1896, Edmund Beecher Wilson, a leading cytologist and embryologist at Columbia University, published *The Cell in Development and Inheritance,* a stunning synthesis and analysis of the universe of the cell. As shown by his title, for Wilson—as for embryologists in general—heredity and devel-opment were intimately linked, and the key to both was cytology. Evolu-tion, the third of the great problems of growth and form, was also part of Wilson's program, because of the continuing analogy of evolution with growth. In 1900, Wilson wrote that "the general problems of embryol-ogy, heredity, and evolution are indissolubly bound up with those of cell structure."[14] They were, but the bonds that united embryology, heredity, and evolution proved to be quite soluble.

Scott Gilbert, an embryologist and historian, has shown that embryologists were the first to link heredity with the chromosomes. Which cellular compartment—nucleus or cytoplasm—directed the intricate processes by which a fertilized egg becomes a complex organism? Mounting evidence pointed to the nucleus. In 1892, August Weismann had proposed a nested series of hereditary elements—ids, determinants, and biophors—that distributed themselves among the cells as development progressed. According to Weismann, a fertilized egg contained a full set. Each time it divided, half of the determinants went into each of the two daughter cells. Cells divided until they had just one determinant, which then decided that cell's fate. Like Aristotle's crystalline spheres in cosmogony, Weismann's theory is beautiful, important—and incorrect. Weismann's answer to how a cell knows what kind of cell to become was to have it winnow out all potentialities save one. In the twentieth century, geneticists showed that all cells retain a full complement of genes. Differentiation is a problem of gene regulation, not segregation. Nevertheless, Weismann's theory explicitly associated heredity with the chromosomes and stimulated much research and thought on heredity.

In 1889, Theodor Boveri removed the nucleus from the eggs of one species of sea urchin and fertilized the empty eggs with sperm, which contribute little more than a nucleus to the embryo, from a different species. The resulting embryos resembled the male parent. Boveri concluded that the nucleus, not the cytoplasm, determined development. In 1903, he and Walter Sutton, one of Wilson's students, independently suggested that Mendel's hereditary elements lay on the chromosomes.

The problem of sex determination was crucial to the establishment of genes on chromosomes. In 1902, Clarence Erwin McClung, another one of Wilson's students, had shown that sex determination seemed to act like a Mendelian element and appeared to be associated with the mysterious "X" or "accessory" chromosome. Nettie Stevens, a student of Thomas Hunt Morgan at Bryn Mawr College, conclusively showed that female beetles fertilized by sperm carrying the X chromosome turned out female.[15]

Meanwhile, cytologists had teased apart and described the dance of the chromosomes. In 1879, Walther Flemming described the process of cell division. In 1881, Wilhelm Waldeyer had described the "chromosomes," or colored bodies, which absorbed pigment when cells were treated with dyes. In the ensuing years, their choreography was worked out. When the cell divides, the chromosomes pair up and align at the center. The chromosomes are in a sense quadruple: there are two copies of each chromosome, one from each parent, and each chromosome has doubled since the previous division, in preparation for dividing again. Poles form at opposite ends of the cell and tiny fibers—the spindle—emanate in arcs from pole to pole like lines of latitude, with the chromosomes lined up along the equator. Now the doubled chromosomes split down the middle. The two halves glide along the spindle to opposite poles. The cell divides at the equator, with one complete set—that is, two of each, in singlets—of chromosomes in each new daughter cell. Flemming called this ordinary type of cell division mitosis.[16]

Just after the turn of the century, the special case of production of germ cells—sperm and eggs—was described. A first division separates the chromosome pairs, with one full complement of chromosomes going to each daughter. A second division splits each chromosome down the middle, as in mitosis. Following the second division, each of the four resulting germ cells has but a single copy of each of the chromosomes. This pair of divisions, the first a reduction in chromosome number, was called meiosis. If a germ cell has n chromosomes (a condition known as haploid), a fertilized egg, as well as other, "somatic," cells have $2n$ (diploid). In 1909, F. A. Janssens looked at chromosomes in the first meiotic division and saw that the arms of homologues often kissed, forming X-shaped figures in which the chromosomes seemed to exchange segments. He called these figures chiasmata; the process became known as crossing-over.[17]

By 1909, few cytologists could think of the chromosomes simply as colored bodies any longer. Mendelism, which at its purest was a statistical formalism, had already begun to impinge on cytology. De Vries, Correns, Bateson, Johannsen, Sutton, Boveri, and others elevated Mendel's principles into laws, fleshed out the language of genetics, began to link them to the chromosomes, and created a new science in opposition to the existing theories of heredity, evolution, and development.

The Mendelian Wedge

The early history of genetics falls conveniently into decades. Mendel's principles were rediscovered in 1900. Between 1900 and 1909, the principles were codified into laws, language was given to the key concepts of Mendelian heredity, meiosis was described, and Mendelian genetics was related, however abstractly, to the chromosomes. In late 1910, Thomas Hunt Morgan integrated Mendelism and cytology.

Morgan came to the study of heredity by a circuitous route. After receiving his Ph.D. from The Johns Hopkins University, where he learned marine biology and embryology from William Keith Brooks and experimental physiology from Henry Newell Martin, Morgan had gone into experimental embryology. He took a year off from his faculty position at Bryn Mawr College and spent a year at the Naples Zoological Station, where he worked with Hans Driesch, Jacques Loeb, and Wilhelm Roux.[18] In 1904, he moved from Bryn Mawr to Columbia University and became a colleague of E. B. Wilson, with whom he shared a Hopkins background. He began to conduct experiments in one of the hot areas of turn-of-the-century biology: experimental evolution.[19]

Born of an effort to bring the hard-nosed rigor of experimental embryology to the study of evolution (both Darwinian and non-Darwinian), experimental evolution signified efforts to produce, manipulate, or otherwise influence evolution under controlled conditions. A center for this new method lay just an hour to the east of Columbia, at the newly founded Station for Experimental Evolution, run by the Carnegie Institution of Washington and sited in 1904 next door to a sleepy marine laboratory at Cold Spring Harbor, Long Island. The Carnegie station cackled, squawked, and bleated like a barnyard. Its scientists studied the inheritance of patterns of coat color in chickens, sheep, and guinea pigs; they probed the biochemistry of plant pigments and the inheritance of alcoholism in rats; they raised salamanders in the dark to try to produce eyeless varieties. It was probably one of the scientists there, Frank Lutz, who, about 1906, sent Morgan his first culture of vinegar flies, *Drosophila melanogaster* (then widely known as *ampelophila*), as a promising organism for experimental evolution research.[20]

The tiny flies had several advantages. They were already domesticated—to collect them, simply leave some bananas to ripen on a windowsill. They were simple to care for and bred easily in the laboratory. And they had a generation time of just ten to twelve days. On the down side, however, they were hard to see and seemed to have few easily distinguishable traits. Lutz got fed up with flies; by 1908, he had forsaken them for crickets and Cold Spring Harbor for the American Museum of Natural History.

At first, Morgan got little of value from them himself. But in early 1910, extensive inbreeding began to produce small, true-breeding variations in color and other traits. Then, in May, he noticed that the pinhead-sized eyes of one male fly were white instead of the usual ruby red. The inheritance pattern of the white-eye trait suggested that it always went with the X chromosome, which, as McClung and Stevens had shown, determined sex. Soon, Morgan would be less timid in his conjectures and state flatly that the white-eye trait lay *on* the X chromosome.[21]

In the preceding years, the question of sex determination had made Morgan a convinced anti-Mendelian; Mendelism's assumptions seemed to him too restrictive and too speculative.[22] The white-eyed fly began Morgan's conversion into one of Mendelism's most ardent advocates. In so doing, it set the trajectory for much of twentieth-century biology. It killed off experimental evolution, by driving a wedge between heredity and evolution. The separation of Mendelism and Darwinism persisted another two decades.

In the fall of 1910, Alfred Sturtevant, a junior at Columbia, and Calvin Bridges, a sophomore, approached Morgan about doing research projects in his laboratory. They, along with Hermann Joseph Muller, a recent graduate, and Morgan crammed into a sixteen-by-twenty-three-foot room on the sixth floor of Schermerhorn Hall. This became known as the "fly room."[23] Over the next two years, several more mutations appeared that were segregated together with sex and with white-eye; these formed a linkage group. By 1912, Bridges had found the first case of autosomal, or non–sex chromosome, linkage. Soon they realized that all mutations clustered into four linkage groups—the same number as Drosophila has chromosomes.[24]

Genetic linkage is the crucial exception to Mendel's rule of independent assortment; the monk would have seen it as contamination. Several of the genes Mendel was working with are in fact linked, one pair sufficiently tightly that if Mendel had reported on this cross the data would have contaminated his remarkably tight correlations.[25] Morgan and the fly boys showed that genes on different chromosomes behave like ideal Mendelian units: independent, autonomous. Genes within a linkage group segregate together. Morgan, Sturtevant, Bridges, and Muller became a scientific linkage group. The three students completed their doctorates under Morgan. Muller later pursued an independent career; Sturtevant and Bridges spent their entire scientific lives with Morgan.

Occasionally, one or a few genes switched linkage groups. The first instance of such crossing-over occurred in 1913. Morgan recognized that breaks in linkage could be explained by Janssens's chiasmata. He extended and specified the Sutton-Boveri hypothesis, developing a model in which Mendelian genes were arrayed linearly along the chromosomes, like a string of pearls. Each genetic linkage group corresponded to the genes on a given chromosome. Linkages were broken at meiosis, when homologous chromosomes exchanged segments. Genetic crossing-over, then, corresponded to chromosomal crossing-over.

Morgan recognized that crossing-over between any two genes occurred with a constant frequency. Sturtevant then realized that comparing crossover frequencies allowed one to order the genes within a linkage group, and even compute relative distances between genes. Crossover frequency thus became a measure of "genetic distance." Sturtevant used this principle to assemble the first genetic map, plotting the genes of the X-chromosome linkage group along a line, according to their crossover frequencies.[26]

The Morgan lab worked out an extraordinarily productive experimental method and philosophy. Within five years, the Morgan group had more mutants than they could analyze and began discarding them.[27] New genes were identified by mutations—it became an easy shorthand to think of a mutation *as* a gene. A chromosomal site identified by mutation and crossing-over was given the Latin name for place: locus. Soon, locus and gene became roughly synonymous. A locus was a site, sites were identifiable by mutations, and mutations occurred at genes. Mutations were analyzed by assigning them first to a linkage group, then to a position within that linkage group.

The product of the lab was genetic maps: schemata drawn on huge sheets of paper in which strings of genes were placed according to crossover frequency with their neighbors. Publications and books poured out of the lab; students and funding poured in. More Drosophila groups sprung up at other labs around the country. Robert Kohler has shown how both the scientists and the animals were tuned to produce this sort of scientific data. The fly folks soon became adept at producing "trick" chromosomes that yielded genetic secrets. Trisomics: flies with extra chromosomes. Attached X, found by Morgan's wife, Lillian (a superb geneticist): female flies with their two X chromosomes attached at the tips. Inversions: in which a segment of a chromosome breaks off and reattaches upside down, reversing the order of the genes. Translocations: in which a chromosome chunk breaks off and reattaches to the end of another chromosome.

Most of these appeared spontaneously and were at first interesting genetic puzzles. The fly virtuosi, however, soon turned them into powerful research tools. Trisomics showed nonstandard Mendelian ratios and dosage effects, in which a third "dose" of a gene intensified a trait. Crossing-over was suppressed in attached-X flies and flies with inversions.

Translocations revealed the effects of transposing genes from one set of gene neighbors to another. Much of the art of genetics lay in selecting strains with precisely the genetic constitution that would reveal the phenomenon of interest.[28]

The Morgan program united the theoretical and mathematical rigor of Mendelism with the empiricism of cytology. In so doing, it split the problems of heredity from those of development. Morgan explicitly differentiated the two sets of questions. In 1910, he could still write of "the modern literature of development and heredity." But seven years later, in a defense of the Mendelian approach, he argued that critics of Mendelism confused "the problem concerned with sorting out of the hereditary materials (the genes) to the eggs and sperms, with the problems concerning the subsequent action of those genes in the development of the embryo."[29]

Maize

Morgan's famous flies tend to cloud the fact that most early genetics was done with plants. Mendel, de Vries, Correns, Tschermak, and Johannsen were all botanists. By the time Drosophila genetics had taken hold, maize, or Indian corn, was the most important plant model for heredity studies. It offered a completely different set of advantages. Flies are small and breed every week and a half; maize is large and produces but one crop a year, two in the subtropics or a greenhouse. Systematic breeding of flies began in the twentieth century and, at least at first, clear traits were hard to recognize. Maize, by contrast, has been cultivated for centuries and has scores of easily distinguishable characters, many of which have several alternate forms.

Maize's biology is complex; though this proved puzzling at first, maize geneticists learned to exploit it as a powerful tool. An individual maize plant produces both male and female gametes. The pollen is produced in the tassel, which sprouts from the top of the plant in high summer, and the female gametes are at the base of the silks, which emerge from the incipient ear arising from the plant stalk. Fertilization involves two sperm cells and three nuclei in the embryo sac. One sperm fertilizes the egg, which forms an embryo. The other sperm fertilizes two "polar bodies," genetic twins of the egg, forming a triploid ($3n$) nutritive tissue called the endosperm—the fleshy part of the kernel that makes corn good to eat.

This complex biology gives maize two of its most important qualities as a genetic model organism. First, the immediate progeny of a maize cross, the kernels, gives a preview of the genetic constitution of the next year's crop. Second, the various maize tissues provide haploid (pollen spores), diploid (embryo and maize plant), and triploid (endosperm) tissue, which can provide important clues to dominance relationships and the effects of gene dosage.

Mendel, in letters to his friend and colleague Carl Nägeli, described experiments on the inheritance of kernel color in maize. In 1870, he mentioned having demonstrated that hybrids of maize as well as other plants "behave exactly like those of Pisum."[30] After the turn of the century, Mendelians quickly discovered the advantages of maize. Most early maize researchers had an agricultural bent. For example, George Harrison Shull, one of Frank Lutz's cohort from the University of Chicago, was one of the first and best-known experimental evolutionists at the Carnegie Institution's Station for Experimental Evolution at Cold Spring Harbor. Between 1905 and 1908, Shull demonstrated "hybrid vigor," the principle that hybrids were often bigger, healthier, and more productive than their pure-line parents. Shull, the first of a long line of agriculturists and breeders who helped frame American Mendelism, invented the "double-cross" technique, which led to the development of modern agricultural hybrid corn.[31]

Not long after Shull began publishing his work, Edward M. East moved from his native Illinois to breed corn at the Connecticut agricultural experiment station, across Long Island Sound from Cold Spring Harbor, and in 1909, he joined Harvard's Bussey Institute.[32] One obstacle to the reconciliation of Mendelism and Darwinism was the Mendelians' insistence that discrete genes produced either the presence or the absence of a trait. Darwinian skeptics pointed to the many traits in nature that varied over a smooth continuum. In 1910, just months before Morgan's white-eyed conversion to Mendelism, East showed that various combinations of the three or four genes then known to contribute to kernel color could render at least fifteen different shades, from pale yellow to deep red. In short, particulate inheritance can produce continuous variation. East concluded by noting that if this hypothesis were true, it would "add another link to the increasing chain of evidence that the word mutation may properly be applied to any inherited variation, however small."[33] East thus reduced the grand de Vriesian mutations to small-scale heritable changes. The change was not fully accomplished until the 1920s, with the pioneering work of Ronald A. Fisher and others, but the association of mutation and Mendelism was a crucial step as genetics became more than mathematical formalism—as it became the science that describes and predicts heredity.[34]

From early on, variegation was a favorite problem of maize geneticists. Variegated plants, in which regions or entire plants are streaked, striped, or spotted, seemed to represent cases of non-Mendelian inheritance, and so were of special interest to geneticists trying to confirm or refute the newly rediscovered principles. Variegated kernels appear in many strains of maize and give Thanksgiving Indian corn its decorative patterns. If variegation is treated as a trait, alternative to the dominant color or recessive lack of color, its inheritance is peculiar. Hugo de Vries called such strains ever-sporting varieties. Carl Correns, who also studied variegation, interpreted them as the result of "sick" genes.[35]

Rollins A. Emerson took a special interest in this problem. Nebraska-raised and corn-fed, Emerson earned a doctorate with East, after having already spent several years as a professor of horticulture at the University of Nebraska at Lincoln. In 1914, he moved to Cornell University to head the plant breeding department. That year, he published a study that interpreted variegation in corn kernels in strictly Mendelian terms.[36] He described several allelomorphs—not distinct genes—for variegation. The mutating gene he named P, for pericarp, the outer covering of the kernel, which it colored. The dominant allelomorph is "self-color," solid red.

Emerson assumed the variegation was an unstable allelomorph that mutated continually between the dominant and recessive forms of the gene. This enabled him to interpret variegation as though it were a simple, recessive Mendelian trait. He designated this allelomorph V, for variegation. The P gene, he later found, also affects the pigmentation of the cob; both may be variegated, and the allelomorphs were independent.[37] Indeed, there were many variegating alleles, each with different properties. Emerson hypothesized that a single factor for pericarp color had mutated several times to give rise to the various alleles of V. Further, V could mutate to S (self-color), thus becoming stable once again, and different alleles of V mutated at different rates. "In fact," he wrote, "the principal difference between certain of the factors is thought to lie in their relative frequencies of mutation."[38] Variegating alleles, then, differed only in their *rate* of mutation. The greater the rate of mutation, the finer the streaks of pigment in the kernel. Emerson suggested that mutation rate was influenced by other genes, as yet undetermined, that modified the effects of the variegating allele.[39]

As late as 1917, then, Emerson could still distinguish between mutation and Mendelism as though they were at odds. He wrote, "Genetic modifications are not the concern of Mendelism but of mutation. The essential feature of Mendelism is the segregation of unit factors without their contamination," or blending.[40] Mutation still connoted deviation from expected inheritance patterns; Mendelism, the absolute constancy of unit factors. By the 1920s, mutation and Mendelism were seen to be on different logical levels: the one, a heritable change in a gene; the other, an idealized model of inheritance. Mutation was therefore no longer anti-Mendelian.

This early maize work was strictly genetic. Maize chromosomes are much smaller and more numerous than those of Drosophila. Cytogenetic analysis could not then be done in corn. The number of maize chromosomes had not been defined, nor could one chromosome be distinguished from another. During the 1910s and 1920s, many mutants of maize were identified and linkage groups were established, but the linkage groups could not be assigned to specific chromosomes.

Challenging the Gene Concept

Though Emerson rescued "ever-sporting" or variegating traits from the morass of non-Mendelian inheritance, variegation continued to puzzle geneticists. By the end of the First World War, the gene was an accepted if nebulous concept. It was conceived as an atom of heredity, indivisible and independent of other genes in its action. During the 1920s, exceptions and complications began to mount; the theory of the gene was challenged on several fronts. These challenges arose out of studies of variegation, in maize and other organisms.

First came the suggestion that the gene, like the atom before it, might be divisible. In 1920, Ernest Gustav Anderson, a student of Rollins Emerson who received his Ph.D. that year, suggested in a letter to Emerson that "genes are not the ultimate units of heredity but are compound structures composed of still smaller units." In 1924, another Emerson student, William Henry Eyster, developed the idea and spruced up the name: the genomere (parts-of-genes) hypothesis. In Eyster's formulation, genes comprised clusters of identical units, each contributing to the gene's expression. For example, the maize pericarp occurred in a graded series of alleles: from cherry through several shades of orange, then yellow, and on to white. Eyster suggested that each darker shade resulted from the addition of one or more subunits of the red gene. The effects were quantitative, additive.

Subgenes, like August Weismann's ids and biophors, are an example of a good wrong idea. By making genes quantitative, the subgene hypothesis suggested many new experiments. It seemed possible to understand the structure of a gene, to estimate its size, to explain aspects of its dynamic action during development. The subgene hypothesis also suggested an explanation of unstable, variegating genes.[41]

Milislav Demerec became one of the biggest advocates of the genomere hypothesis. Demerec emigrated from Yugoslavia to the United States in 1919 and took his Ph.D. at Cornell under Emerson in 1923, the year McClintock received her bachelor's degree. He knew Eyster and Anderson and discussed and coauthored articles with them on the subgene hypothesis. Subgenes seemed to offer an explanation of the mottled pigmentation patterns in variegated corn kernels. The depth of color of the spot could be explained by exchanges of gene subunits during crossing-over. Spots of different sizes could be the result of differences in the timing of gene action, with large spots originating early in the development of the ear and small ones originating later.[42]

Emerson's *P* gene system was good for exploring unstable genes, but when Demerec finished his degree he decided to look for unstable genes in other species, perhaps with faster generation times or presenting other advantages. In 1923, he took a job as a scientist at the Biological Laboratory at Cold Spring Harbor, sister institution to the Carnegie Institution's Department of Genetics (as the Station for Experimental Evolution had been renamed). There, he began working with larkspur (*Delphinium* spp.), where he indeed found unstable genes—three of them. On the suggestion of Charles W. Metz, a Carnegie colleague, Demerec began working with a Drosophila species called virilis. Between 1924 and 1926, he found three unstable genes in virilis: *reddish,* a mutable allelomorph of the *yellow* locus; an unstable *miniature wing* mutation; and an unstable form of the *magenta* eye color gene.

Complexities of the Delphinium and Drosophila genes torpedoed the subgene hypothesis. First, the Delphinium *lavender* gene showed differences in mutation rate at different times in development. Demerec then found one allele (allelomorph had by now been truncated) of the Drosophila *miniature* gene that was stable in the early embryo but unstable later in development; another was unstable both in germ cells and in somatic tissues (wings, legs, guts, and the like); yet another was unstable only in somatic tissues. Subgenes could not account for such behavior, unless one added numerous ad hoc assumptions. Instead, Demerec opted for but a single ad hoc assumption. If mutations were or were like chemical reactions involving gene molecules, he reasoned, instability could be modeled as a dynamic equilibrium, going first one way and then the other about a stable center.[43]

While subgenes challenged the unity of the gene, position effects emerged to challenge its autonomy. The first position effect was described in the Drosophila *bar* gene. From its initial detection in 1913, *bar* was a peculiar gene. *Bar* flies' eyes are reduced from the normal round form to a narrow vertical band. The *bar* gene showed a high frequency of reversion, or back-mutation from the recessive to the dominant, wild-type form. Numerous alleles of *bar* were soon discovered, revealing a spectrum of narrowing, from *ultrabar* to *infrabar*. In 1923, Alfred Sturtevant suggested that the various *bar* alleles arose from unequal crossing-over—that is, crossing-over between nearby but not identical sites on homologous chromosomes—which led to duplications of all or parts of the *bar* alleles. Two years later, Sturtevant showed that two *bar* alleles next to each other on the same chromosome gave a stronger effect than when they lay on different chromosomes. Thus, the second allele's position relative to the first had an effect on *bar* expression.[44]

In 1936, Jack Schultz, another geneticist from the Morgan group, modified the position-effect concept to describe variegation or mottling. He did so by uniting position effects with a mysterious substance called heterochromatin, a thick, dark-staining chromosome material first described by Emil Heitz in 1928 and 1929. Whereas most regions of the chromosomes spread out into thin strings, becoming condensed only prior to and during cell division, these regions never decondensed. They also seemed devoid of genes. Gene chromatin became known as "true" or euchromatin, in contrast to the dark and geneless heterochromatin. Heterochromatin composed the spindle-fiber attachment region, the ends of chromosomes, and occasional tracts within the chromosomes. A large section of the Drosophila X chromosome consists of heterochromatin.

Crossing-over, inversion, or translocation at the border between heterochromatin and euchromatin could bring a gene normally in the middle of a stretch of euchromatin into contact with heterochromatin. Schultz found that in such cases, a normally active gene fell silent. He proposed that intermittent contact with heterochromatin caused a type of variegation. He called this position-effect variegation. This mechanism did not explain *bar* eyes or many other kinds of mutations, but it was suggested that position effects could simulate true gene mutations—defined as chemical changes that inactivated or otherwise altered the gene itself. Position effects, then, provided some of the first evidence that chromatin interactions—where genes lay on the chromosomes, which genes and nongenic materials were their neighbors—affected gene action.[45]

The concept of the stable, autonomous gene found new support when geneticists discovered how to damage genes with ionizing radiation. Morgan had tried unsuccessfully to produce mutations with radium as early as 1914. In 1927, Hermann J. Muller, who had moved to the University of Texas, successfully produced mutations in flies using X rays. Muller turned up the contrast for detecting mutations resulting from chromosome breakage by using strains of flies that carried chromosomal inversions. The inversions suppressed crossing-over and therefore maximized the chance of recovering a broken chromosome. Muller found a large number of dramatic mutations, especially translocations. The next year, Lewis J. Stadler, a plant geneticist at the University of Missouri, reported producing mutations in barley with X rays and radium.[46]

X-ray studies greatly facilitated the search for new genes, challenged aspects of Darwinian and Mendelian theory, bolstered the string-of-pearls gene theory, and raised philosophical questions about humans' "load of mutations." They also concentrated attention on the chemical nature of the gene. Muller became the chief spokesman for the view that mutations were chemical changes in gene molecules. X-ray studies, for example, convinced Demerec that unstable genes were the result of chemical reactions rather than changes in genomeres, and they helped put position effects on the back burner as an unsolved problem, probably of limited importance.

The Premodern Developmental Synthesis

Such was the status of genetics three decades after the rediscovery of Mendel. The string-of-pearls model of the gene remained a powerful tool for generating new experiments, though it had been challenged by messy problems such as variegation, position effects, and heterochromatin. Drosophila was the queen of genetic model organisms, but maize was a close second, hampered only by the apparent impossibility of distinguishing among its chromosomes.

No serious biologist in the first third of the twentieth century could have doubted that the problems of growth and form—evolution, heredity, and development—must eventually be resynthesized. Yet the questions of when and how to synthesize them created the dynamic tension that propelled much of biology during the first two thirds of the twentieth century. Mendelism united with cytology became cytogenetics. By the 1920s, this was clearly the fastest growing, most rapidly progressing of the three fields. In order for their discipline to continue to thrive, evolutionists and embryologists had to explain their biological problems in terms of heredity. Uniting those fields with genetics thus became a matter not only of intellectual but also of professional interest.

With its unit characters and its emphasis on discontinuous variation, early Mendelism was a direct challenge to the slow evolutionary change required by Darwin's natural selection.[47] A true synthesis of Darwinism and Mendelism did not occur until the 1930s. The breakthrough was the development of a science of changes in gene frequencies in populations, as opposed to gene changes in individuals. The modern evolutionary synthesis was therefore unconcerned with the question of gene action, and was consequently unruffled by the challenges to the gene concept.

An overlapping effort to synthesize the problems of development and heredity confronted directly the problem of gene action. Because embryologists, unlike evolutionists, were concerned with changes in individuals, they were forced to contend with the chromosome theory of heredity, with its independent, autonomous strings of genes. They recognized that current genetics could not account for the elegant and mysterious ways in which the fertilized egg becomes a plant or an animal. If all cells in an organism have the same genes, what tells one group of cells to become a liver, another an eye, another a toe? Such a question requires a theory of the action of genes, which the chromosome theory did not provide. The embryologists therefore looked outside the nucleus, to the cytoplasm.

The center of this research program lay in Germany, where it had its origin among scientists exploring white and green variegation in plants. In 1909, at the University of Berlin, Erwin Baur showed that the segregation patterns of white-green variegation in chloroplasts, the organelles that contain chlorophyll, could not be explained by Mendelian mechanisms. He postulated a separate mechanism of heredity operating through the chloroplasts themselves. Carl Correns, one of Mendel's rediscoverers, disagreed and proposed that the cytoplasm itself impressed upon the chloroplasts their final character. In the 1920s, German botanists suggested that the cytoplasm contained its own genes, independent of the nucleus. From this work grew a diverse but coherent school of cytoplasmic inheritance studies.

One of this school's brightest and most controversial students was Richard Goldschmidt, head of the Kaiser Wilhelm Institute for Biology at Dahlem, in western Berlin. Goldschmidt studied intersexes (hermaphrodites) in the moth Lymantria and developed a brilliant, synthetic, and incorrect "balance theory" of sex determination that involved nuclear male-determinants and cytoplasmic female-determinants.[48] In the United States, few geneticists were concerned with cytoplasmic inheritance until after the Second World War. Tracy Sonneborn became the leading American advocate of the influence of the cytoplasm on heredity. He worked with a model organism unfamiliar to most geneticists: the unicellular pond animal Paramecium. Beginning in the 1930s and continuing for decades, Sonneborn produced a string of surprising findings that suggested a relationship among nucleus, cytoplasm, and environment more complex than anything known in multicellular plants and animals.[49]

In short, by the time Barbara McClintock emerged as a scientist in 1927, the effects of the Mendelian wedge were being played out. The string-of-pearls model of the gene was still intact but seriously questioned. The isolation of genetics from the rest of biology, the separation of heredity from development and evolution, was beginning to erode, and it was not yet clear what would become of the theory of the gene, or which integrations would succeed and which would fail.

Wizard of the Microscope

Every course involving microscopy seems to have one student who is a natural. She always wins the cellular treasure hunts; she finds the chromosomes and chloroplasts and mitochondria before anyone else. Her specimens are always better prepared, better lit, better focussed than one's own. Through her eyepiece, structures appear as in the textbook, while one's own reveals only eyelashes, bubbles, and dust. It is around her microscope that the instructor herds the class, squinting in turn through her ocular to see at last what one is supposed to see. Barbara McClintock would have outshone a laboratory full of such naturals. "If at any time she came to look at a slide on your microscope and did not immediately make some adjustment, you felt triumphant relief," recalled Harriet Creighton, who arrived at Cornell in 1929.[50]

McClintock received her bachelor's degree in 1923, then stayed on for a master's degree and a doctorate. From the start, her primary interest and greatest talent was looking at chromosomes. Amid professional colleagues of vast experience and reputation, McClintock was so good that she often found herself not only teaching her peers and elders but solving their problems for them. This did not always endear her to them.

"When I was first a graduate student, second year, or maybe second term of the first year, I was made the assistant to a man who (this was for money purposes) was working on maize chromosomes," McClintock recalled in a 1978 interview.[51] The man was Lowell "Fitz" Randolph, a former student of Sharp's who had recently been promoted to the faculty. Randolph, no dullard, "was more methodical and less gifted" than his "quick, imaginative, and perceptive" young female assistant, said Marcus Rhoades, a graduate student there.[52] Randolph recognized the problem of the maize chromosomes. In 1926, while McClintock was still a graduate student, she and Randolph published a short note on a triploid maize plant, a plant with three copies of every chromosome rather than the usual two. The cytology and genetics of this plant, the first triploid known in maize, became McClintock's Ph.D. thesis, which was published in 1929 as a long article in *Genetics*, the primary journal of her field.[53]

Downstate at Cold Spring Harbor, Albert Blakeslee had for years been crossing jimson weed *(Datura)* triploids to diploids to obtain trisomics, individuals with a single extra chromosome $(2n + 1)$. Trisomics had interesting genetic properties and, because they disrupt normal Mendelian ratios, were powerful tools in locating known genes on chromosomes. The trick does not work for all chromosomes—some trisomies are lethal. In humans, most trisomies cause the death of the embryo; one that survives is trisomy of chromosome 21, which results in Down's syndrome. McClintock set about searching for viable trisomics in maize and associating them with genes in known linkage groups.

The project raised the question of visually distinguishing the maize chromosomes. Randolph had been trying for years to do this. As was standard practice, he used chromosomes in metaphase, a stage in which the chromosomes are condensed, appearing as little sausages. He sliced tissue very fine, stained and fixed it on slides, and pored over the sections in the microscope. Yet he could not discern distinguishing features in them. In bounced McClintock. "Well, I discovered a way in which he could do it, and I had it done within two days or three days—the whole thing done, clear, sharp and nice."[54]

She pulled it off with two innovations. The first was to look at cells in an earlier stage, pachytene, rather than the metaphase chromosomes used conventionally. Pachytene chromosomes are much longer and thinner than the dense sausages of metaphase chromosomes. Moving to the long threads of pachytene revealed rich chromosomal landscapes, as slowing down a fast piece of music reveals nuances of harmony and texture.

Immediately, differences in length among the chromosomes became apparent. As she trained her eye, the spindle-fiber attachment region (also known as the centromere) became discernible as a "constriction" in the chromosome. In some chromosomes, it was situated in the middle; in others, toward one end. The relative length of the arms was thus another useful marker. Some chromosomes had dark blobs she called knobs, often at one end but sometimes in the middle of an arm. The fifth largest had a "satellite," a piece of chromatin connected to the main chromosome by a tiny thread. Here—in the trisomics, the centromeres, the satellite, the knobs—were the materials of much of her scientific career. In 1929, she published the haploid chromosome number for maize (ten) as well as the first idiogram, or schematic diagram of all the chromosomes, of maize as a short note in *Science*—a journal of unusually high visibility for such a young scientist.[55]

Her second innovation was technical. Randolph, she said, had been "just following the path that had been laid down by the early people." As a graduate student, assisting in the cytology course taught by botany professor Lester W. Sharp, she studied the history of microscopes and tissue staining, allowing learning to sharpen and give depth to her skills. "My background in the staining part sure gave me freedom that Randolph wouldn't necessarily have had."[56] She adapted to maize a technique recently developed by Blakeslee's student John Belling, the "squash" technique, in which cells are spread on a slide, stained, and flattened on a coverslip with the thumb. This procedure allows the cytologist to view whole chromosomes, thereby eliminating the ambiguities and distortions produced by making serial sections through the cell. McClintock's modification of Belling's technique was one of many technical innovations she made that have become mainstays of cytogenetics.[57]

"That was the beginning and end of a friendship," she recalled in 1978. Randolph "was furious at me," she said. "I never thought I was taking anything away from him," she said. "It was just exciting that here we could do it, here we could tell one chromosome from another without any difficulty, and so easily."[58] If she had at this early stage anything like the intimidating personal style she later showed, she may not have been as innocent of offense as she made out. Nevertheless, McClintock left Randolph's group and completed her doctorate under Sharp. "Sharp said I would do better to work for myself than for him, so he gave me my freedom except that I helped in his cytogenetics course."[59]

After being awarded her Doctor of Philosophy in 1927, McClintock stayed on at Cornell as an instructor. She soon assembled a small maize cytogenetics group. McClintock and two graduate students, George Beadle and Marcus Rhoades, formed the core of the group. "We were considered very arrogant," she recalled. "We were way ahead of all of these other people, and they couldn't understand what we were doing. But we knew, and we were really a very united, integrated group."[60]

George Wells Beadle was a farmer's son from the sober Nebraska town of Wahoo. He received his bachelor's degree from the University of Nebraska in 1926 and stayed on another year to do a master's, on hybrids of wheat. He came to Cornell to work with Emerson in 1927. He had a field hand's tan, bristly hair, and fine, strong features. McClintock recalled that he stopped by her "cubbyhole" in the botany department. Beadle wanted to examine male sterility, a research problem stemming back to former Emerson student W. H. Eyster. He found that male sterility—inviable pollen—was inherited as a single gene. But what did that gene do? "He needed the cytology," McClintock recalled. "I think he came to me and asked me if I would look at these things. I said, 'You can do it because it's very simple.'" She showed him how to prepare and identify the chromosomes. "He was very good, very fast. So then he was off on his own." Beadle, however, recalled being not quite so on his own. "It was difficult to dissuade her from interpreting all my cytological preparations. Of course she could do this much more effectively than I."[61]

The cytology revealed that the pollen was sterile because the chromosomes often failed to pair up in meiosis, resulting in plants with too many or too few chromosomes. Most of them were not viable. In the fall of 1928, Beadle and McClintock sent off a note to *Science* describing their findings.[62] McClintock's friendship with Beadle continued for the rest of Beadle's life, though after he left Cornell he abandoned maize to work on Drosophila and then the bread mold Neurospora. He went on to head the biology division at the California Institute of Technology (Caltech) and win a Nobel Prize for his and Edward Tatum's one gene–one enzyme hypothesis of gene regulation.

That fall, Marcus Morton Rhoades arrived. Born in 1903, the same year as Beadle, Rhoades grew up in Graham, Missouri. Lanky, with a large nose and deep-set eyes, Rhoades had a gentle disposition and a diplomatic nature. He attended the University of Michigan, receiving a bachelor's degree in 1927 and a master's the following year, both in botany. While there, he met Ernest Gustav Anderson, who pushed him toward Emerson and Cornell for his doctorate. After a year at Cornell, he took a teaching assistantship at Caltech in 1929–30 in order to learn fly genetics in Morgan's laboratory.[63]

In McClintock's telling, Rhoades entered her life the same way Beadle had: he popped into her cubbyhole, "asked me what I was doing, and I told him. Well he was very excited and he said, 'Can I join you?'"[64] Rhoades went on to become one of the country's leading maize geneticists. He served as vice president and then president of the Genetics Society of America and as editor of the journal *Genetics*. He was elected to the National Academy of Sciences in 1946. He and McClintock remained close friends and colleagues for the rest of his life. He was her confidante and advocate and, during the late 1940s and early 1950s, her sounding board for her developing ideas on gene regulation.[65]

After graduate school, McClintock "wanted to do something very special," she said. "I had discovered which chromosome was which and could tell one from another. What I wanted to do was to link up the linkage group with the particular chromosome."[66] Drosophilists had done this years before. The lack of cytological underpinning was clearly hampering maize genetics. Randolph had tried for years to distinguish the maize chromosomes from one another, which is only worth doing in order to associate them with linkage groups. And by this time other maize geneticists, including Royal Alexander Brink at the University of Wisconsin, were also trying to associate linkage groups with specific chromosomes.

Yet in interviews, McClintock insisted that no one could see why she thought the problem important. "It was so new at that time that I was ostracized."[67] The only person who did understand, she recalled, was Rhoades. In her telling, Rhoades went to Caltech *before* coming to Cornell; this provided her with an explanation for why he alone saw the importance of what she was doing. Rhoades "had been in a different environment earlier to get his master's degree, where this was known." He had "been involved with the Drosophila people."[68] Everyone else, she said, "thought I was crazy for doing what I was doing at the time I was doing it, and it was only Marcus who made it possible for me to be accepted."[69] It is impossible that no one saw the importance of her work. The rejection was part of her private myth.

Trisomics were the key to her solution. In a viable trisomic, the classic Mendelian 3:1 ratio of a hybrid cross is distorted by the addition of a third copy of the gene. McClintock crossed trisomics to normal diploid plants and pored over her crosses, studying the ratios of known traits till she found one that deviated from the diploid ratio in the expected way. The linkage group to which the deviant gene belonged could then confidently be assigned to the tripled chromosome. McClintock assigned six of the ten linkage groups in this way. The remaining four were assigned—by McClintock, Brink, and Charles Burnham—by 1931.[70]

*   *   *

Within two years of receiving her Ph.D., McClintock had six publications, most in major journals, four of them single-author. She had identified the correct haploid chromosome number for maize, developed new techniques, distinguished the chromosomes from one another, and assigned most of the linkage groups to a chromosome. Her discovery of the pachytene stage for cytology gave maize geneticists microscopic superiority over the drosophilists; the pachytenes revealed details fly geneticists could only dream about. Her work ushered in the golden age of maize genetics, which lasted until about 1935. In 1933, Theophilus Painter, at the University of Texas, showed how the giant chromosomes found in salivary gland cells of Drosophila and other flies revealed new detail in banding patterns. (In these cells, the chromosomes reduplicate many times, swelling to one hundred or more times the size of other chromosomes.) This discovery sparked new mapping efforts in flies, and within five years Drosophila was once again the premier genetic model system.[71]

In 1931, the year after Rhoades returned from Morgan's Caltech laboratory, Beadle finished his doctorate and went to Morgan's lab himself, on a National Research Council (NRC) fellowship. The year before, however, two new members had joined the group: Charles Burnham, who had done his Ph.D. with Brink at Wisconsin, came on an NRC fellowship in June 1929; and Harriet Creighton, who arrived as a new graduate student that fall. Rhoades completed his dissertation, *The Cytoplasmic Inheritance of Male Sterility,* in 1932. But in these early years of the depression, jobs were scarce. Rhoades stayed on at Cornell for three more years as a researcher in the plant breeding department. None of the stellar trio of McClintock, Beadle, and Rhoades obtained permanent jobs right out of their doctoral program. The National Research Council fellowships were a godsend to bright young scientists in the 1930s.

The Cornell cytogenetics group was an ideal place for Burnham. Beadle and Rhoades shared his interest in pollen sterility. But it was McClintock who helped him solve his problem. Once again, it was by bringing her dazzling cytological skills to bear on another's genetic analysis. Three mutants of *semisterile* showed peculiar linkage relationships. What was more, in some of the plants, they could see a ring of four chromosomes. Burnham had suggested that a translocation could account for the rings and the ratios. McClintock demonstrated that indeed there was a reciprocal translocation and published a paper on it the same year.[72] But it only partly explained the problem. Whereas some pollen should have been normal and some should have been 50 percent sterile, there was a class of pollen with intermediate sterility, 25 to 30 percent.

McClintock recalled in 1978 that Burnham "told me this in the field, and he was disturbed." In a 1992 memoir, Burnham bristled at this. "Plants with intermediate sterility did not 'disturb' me. They were something to be explained." But to McClintock, this was exciting disturbance. "This challenge was the kind of challenge that was disturbing, the fact that it would be answered and you recognize that it can be answered, that it must be answered, that it's part of a system or it wouldn't be there. Therefore, I could answer it." She was disturbed too. "I was so disturbed that I left the field, which was down in a hollow, and I walked up to my laboratory and I sat in my laboratory for about thirty minutes. Just sat there thinking about it, and I suddenly jumped up and ran down to the field. I was at the top of the field and everybody was down at the bottom, and I was saying, 'Eureka, eureka, eureka, I have it!'" It was a trisomic, an extra chromosome. She joined the others down in the hollow, and Rhoades turned to her and said, "Prove it." So she sat down with a brown paper bag, which they used to cover tassels to prevent inadvertent fertilizations, and began to sketch out her solution." She was right.[73]

With Harriet Creighton, McClintock put the capstone on the establishment of maize as a cytogenetic model system, by at last demonstrating conclusively the equivalence of genetic and cytological crossing-over, assumed since 1910 but never demonstrated. Their proof consisted of taking two known genetic markers and two known cytological markers and showing that when crossing-over occurred, they traveled together. The paper shows this is referred to simply as Creighton and McClintock 1931.

They used chromosome 9. At the tip of the short arm was the knob she described in 1929. On the long arm was the reciprocal translocation (or "segmental interchange") described in her 1930 paper on Burnham's *semisterile*. She produced plants in which chromosome 9 carried both the knob and the translocation, giving her a chromosome visibly marked at each end. In between lay two known genetic markers, red kernel color *(C)* and waxy kernel starch (*Wx* = wild type, starchy; *wx* = waxy). Starchy pollen stains blue with iodine potassium iodide; waxy pollen stains brick-red. In a rarely cited paper in the *Proceedings of the National Academy of Sciences*, McClintock established the order and crossover frequencies of the genes *C* and *Wx* and of a third, shrunken kernels *(Sh)*. The next article in the issue is Creighton and McClintock 1931.[74]

The crux of the experiment is the setting up of a clever cross. McClintock and Creighton made plants that carried knob-*C*-*wx*-interchange on one chromosome and knobless-*c*-*Wx*-no interchange on the other. They crossed them to plants with no knobs, no interchange, recessive for the *C* gene, and heterozygous for *waxy*. The analysis of the cross was simple. If crossing-over occurred between *C* and *wx*, the knob should go with the *C* locus and the interchange with *wx*.

They showed that plants grown from red kernels that had a knob but no interchange produced blue-staining pollen. Thus, when the *waxy* gene was lost by crossing-over, so was the interchange. The experiment yielded other crossover possibilities, which added to the evidence, but the analysis was the same.[75]

The experiment was elegant and clear-cut, but the data were scant. The crop in the summer of 1930 had been poor, and so McClintock and Creighton had many fewer plants than they'd planned. Creighton recalls Thomas Hunt Morgan, visiting Cornell in the spring of 1931, urging them to publish right away. The German drosophilist Curt Stern had been doing similar experiments and if the maize geneticists did not publish now they would be scooped. The two young scientists protested that they had insufficient data; next year they would have more. But in the end, "the Boss" convinced them to publish. Harriet Creighton recalled Morgan saying, "I thought it was about time that corn got a chance to beat Drosophila!" The paper appeared in August. Stern published soon after.[76]

Already by 1931, McClintock was constructing an intricate scientific edifice, with each new publication building on several previous ones. This trend contributed to her growing reputation for producing brilliant but difficult articles. James A. Peters, introducing the Creighton and McClintock paper in his 1959 compilation, *Classic Papers in Genetics,* calls it a cornerstone of the science, but "not an easy paper to follow, for the items that require retention throughout the analysis are many, and it is fatal to one's understanding to lose track of any of them. Mastery of this paper, however, can give one the strong feeling of being able to master almost anything else he might have to wrestle with in biology."[77] This quality increases with almost every paper McClintock published over the next fifty years.

Rings and Broken Ends

In 1931, McClintock herself won a National Research Council fellowship. Although within her small Cornell cohort receiving one of the coveted fellowships was standard, among women scientists it was a signal honor; in botany in the 1920s and 1930s, men were three times more likely than women to win one.[78] McClintock spent hers traveling among Cornell,

Caltech, and the University of Missouri at Columbia. Her two-year tenure as an NRC fellow was one of the most productive periods of her career.

Lewis J. Stadler was the reason she went to Columbia, Missouri. Stadler received his Ph.D. in 1922 from Missouri, nominally under William Eyster; his real training, however, had come from Edward East at Harvard. In Stadler's case, Drosophila had beaten maize. In 1927, Hermann J. Muller had scooped him on the demonstration of X ray–induced mutations. Muller went on to win a Nobel Prize in Physiology or Medicine in 1946 for this work. In the late 1920s, Stadler began traveling regularly to Ithaca to visit and work with the Cornell maize cytogeneticists. McClintock befriended him and was fascinated by the experimental prospects of X rays. She spent the summer of 1931 in Columbia, looking at various mutations in plants Stadler had grown from irradiated pollen. In December she published the results as a bulletin of the University of Missouri's Agricultural Experiment Station.[79]

Her deep understanding of chromosome behavior led her to astute observations and predictions. In one case, the chromosome broke twice, on either side of the centromere, forming one small fragment with a centromere and two large fragments lacking a centromere. The broken ends of the small fragment fused, forming a ring. Because the ring possessed a centromere, it lined up with the other chromosomes at mitosis, synapsed with its homologue, and persisted through future cell divisions. The two long segments, lacking a centromere and therefore unable to attach to the spindle, were lost at cell division.

Another result of X-raying was an inversion, in which a chromosome broke and the fragment reattached upside-down. McClintock predicted that a translocation within this inversion could produce one chromosome with no centromere and another with two centromeres. Such a case, she wrote, "would produce a chromosome with two insertion regions [centromeres] . . . It is probable that in one of the early mitoses the two insertion regions on the single chromosome would pull toward opposite poles, causing a break to occur in the chromatin thread between the two insertion regions."[80] A few years later, she indeed found such a case. It led to one of her most important discoveries: the breakage-fusion-bridge cycle.

In late August 1932, the Sixth International Congress of Genetics convened at Cornell. It was a landmark gathering, especially for maize geneticists. Demerec and Rhoades organized a series of exhibits, including a living chromosome map of maize, consisting of ten rows of plants, one for each chromosome, with plants demonstrating a mutation of each known gene planted in sequence. In a session chaired by Ralph Cleland, a geneticist at The Johns Hopkins University, McClintock gave a talk on maize cytology that featured her new results on the association of nonhomologous chromosomes, and she served as vice chairman of a session on cytology. In the published proceedings, her work was cited by Emerson, Muller, Stadler, Stern, and others. Karl Sax, surveying recent work on the cytological mechanism of crossing-over, noted that the equivalence of cytological and genetic crossing-over, long assumed, "has been confirmed by the brilliant investigations of Creighton and McClintock (1931) with Zea, and of Stern (1931) with Drosophila."[81] McClintock, just five years out of graduate school, received attention and respect usually reserved for much older, more distinguished scientists.

McClintock had gone from Columbia, Missouri, to Ithaca, where she spent the summer with some of Stadler's plants. She enjoyed telling the story of why. The previous winter, she had received a paper from Susumu Ohno at Berkeley that reported unusual variegations in their plants. "They'd looked at the chromosomes and found a little chromosome that was apparently being lost." She was filing the paper—in one telling she was standing on a stool—when "it suddenly occurred to me that this must be a ring chromosome." The insight was topological, a flash of understanding of the kinds of knots chromosomes could form by crossing-over in different configurations and how those knots would sort out at cell division. She recognized that the variegation must have arisen from sister-strand exchanges, reciprocal translocations between equivalent arms of a chromosome. Rings were known to occur occasionally when chromosomes broke; the ends would find each other and join. But rings had never been implicated in the formation of variegated patterns.[82]

Here is how it worked. A ring chromosome had formed early in the development of the plant. The ring carried the dominant form of a pigment gene, while its homologue, a normal chromosome, carried the recessive form. As the plant developed, the ring was propagated normally in most cells. But occasionally, exchange occurred between sister strands, or chromatids, producing a small, acentric ring and a dicentric ring. If the dominant allele ended up on the acentric ring, it would be lost; the resulting cells would be recessive. These occasional losses would result in a variegated plant.[83]

McClintock sent her idea to Ohno. She recalled that he replied that it was "a crazy idea, but it's the only one we've had." She went back to Missouri the next summer to look for rings herself. Charles Burnham was in

Columbia as well. He and Stadler met her at the train. "We went out to dinner and I talked about ring chromosomes and they began to kid me about ring chromosomes." When she started walking the fields, she was already calling the variegated plants ring chromosome plants. "I went through the field, before examining the chromosomes, and made my guess for every plant as to what kind of rings it would have." She began to worry that she had been overconfident, but her fears proved groundless. "When the material was ready I examined the first plant and saw a ring chromosome and called Stadler on the phone and told him, 'Yes, it's rings.'" Visiting Caltech the following winter, 1932–33, she made a trip to Berkeley to visit Ohno. "He brought me to his lab and there was a microscope set up, and he said, 'Look down,' and I looked down and there was the little ring."[84]

Getting Organized

In the fall of 1933, still on her National Research Council fellowship, McClintock returned to Caltech. Thomas and Lillian Morgan were there of course, and George Beadle, and maize geneticist E. G. Anderson. Looking through Anderson's X-rayed material, she found an unusual translocation involving chromosome 6—the chromosome with the satellite region connected by a threadlike stalk. Emil Heitz had recently argued that such stalks were the site of formation of the nucleolus, a poorly understood, deep-staining body that disappears at cell division. Some argued that the nucleolus had no genetic significance, yet others pointed out that it was associated with chromosomes and in some species seemed to communicate with the cytoplasm, perhaps by sending or receiving important materials.[85]

For McClintock, the nucleolus was part of the landscape of the chromosomes. She had observed that it formed not at the stalk but at a special, deep-staining region on chromosome 6 near the base of the stalk. In Anderson's material, she found a case of a reciprocal translocation between chromosomes 6 and 9 that divided this region. The result was that two chromosomes, one mostly 6 plus a chunk of 9, the other mostly 9 with a chunk of 6, had a portion of the nucleolus-associated region. In the resulting cells, two nucleoli formed, one associated with each of the hybrid chromosomes, right at the site of the translocation. Cytologically, McClintock could discern that the break had not split the region evenly in half; one chromosome had a larger portion of the region. The two resulting nucleoli were unequal in size. The larger nucleolus formed out of the smaller chromosomal region. Yet when the larger region was present alone in a nucleus, it formed a nucleolus of normal size.

According to the leading theory of the day, chromosomes were embedded in a gooey matrix. When the chromosomes condensed prior to division, this matrix was said to fill in the gaps that must be present when long stringy chromosomes came to appear like solid sausages. After division, when the chromosomes decondensed, the matrix spread through the nucleus. The nucleolus, it was believed, formed from this matrix, as did the nuclear membrane surrounding the chromosomes. Something pulled the matrix together and wadded it up to form the nucleolus. McClintock concluded that this region of chromosome 6 was what accomplished this. She called the region the nucleolar organizer, she said, "because there were materials in the nucleolus that had to be taken from where they were and put through this body to get into the nucleolus." In the published paper on the nucleolar organizer, she wrote that without the organizer, "droplets" of nucleolus substance would simply accumulate along the chromosome without coalescing into a functional nucleolus.[86]

In 1933, the remarkably productive, free-ranging years of her NRC fellowship came to a close. At the suggestion of Ralph Cleland, she applied for a Guggenheim fellowship to study abroad. The three giants of her various institutions, Emerson, Stadler, and Morgan, as well as Sharp, Cleland, and Madison Bentley, head of the Cornell psychology department, all wrote letters of support for her. Out of 871 applicants, she won one of 38 fellowships.[87] She intended to spend a year in Curt Stern's laboratory—clearly Stern did not harbor resentment at being scooped by McClintock on the crossing-over story—at the Kaiser Wilhelm Institute for Biology in Dahlem.

She could not have picked a worse time to go to Germany. Hitler had just become chancellor. Jews in academia were being removed from distinguished positions. Stern never returned to Germany: from Caltech, he got a job at Berkeley, where he remained for the rest of his career. Stateside, it still was not obvious how bad the situation was. The Guggenheim Foundation urged McClintock to keep her plans and go to Berlin. One reason was Richard Goldschmidt, the head of the Kaiser Wilhelm Institute for Biology. Goldschmidt was Jewish but had not yet been purged from the academy. Though nationwide the situation was deteriorating rapidly, in the fall of 1933 "general conditions for the non-Aryans were still rather good," Goldschmidt wrote in his memoir. "It was not pleasant to return from a concert at night and to ride in the suburban train with a gang of storm troopers who were singing the beautiful song, 'If the Jew's blood drips from our knives, we feel happy, oh so happy.' But these things still were mere pinpricks compared with what was to come later after we had left."[88]

McClintock liked Goldschmidt. "When I came we immediately became intimate friends because he had nobody to turn to but me," McClintock recalled. "We stayed that way until he died. We understood one another, we liked one another." In some ways, they could hardly have been more different. McClintock tended to detach herself from "the body." Goldschmidt, however, "had an enormous ego," McClintock said. "It was one where he had to show off a little bit . . . He had to expose his skills. I found that very entertaining. Those things never bothered me."[89]

She liked Goldschmidt's freedom from scientific dogma. He argued against Darwinian gradualism, taking a line similar to de Vries's, that species were formed suddenly, by large "macromutations." He argued against the gene: "There are no genes, no gene mutations and no wild type allelomorphs," he wrote in 1937.[90] He enjoyed proposing farfetched theories; these were usually based on scant evidence and have proved to be wrong literally, but some have had a kind of poetic truth to them.[91] "I approved of his way of thinking," McClintock said. "I approved of his freedom."[92]

Yet Berlin was terrible. In interviews, McClintock would not delve into this period; all she would say about Berlin was, "Nobody smiled." But she was so miserable there that after a few weeks, Goldschmidt suggested she leave. (Goldschmidt himself left Germany in 1936.) His colleague Friedrich Oehlkers was in Freiburg; she could stay there, "where there was not so much Hitlerite business," as she told the Guggenheim Foundation. Oehlkers was firmly part of the German school of developmental genetics. He favored theories of cytoplasmic inheritance, which in the United States had little support. His own work was on what he called the physiology of meiosis, which consisted of studies of environmental influences on the formation of germ cells.[93]

The lion of Freiburg was Hans Spemann. Spemann came to Freiburg after the First World War and remained until his retirement in 1938. Between 1901 and 1906, he had developed the concept of embryological induction, by which one tissue induces, or causes, the formation of a new kind of tissue by contact with it. In the 1920s, Spemann developed his concept of the organizer, a special tissue that caused induction of another tissue, in this case the neural plate, a primordial tissue of the nervous system. To Spemann, the organizer did more than simply secrete chemicals that changed the overlying tissue; it created a "field" that influenced neighboring tissue. Spemann's student Johannes Holtfreter said Spemann insisted on a "metaphysical concept of the organizer as being a vitalistic agency, a sort of planning and disposing manager who instructs and organizes the adjacent, as yet undetermined, tissue." Despite the tinge of vitalism, his work greatly influenced other embryologists and was a major force in the development of embryology away from genetics. He won a Nobel Prize in 1935.[94]

While she was at Freiburg, McClintock wrote up her paper on the nucleolar organizer. Its inconsistent use of terms and poor organization testify to her stress and difficulties while there. Years later, she complained that scientists still did not grasp its message. Today, the nucleolus is understood as the site of synthesis of ribosomes, organelles made of ribonucleic acid (RNA) that are shipped out to the cytoplasm, where they serve as sites where proteins are synthesized. The nucleolus organizer is a stretch of ribosomal genes. But as late as 1979, McClintock thought the nucleolus organizer was more than that. "I think it will come out later that this *is* an organizer, that it not only has the ribosomal parts in it, but it also has some other function that it's performing."[95]

As early as 1938, though, she thought her nucleolus organizer had not been accepted. On 15 November of that year, in response to plant geneticist Edgar Anderson's request for a reprint of that paper, she replied, "With the full conceit of the author I have always felt that this investigation has not been really understood. References I have seen concerning it have been muddled in their understanding—very possibly my own fault in not making the real points clear."[96] As with her early cytogenetics and, later, her controlling elements, it seemed to her that the full significance of this discovery failed to be grasped by her colleagues.

Germany devastated McClintock. She returned early, in April 1934. The Guggenheim Foundation allowed her to continue her fellowship in the United States, providing a few more months of funding, but after that she knew not what would come.[97] This began a low period for her. She had run out of fellowships. Emerson was glad to provide her with space, but he had no money to support her. The university was not forthcoming with a job. Harriet Creighton was still at Cornell, but she soon left for a teaching position at Connecticut College.[98] Rhoades left in 1935 for a job at an outpost of the Department of Agriculture in Ames, Iowa. Though hardly a plum academic research position, in the depression a job was a job. McClintock had nothing. Her nucleolus paper came out in 1934. In 1935, she published a masterly review article with Rhoades and a brief paper with Harriet Creighton corroborating their 1931 findings on crossing-over, answering R. A. Brink's criticism that they did not have enough data.

Her situation brightened in 1936, when Lewis Stadler invited her to join the faculty at Missouri. Stadler headed a Department of Agriculture research group at the Agricultural Experiment Station in Columbia and had a courtesy appointment at the university. He was proving to be a good fundraiser; he had obtained a major grant from the Rockefeller Foundation to build a center of genetics. McClintock joined the botany department as an assistant professor, supported by Stadler's grants.[99]

The job, however, was less than ideal for her. It came with the usual trappings of academic life: teaching, students, collaborations. She found it overwhelming, and her research productivity remained low. In 1936, she published nothing at all, and the following year only an abstract for the Genetics Society meetings. She made few friends and chafed at the conventions of professorial life. She became something of a troublemaker. For one thing, she never seemed fully to settle in. Her winter trips to Pasadena had become a habit; her first winter as a professor she was back at Caltech, where she met, among others, Jack Schultz, who had just described variegated position effects in Drosophila, and the dashing young French bacterial geneticist Jacques Monod. Summers, she returned to Ithaca to plant her corn, sometimes arriving back in Columbia late for classes. She shocked the faculty and administration with her flouting of procedure, decorum, and authority. In one story, which circulated on the graduate student grapevine, she climbed in through her laboratory window after hours when the building was locked.[100]

Increasingly during these hard times, she confided in Marcus Rhoades. Rhoades knew not to take offense at her sometimes abrasive style. She told him of her science, but also of her loneliness and "evil thoughts."[101] One formidable obstacle to her happiness seems to have been Mary Jane Guthrie. A protégée of Winterton C. Curtis, a biologist and the dean of Arts and Sciences, Guthrie was as intimidating as McClintock. The two clashed. Helen Crouse, a student of Guthrie's who also worked with McClintock, noted in a memoir that Guthrie once lectured for an hour on the nucleolus without even mentioning McClintock's 1934 organizer paper.[102] In a letter to Rhoades describing a tiff with Guthrie, McClintock wrote sarcastically, "Me and Miss Guthrie are apparently just like that

(Fingers crossed!)."[103] Bentley Glass, a drosophilist whose wife, Suzanne, worked in Guthrie's lab during this time, told me Guthrie was very bright, "but very acid in temperament and very jealous. She made life miserable for a good many of the graduate students in zoology. My wife barely escaped. Her roommate did not." When Guthrie took a dislike to someone, he said, "she'd move heaven and earth to do that person injury. And from the day that Barbara got there Mary Guthrie set out to destroy her, I am convinced of that."[104]

McClintock also began to have troubles with Stadler. She became suspicious that the bargaining and politicking he did as head of the genetics group was being applied against her behind her back. She occasionally argued with him, and she fired off blistering letters to Rhoades about him. On 28 April 1941, she wrote vaguely but bitterly of his "machinations;— or machinations as I call them. One simply never knows where one actually stands. He always plays for your good favor with the most trusting expressions. It makes one boil with resentment and anxiety, for during all he may be working entirely in the opposite direction with the powers that be."[105] She did not specify any incidents, nor have I turned up examples or corroboration that Stadler in any way double-crossed her.

In fact, Stadler remained a staunch advocate for McClintock. In March 1941, he wrote to Rhoades, concerned. McClintock had just informed him she was quitting her position at Missouri. In McClintock's telling, she had been told by the dean that if Stadler ever left, she would be fired, so before they could do so, she quit. She told Evelyn Fox Keller that at Missouri she had always felt her days were numbered. She was always alone, had no chance of being promoted.[106] (She had, after all, only been there five years—one rarely comes up for tenure in fewer than six.) Stadler's letter gives a striking new perspective on McClintock's departure from Missouri.

In strictest confidence—is there any chance of a job for Barb at Columbia, or at any of the neighboring institutions? As you know, she has never been very happy here, and during the past winter she has been feeling especially low. Finally last week she told me that she had definitely decided to quit at the end of this term. She talks of quitting the field altogether, but I don't think she really wants to do this. I think she is just completely fed up with the job here, but would get her old pep back completely in more congenial surroundings.

There is nothing in the local situation except her own feeling to make it necessary for her to move. She is definitely slated for a promotion this spring, and Tucker (Botany dept. chairman) has told her so.

Her own feeling as she expressed it to me is simply that the job here is not permanent, since pure research jobs are a luxury to a University of this level. She thinks the genetics project here might be closed out in some time of stress, and that if she can't get established in a research job that she considers permanent, she'd rather just quit now and go into something else.

God knows no one can guarantee permanence in times like these, though I think the job here is pretty permanent as jobs go. The University at the time she and [illegible] were appointed, gave official assurance that the research jobs would be just as permanent as teaching appointments. Presumably her promotion this year would make her an associate professor, which is the grade here at which permanent tenure becomes automatic. [107]

Although it is possible Stadler's letter was calculated and insincere, this seems unlikely. Rhoades could have verified the story with Tucker, the botany department chairman, and Stadler was, after all, writing to help find McClintock security. McClintock's version of the story was given decades afterward. The most plausible resolution of the two versions of the story is that McClintock left Missouri for personal reasons, and that after the fact she massaged the events into her version.

In May, McClintock told Rhoades she was thinking about leaving genetics, possibly "going over into meteorology."[108] Rollins Emerson, desperate to keep her in the field, nominated her for the National Academy of Sciences.[109] In 1944, McClintock became the third woman to be elected in the National Academy's eighty-one-year history. Moreover, at age forty-one she was young for this honor. "I knew then I was caught," she reflected in 1980. Caught? "You see, I had all this freedom. Now I figured I couldn't let the women down." Election bound her to the community of science. She recognized that other women in science needed role models and leaders, yet she chafed at any obligation.

In May 1941, McClintock wrote to Rhoades wondering whether he could arrange a year's visiting professorship for her at Columbia. Rhoades thought the idea splendid, and within a week all was settled but the details. Meanwhile, the Carnegie Institution was looking for a new director for the Department of Genetics at Cold Spring Harbor. McClintock had been asked for suggestions and thought of Milislav Demerec, the mutable-genes man who had been at Cold Spring Harbor since 1923. But Stadler, she said, was angling for the post. "He plans to try and get the job and then try and get a job for me there (at least that is what he says)," she told

Rhoades, ever suspicious now of Stadler's motives.[110] Demerec got the job, which made him director of both Cold Spring Harbor laboratories: the Carnegie department and the Biological Laboratory, run by the Long Island Biological Association. He invited McClintock and many other maize geneticists, including Rhoades, Stadler, Emerson, and Creighton, to Cold Spring Harbor for the summer. They all planned to attend the annual symposium there in June.

McClintock got the appointment as a visiting scientist in botany at Columbia, but that summer Demerec offered her a position with the Department of Genetics. He could make a temporary appointment beginning in September, he told her; he would then work to make it permanent. The pay was low, but that did not mean he undervalued her skill. He praised McClintock to Vannevar Bush, president of the Carnegie Institution, calling her one of the top ten cytogeneticists in the world, someone who "has the reputation of doing her work so thoroughly and competently that there is no necessity for someone else to do additional work on the same problem." Personally, she was an "individualist," who would do best in a setting where she would be free of teaching duties. In retrospect, McClintock said she wasn't sure she wanted the job. "I didn't want to commit myself to anything, because I enjoyed the freedom, and I didn't want to lose my freedom." She became a permanent staff member of the Department of Genetics on 1 April 1942. When E. Carleton MacDowell, another scientist at Carnegie's Cold Spring Harbor department, heard the news, he leaped in the air and exclaimed, "We should mark today's date with red letters in the Department calendar!"[111]

The Carnegie department was the best place for McClintock. Writing from Cold Spring Harbor in April 1942, she told Edgar Anderson, a geneticist at the Missouri Botanical Garden in St. Louis, that "it was a hardship to turn down the excellent equipment and conditions for the meager conditions here. However, all-in-all, I believe remaining here is the wisest thing to do—being a woman!"[112] Further, teaching frustrated her. The student population was far too coarsely filtered for her taste. She could not tolerate substandard minds. When Harriet Creighton left research for a teaching position, McClintock told her, that's fine for you, "you like people more than I do." But Creighton told me she knew McClintock had "really meant, 'You don't mind teaching people, dumb as they may well be.'" In 1980, she gave the Carnegie Institution her highest praise: "That's what the Carnegie's for, to pick out the best people they can find and give them freedom."[113]

"Suddenly You Know the Answer"

By the time McClintock arrived at Cold Spring Harbor, she had already developed the extraordinary mental process she called integration. She had many examples of solving puzzles this way: distinguishing among the maize chromosomes; Burnham's semisterility problem; ring chromosomes; and, later, the cytology of Neurospora and genetic transposition. She could not say where these solutions came from. "I've had so many experiences in my life of getting these signals from my subconscious that I cannot tell you necessarily where they come from, but the whole thing is solved suddenly."[114]

The experience McClintock named integration seems to be reported most often by mathematicians and physicists. The most famous integrator of all was Albert Einstein. Einstein once explained that when he worked problems, it was not with numbers or words but with images, "certain signs and more or less clear images which can be voluntarily reproduced and combined . . . The above mentioned elements are, in my case, of visual and some of muscular type. Conventional words or other signs have to be sought for laboriously only in a secondary stage."[115] For McClintock, the solution came too fast for pictures. "It doesn't have time to come totally in pictures . . . It's faster than I can recognize. Suddenly everything seems clear. Then you are able to put it in steps: 1, 2, 3, 4 and tell somebody about it and it seems perfectly logical. But that's not how it's arrived at, it's arrived at in some complex way that I have no way of stating. Suddenly you know the answer."[116]

Einstein said that he reached his theory of relativity by imagining he was riding a light wave and then looking around and describing what he saw. His famous *Gedankenexperimenten,* or thought-experiments, were ways of getting down into the phenomena he was contemplating. When McClintock could not solve a problem, she said it was because she had not "oriented" herself properly to see the relationships among the pieces of her puzzle. Once oriented, she said, "I could integrate whatever I saw immediately."[117]

Other mathematical geniuses integrated as well. Srinivasa Ramanujan Aiyangar, the self-taught Indian mathematician discovered by G. H. Hardy, produced fantastic conjectures; Hardy then worked with him to derive formal proofs. Ramanujan was said to be "personal friends" with every positive integer, producing spontaneous observations such as that 1729 is the smallest number expressible as the sum of two cubes in two different ways.[118] John Nash, the creator of game theory, also integrated. According to Sylvia Nasar, his biographer, "Not very long after he started thinking about a problem, he would have just a very clear vision of where the solution lay. And he wouldn't know how to get to it and it might take a year or two to get there, but he had this vision."[119] Physicist Richard Feynman was a famous integrator. His colleague Murray Gell-Mann once described the Feynman problem-solving method: "You write down the problem. You think very hard. Then you write down the answer." Like McClintock, Feynman could also chagrin colleagues by instantly solving their problems or, worse, more general theorems that encompassed their problems. With both, it was often not clear whether the answer sprang from inspiration or prodigious learning and memory.[120]

McClintock described integration not as a single holistic leap of intuition, but rather as a form of computation, a rational, rapid process of working out connections and logical steps. When describing how she solved Burnham's semisterility problem, she said, "Something was going rapidly, integrating, backtracking and making sure everything was all right because it was a very intricate process."[121]

When she reflected on how she had solved a given problem, decades later in some cases, she nearly always could reconstruct the reasoning. It usually involved the synthesis of many bits of knowledge about chromosomes and gene action, combined with an extraordinary ability to visualize chromosome behavior. There was nothing mystical about it. When other scientists failed to solve problems, often it was because they had gathered all the data but had not integrated it. She illustrated this difference with another story. The details of the science do not matter; the significance is in her interpretation. She had read a paper on Drosophila, she recalled, "and all of this data was giving nothing integrated. I read the paper and when I put it down I said, 'This can be integrated.' My subconscious told me that. I forgot about it, and about three weeks later I went into the laboratory one morning at the office. I said, 'This is the morning I'll solve this.' It turned out to be the most elaborate solution . . . It was perfect."[122]

For McClintock, integration was an internal process, one of self-control and awareness. Describing another, later problem and her temporary inability to solve it, she said, "Everything was there, but I wasn't integrating properly." In explaining how she got into the integrating frame of mind, she could only repeat, "You do something with yourself. You do something with yourself."[123]

Sara McClintock with the family dog, Mutty. Reprinted with permission from the American Philosophical Society Library.

Left to right: Mignon, Malcolm Rider (Tom), Eleanor (Barbara), and Marjorie McClintock, as children and in the 1950s. Reprinted with permission from the American Philosophical Society Library.

McClintock in the 1920s. Reprinted with permission from the American Philosophical Society Library.

The "Animal House" in 1937. Four years later, McClintock moved into the laboratory in the front upper left corner. Her cornfield would have been just to the photographer's left. Reprinted with permission from Cold Spring Harbor Laboratory Archives.

McClintock in 1947, at Cold Spring Harbor. Reprinted with permission from Cold Spring Harbor Laboratory Archives.

McClintock in her cornfield in 1951. Reprinted with permission from Cold Spring Harbor Laboratory Archives.

Richard Goldschmidt, at the 1951 Cold Spring Harbor Symposium. Reprinted with permission from Cold Spring Harbor Laboratory Archives.

Tracy Sonneborn and McClintock, probably at the 1951 Cold Spring Harbor Symposium. Reprinted with permission from Cold Spring Harbor Laboratory Archives.

McClintock relaxes on the patio at the 1951 Cold Spring Harbor Symposium.
Reprinted with permission from Dr. Norton Zinder.

McClintock and Harriet Creighton at the 1956 Cold Spring Harbor Symposium.
Reprinted with permission from Cold Spring Harbor Laboratory Archives.

The races-of-maize group, circa 1961. *Left to right:* Almiro Blumenschein, T. Angel Kato Yamakake, and McClintock. Reprinted with permission from the American Philosophical Society Library.

By 1986, fame and fortune drove McClintock to attend lab functions incognito. Reprinted with permission from Cold Spring Harbor Laboratory Archives.

4

PATTERN

In large measure, her ability to anticipate and exemplify new directions has come from an appreciation of the meaning of pattern. Pattern always indicates an underlying system of control over the material under observation.

—JAMES SHAPIRO

At first glance, the colors on an ear of Thanksgiving Indian corn seem chaotic. A closer look reveals regularities. There may be four kinds of kernels: solid red, solid purple, solid yellow, and variegated. The variegated kernels are striped with red on a yellow or creamy white background. Some are striped all over. Others are divided into sectors: part of the kernel may be plain yellow or white, with the remainder shot through with streaks of deep red. Some kernels may be half-variegated and half-solid; others may be mostly variegated, with only a small nonvariegated sector; while others may be mostly solid, with a pie-wedge of striped tissue. The red streaks may be few—mere threads lacing an otherwise pale kernel—or so numerous that they crowd together, leaving only a few cracks for the white to show through.

These color variations are the result of the P gene, first described by Rollins Emerson.[1] The P gene encodes an enzyme that makes the red pigment anthocyanin. The pigment appears in the pericarp, or outer covering of the kernel (and, as it turns out, in the cob). The P gene occurs in several alleles. With one, P^r, the enzyme is made continuously, resulting in solid red color. With the P^w allele it is not made at all, yielding white or yellow kernels. Emerson discovered a third form of the gene, P^v, for variegated. The stripes occur because the P gene turns on and off as the kernel grows: some cells make anthocyanin while their neighbors do not. P^v is an unstable, or mutable, allele. Emerson described many forms of P^v, which he distinguished from one another by the rate at which they caused mutations to occur: a slow rate meant few, thick stripes; a fast rate yielded many fine streaks.

Under Emerson's model, the patterns on Indian corn are easily understood. Those that are striped all over have P^v in all cells of the kernel. The first cell of that kernel must have been P^v; when it divided, both daughter cells were P^v, and when they divided all four cells were P^v, ad infinitum. More interesting are the variegated sectors. In those, the P gene changed either from stable to unstable or from unstable to stable. The histories of some kernels are easy to read: in a half-white, half-variegated kernel, the alteration must have come at the first cell division, so that the descendants of one of the first two cells were stable white and those of the other were unstable. A kernel that is three-quarters variegated with one quadrant white must have had P^v to begin with, but in one of the cells of the second round of division the gene must have stabilized. The smaller and more numerous the streaks, the later during kernel development the change must have occurred. As the patterns become more complex the analysis becomes trickier, but the principles are the same. It is a kind of puzzle, the solution of which requires knowledge of genetics, the behavior of chromosomes during cell division, and the growth patterns and life cycle of maize.

From her discovery of ring chromosomes until the end of her career, McClintock's science was about solving such puzzles. Each innovation came from seeing a new and unexpected pattern and making an inductive guess about the genetic events that produced it. When she found an anomaly of pattern, she plucked out the kernel that showed it, grew it up, and began experiments to verify her hypothesis. The experiments were of two basic types. She did breeding experiments, either self-fertilizing the plant or crossing it with other "tester" stocks whose genetic constitution she knew, to determine whether the gene or genes involved behaved the way she predicted. She also examined the chromosomes under the microscope, looking for breaks, inversions, rings, or other physical changes she thought underlay the anomaly of pattern. Cytogenetics is in essence the study of the cellular basis of heritable patterns.

In the late 1930s, unusual patterns led her to a new kind of chromosomal behavior: the breakage-fusion-bridge cycle. Twisted, rearranged chromosomes, formed by X-raying corn, began to break and rejoin on their own, creating spontaneous mutations. At Cold Spring Harbor in the early 1940s, she explored the effects of this repeating pattern of chromosome breakage and re-fusion. In one experiment in 1944, she unwittingly disrupted the normal system of producing color patterns in her corn. She recognized it immediately. Scores of new patterns emerged in her plants, including many of the variegated type, similar to those seen by Emerson decades earlier. Among these, McClintock discovered patterns that led her to controlling elements.

Pattern, then, is the starting point for understanding McClintock's discovery of controlling elements and, beyond that, her unique vision of nature. She was particularly attuned to nuances of pattern; it was her key to the behavior of the chromosomes, and it is our key to her visual style.[2] Through pattern, one can begin to see like McClintock.

Breakage-Fusion-Bridge

McClintock's discovery of ring chromosomes in 1932 was a model of cytogenetic reasoning and observation. The genius of the discovery lay not in the prediction that rings would form—that had been done—but in her ability to recognize the variegation patterns that resulted when the rings were lost. In 1938, while she was at Missouri, she returned to ring chromosomes. In this much more extensive study, she described plants that showed brown stripes, or variegation, in the stem and leaves. The gene in question was *bm1, brown midrib,* which in the recessive state produces a brown color on the central longitudinal ridge of the leaf. She found in X-rayed plants two chromosome breaks, one of which split the centromere (much as she had shown in 1934 that the nucleolar organizing region could be split). The result was a rod-shaped chromosome and a ring chromosome, both of which possessed part of a centromere, and a fragment, which lacked a centromere. The *Bm1* locus was located within the ring; the rod was *bm1,* the recessive allele. Both ring and rod were capable of lining up at mitosis and persisting through cell division, which meant that both partial centromeres were functional. In such cells, normal *Bm1* (not-brown) resulted. The ring, however, was sometimes lost at cell division, resulting in a cell that was *bm1/bm1* and expressed the brown color. Through various kinds of crossing-over events, the rings often increased or decreased in size during plant development. Decreases were associated with loss of *Bm1* and consequently with variegation. The larger the initial ring, the more frequent were these events. This led McClintock to conclude that the extent of the variegation in the mature plant was proportional to the size of the ring that was lost.[3]

In a companion paper that same year, McClintock explored further the behavior of broken chromosome ends. She found that two broken ends in the same meiotic nucleus tended to find each other and fuse. This formed a new dicentric chromosome. At cell division, one centromere migrated toward one pole and the other to the other, so that a thread connecting the halves of the double-waisted chromosome spanned the pinching middle as the cell split. McClintock called this a bridge. The bridge ultimately broke, leaving one broken chromosome in each new daughter cell. The cycle repeated in the next cellular generation.

She explained this by taking sides in a debate over chromosome structure in meiosis. Cytologists agreed that at the beginning of the first meiotic division, the chromosomes behave as though they were single. They also agreed that by late in this first division, they behave as though double. This is where crossing-over occurs between sister chromatids, the twinned arms of a chromosome. What happens in between, when the chromosomes took on this double character, was open to debate. McClintock recognized that the cytology could be explained if it were assumed the chromatids were double. When the chromatids separated, each would have a broken end. These broken ends would fuse, forming a dicentric chromatid, leading to another bridge at cell division, leading to two new cells containing broken ends and beginning the cycle again: breakage, fusion, bridge (see Figures 4.1, 4.2).[4]

In 1939, she followed the broken chromatids through successive cycles of cell division. How long did the breakage and fusion continue? She had found in her X-rayed material a nucleus with a rare and complex rearrangement. Three breaks had occurred on chromosome 9, with two of the resulting segments swapping position, one of them in reverse orientation. One of the breaks occurred in the knob at the end of the short arm. If one could sit on this partial knob and look down the chromosome, one would see several long arm genes in reverse order; the rest of the knob; a heterochromatic region and the centromere; the short arm genes in correct order—but now on the long arm; and the rest of the long arm.[5] This chromosome took McClintock into a new realm of cytogenetic reasoning. The trick chromosomes of the Morgan group were children's toys by comparison.

Tracking the behavior of this monstrous chromosome through meiosis and crossing-over was a remarkable feat in itself, but it permitted her many complex and revealing experiments. Soon after she found this rearranged chromosome 9, she wrote to Marcus Rhoades, "I have just run across something that I hope will give a clue to whether broken chromosomes will continue to fuse and then break and fuse again etc. I have a *perfect* set-up for testing it and the evidence looks very promising" (emphasis hers).[6]

Crossing-over between the rearranged chromosome 9 and a normal chromosome 9 resulted in a variety of monster chromosomes, simultaneously carrying inversions, duplications, and two centromeres. One was like Chang and Eng, the Siamese twins who were joined at the head. The

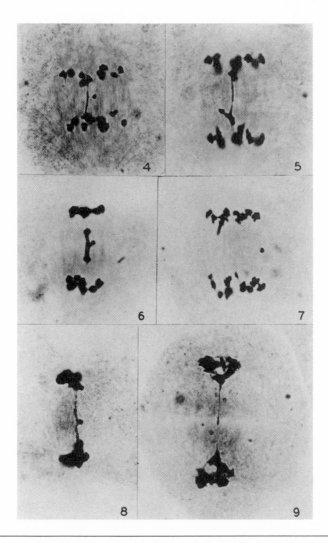

Figure 4.1. Breakage-fusion-bridge cycle. These images of McClintock's from the late 1930s show a cell preparing to divide. The chromosomes have duplicated and separated; the new cell wall will form horizontally at the middle of each image. In all but part 7, a chromatin "bridge" can be seen connecting one chromosome from the upper set to one from the lower set. Reprinted with permission from the American Philosophical Society Library.

entire short arm of chromosome 9 was doubled, with the tip of one arm attached to the tip of the other. If the genes were labeled *a* through *e* from the tip, this chromosome had centromere-*e-d-c-b-a-a-b-c-d-e*-centromere. Breakage might occur anywhere, but when it occurred between the *a*'s, the result would be two broken chromosomes with a full complement of genes. A nondeficient chromosome causes no problems and can form a

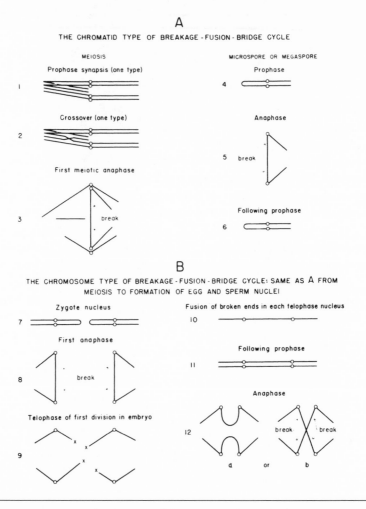

Figure 4.2. McClintock's diagram of the breakage-fusion-bridge cycle. Reprinted with permission from McClintock, "Chromosome organization and gene expression," Cold Spring Harbor Symposia on Quantitative Biology 16 (1951): 14, fig. 1.

viable embryo sac. The broken chromosome thus makes it into the endosperm, or kernel, yielding her *"perfect* set-up" for testing whether a broken chromosome would continue to fuse and break during development.[7]

Curiously, she found that breakage continued in the kernel but not in the maize plant. The short arm of chromosome 9 contained four useful genes. From the terminal knob in, they were *yg*, yellow-green leaves; *C*, red aleurone in the kernel; *sh*, shrunken endosperm; and *wx*, waxy endosperm. McClintock obtained several types of variegation patterns in the kernels, spots of various sizes containing tissue recessive for one or more of the kernel genes. But when she planted those variegated kernels, the resulting plants were not variegated for *yg*, the outermost gene. When she looked at the chromosomes of root tips of these plants, both chromosomes 9 lacked the terminal knob, indicating that they had indeed broken once, but the lack of variegation meant that the cycle of breakage, fusion, and bridge had not persisted. McClintock concluded that, although in the germ cells and endosperm the cycle continued, in the embryo and resulting maize plant, the broken chromosome "healed."[8]

The breakage-fusion-bridge cycle produced nested variegation: variegated spots within variegated spots. If three genes, *A, B, C*, were involved, with *A* near the knob and *C* near the centromere, she might see patches where only *A* had been lost *(aBC)*. If there were dots of variegated tissue inside that patch, they could be *abC* or *abc*, but not *aBc*. Thus, the elaborate variegation patterns of the BFB cycle could be explained by a progressive series of breaks, with each pattern in the nested series resulting from a new break as the kernel developed.[9]

The BFB cycle resulted from fusions between sister chromatids during meiosis. In the germ cells and endosperm, the breakages and fusions continued through development. But in the embryo and resulting maize plant, the ends healed and the cycle ceased. In 1942, McClintock discovered a variant on this form of BFB. If two homologous chromosomes, for example, both chromosomes 9, were broken in the same nucleus, such that both chromatids of both chromosomes were broken, four broken ends would result. These four broken ends could find one another and fuse. At mitosis—not meiosis, this time—the chromatids separate and migrate to opposite poles. This forms *two* bridges, which break and result in two broken ends in each daughter nucleus. The two broken ends then find each other and fuse, beginning the cycle again. McClintock called this the *chromosome* type of BFB cycle, in contrast to the *chromatid* type.

Because chromosome BFB continued in the sporophyte, or maize plant, while chromatid BFB occurred only in the germ cells and endosperm, McClintock now had a way to produce variegation in different plant tissues selectively. To get variegation in kernel genes only, she could introduce a single broken chromosome and look in the spores for chromatid BFB. To get variegation in leaf genes (as well as kernel genes), she could introduce two broken chromosomes and use chromosome BFB.[10]

The BFB cycle gave McClintock a powerful new set of tools. Out of Lewis Stadler's X-rayed maize stocks, she had developed three related systems for generating broken chromosomes and accompanying duplications, deficiencies, and other mutations—each of which persisted *without* further X rays. The initial inversions and rearrangements had been produced by the radiation blasts, but the natural cycles of chromosome fusion and division—in the trick chromosomes, that is—perpetuated them. At Cold Spring Harbor, McClintock began to exploit the BFB cycle the way Stadler, Hermann Joseph Muller, and she had exploited X rays.

Cold Spring Harbor

Cold Spring Harbor in the 1940s was a bucolic, unself-conscious summer camp of science. When McClintock arrived there in 1941, the laboratory buildings rambled over more than fifty acres of prime coastal Long Island real estate. The lab grounds were bordered to the north by Fitzgerald's "great wet barnyard," Long Island Sound, to the east by the narrow inlet of Cold Spring Harbor, to the south by a highway running out to Long Island's east end, and to the west by the staid luxury of Gold Coast estates belonging to Louis Comfort Tiffany, E. H. Harriman, and John D. Jones. In the nineteenth century the site had been a whaling village, nicknamed Bungtown for the barrel-plugs that had littered the cooperage there. Most of the laboratory buildings huddled at the head of the harbor, a farrago of dilapidated nineteenth-century whaling warehouses, graceful turn-of-the-century laboratories, and boxy poured-concrete structures. A central thoroughfare, Bungtown Road, ran along the water's edge.

The road is anchored at the highway by a stately four-story Victorian that was, in 1941, home to the aging Charles Davenport. Cold Spring Harbor's reputation rested heavily on Davenport's career. Science at Cold Spring Harbor had begun in 1890, when the Brooklyn Institute of Arts and Sciences established there a summertime teaching laboratory for marine biology, a Long Island rival to the Marine Biological Station at Woods Hole on Cape Cod, Massachusetts, which had been founded two years prior. Since 1873, when the Naples Zoological Station was built, biologists have summered at seaside laboratories; these institutions have played a crucial role in the development of American biology.[11] The Woods Hole lab was the first American marine biological station. The original Cold Spring Harbor station was titled simply the Biological Laboratory and was known as the Bilab. In 1898, Davenport became director of the Bilab, which was then little more than a small lab and a few decrepit warehouses. Over the next dozen years, he built a small scientific empire at Cold Spring Harbor.[12]

Fifty yards north along Bungtown Road, down a slope to the water, lay the original laboratory, built in 1894 and named after the Bilab's first benefactor, John D. Jones. Once outfitted with sea tables, aquaria, and collecting gear, the Jones lab had since been converted to laboratory space for visiting Drosophila and bacterial geneticists.

To the east, almost at the water's edge, stood the original Carnegie laboratory, by 1941 the library. In December 1901, Andrew Carnegie transferred $10 million in U.S. Steel Corporation bonds to establish the Carnegie Institution of Washington. The institution's aims were to strengthen American scientific research and education and to discover the "exceptional man," wherever he might be found. John Shaw Billings, director of the New York Public Library, and Charles Walcott, chief of the U.S. Geological Survey, became two of Carnegie's closest advisors. They would establish a series of departments, each dedicated to a particular branch of science, independent from universities, with no teaching and negligible administrative obligations imposed on scientific staff. On 16 January 1902, within twelve days of the incorporation of the Carnegie Institution, Davenport submitted a proposal. He lobbied Billings for the Carnegie Institution to establish a Station for Experimental Evolution on grounds adjacent to the Biological Laboratory, which it did in 1904. The other Carnegie departments included a desert laboratory, a department of marine biology, and, later, a nutrition laboratory and department of embryology.[13]

Looping back north along the water's edge, one came to the "Animal House," an Italianate structure built in 1912 and modeled after the Naples Zoological Station. Built in what turned out to be the waning days of experimental evolution, the Animal House was once fitted with cat cages and dirt-floored sheep pens. In 1921, the Station for Experimental Evolution was renamed the Department of Genetics, and the sheep pens gradually gave way to lab benches, refrigerators, microscopes, and other trappings of genetics research. McClintock was installed in a lab on the southeast corner of the upper floor. From her laboratory window, she could look out on an acre plot where she would grow her corn, and beyond that to the water and the village of Cold Spring Harbor. Just south of that field lay the original Carnegie Laboratory.

Behind the Animal House, back up on Bungtown Road, lay Blackford Hall, which housed upstairs a few spartan rooms for visiting women scientists (men boarded down the road) and downstairs the dining hall and, in a room no larger than thirty by fifty feet, the meeting room for the annual symposium. The Cold Spring Harbor Symposia on Quantitative Biology were founded by Reginald Harris, Davenport's son-in-law, who, with his slicked-back hair and pencil mustache, could have walked straight out of Fitzgerald's fictional town of East Egg. In 1924, wishing to focus on genetics and eugenics, Davenport gave his first kingdom, the Biological Laboratory, to Harris. In 1933, Harris organized the first meeting as a five-week "conference-symposium," in which researchers came to Cold Spring Harbor for most of the summer to discuss science and do experiments. The proceedings were published in slim oxblood volumes at the end of the year. Though some, such as the 1938 meeting on protein chemistry, featured important new results, overall the topics were as sprawling and unfocussed as the schedule. But in time, the summer seaside meetings and published volumes became some of the most important intellectual and economic capital of Cold Spring Harbor science. Harris's father-in-law remained treasurer of the Bilab and kept the directorship of the Department of Genetics until his mandated retirement at age sixty-five, in 1934.[14] Harris died two years later. Lab lore has it that it was the stress of publishing the symposium proceedings in the same year as the conference that killed him. Subsequent directors have not taken any chances: not since Harris have the proceedings come out in the same year as the meeting.

Further down the road, heading north, spread a motley cluster of buildings. Harris, anxious to distinguish himself, mounted an effort to establish a year-round research program in biophysics. The program foundered after a few years, but it left a lasting mark on the Cold Spring Harbor landscape. Harris commissioned several new laboratories, which were built at what was then the far northern end of the campus, sprinkled among cottages for staff scientists and whaling warehouses rehabilitated for dormitory and administrative space. Visiting scientists stayed in tents pitched on the lawns near Blackford dining hall. Soon after McClintock arrived, several of Harris's laboratories were turned over to the federal War Department, which set up secret research projects on chemical warfare, aerosols, and penicillin production.[15] After the war, many of these buildings became teaching laboratories, especially for summer courses on bacterial and viral genetics.

The Bilab buildings gave way to a tennis court, woods and meadows, and, just beyond, the Sand Spit, a fat sandbar that pokes into the narrow inner harbor. Summers, children played in the waveless water while scientists diagrammed experiments with driftwood in the sand.[16] McClintock soon discovered these woods and the little beach. Walks around the grounds soon became a major part of her Cold Spring Harbor routine.

The last piece of Davenport's empire lay a few hundred yards uphill to the west. In 1910, Davenport petitioned the newly widowed wife of railroad magnate E. H. Harriman, a close neighbor, to support a Eugenics Record Office. Mrs. Harriman funded the office entirely until 1918, when the Carnegie Institution assumed control of it. The Eugenics Record Office was the American headquarters of the eugenics movement through the 1910s and 1920s. In the late 1930s, it became an embarrassment, and finally the Carnegie Institution shut it down at the end of 1939. By that time, scientists at the Bilab and the Department of Genetics had essentially no contact with the anachronistic office.[17]

Cloven in the late 1920s and the 1930s, the Bilab and the Carnegie department grew together again in the early 1940s. The labs were once again united under a single director, Milislav Demerec. McClintock had scarcely known Demerec at Cornell—he left graduate school the year she began it—but he became a major figure in her life. Demerec was built like a fireplug, with bristly hair, a high forehead, and thick horn-rimmed glasses. He had a rare ability not to let his organism limit his research questions. After two decades of work in mutable genes in plants and fruitflies, during the Second World War he learned bacterial genetics in order to improve penicillin yield for the army, which he did by orders of magnitude, and stayed with bacteria for the rest of his career. Demerec became director of the Bilab and was named acting director of the Carnegie department in 1941, becoming director in 1942. Over his tenure, he gradually integrated the two labs, and two years after his retirement in 1960, they formally merged into the Cold Spring Harbor Laboratory of Quantitative Biology.

When McClintock arrived at Cold Spring Harbor, then, it was in transition. Many elements—structural, administrative, intellectual—remained from its first phase as a pair of laboratories; one a summertime teaching lab, the other a respected year-round research institute for genetics. Yet, with Demerec's directorship, it was taking its first steps toward becoming a world-class center for research and teaching in microbial genetics.

Cold Spring Harbor's annual cycle of activity was the reverse of that of a university. Summers, when a university is the quietest and a maize geneticist busiest, the lab grounds became a bustling scientific resort. Researchers from around the country and abroad attended the month-long annual symposium in June; others came with families in tow for a summer of rest, relaxation, and research. But from September to May, the maize geneticist's time for analysis and discussion, the country club shriveled into a ghost town, with a skeleton crew of year-round researchers and staff. With no students and a small and hierarchical administrative staff, there were no committees to be drafted onto as in a university department. What's more, hugging the shore of Long Island Sound and tucked in amid the grand estates, Cold Spring Harbor could offer only small plots of land for growing crops.

Yet for McClintock, what for many would have been inconveniences presented no hardships—and some were positive attractions. She loved the natural beauty of the lab grounds and was only too glad to escape departmental politics, teaching loads, and especially funding worries and tenure reviews (or lack thereof). She preferred to work alone, without inexperienced students or indifferent field hands. The paucity of arable land, which made Cold Spring Harbor impractical for many corn geneticists, was for McClintock a small imposition. Other corn geneticists planted thousands of seeds each year, deploying platoons of students and field hands to tend the acreage. McClintock grew but a couple hundred plants at a time and personally tended each one. An acre or two would be ample.

True, she had no maize colleagues at hand year-round. At times over the years, this distressed her. But Rhoades was just an hour to the west, at Columbia University. In 1936, he had published the first of several papers on *Dotted (Dt)*, a new mutable gene in maize. *Dotted* affected the *A1* locus, a site on chromosome 3 that governs the production of the kernel pigment anthocyanin. In the presence of *Dotted*, *a1*, the recessive form of *A1*, occasionally mutated to the dominant form, leading to dots of color. By 1938, Rhoades had located *Dt* to the very tip of the short arm of chromosome 9. It was adjacent to, possibly even in, the heterochromatic knob at the end of the arm. By the 1940s, cornfields were vanishingly rare in uptown Manhattan. Rhoades grew his corn at Cold Spring Harbor in the summer. This arrangement suited McClintock. She liked to work out her analyses on her own, on paper, then bounce them off Rhoades, either in writing or in person. A trip in to the city was easily arranged.[18] In 1948, Rhoades moved to the University of Illinois at Champaign-Urbana. McClintock then made regular visits to confer with him and his students.

During the war, the rustic grace of Cold Spring Harbor took on a somber cast. The annual symposium was abbreviated in 1942 and canceled the following three years, stemming the flood of summer visitors. Many Europeans were prevented from traveling, and American-based scientists had restricted mobility as well. Scientists at the two Cold Spring Harbor laboratories pooled their resources, sharing dining facilities and laboratory equipment. The government wartime researchers all had security clearances and were forbidden to talk science with the others, contributing to an air of secrecy antagonistic to the openness once characteristic of summers at Cold Spring Harbor. Demerec prowled the grounds, unscrewing unneeded light bulbs. Scientists and their families planted "victory gardens" at the north end of the campus. The laboratory grounds were isolated—the nearest pub was a thirty-minute walk into the village of Cold Spring Harbor and the nearest store or movie theater was three miles away in Huntington. With few visitors and rationed gasoline, there was little to do but science.

Some scientists were even more isolated than most. McClintock did not serve, and her research materials were not valuable to the war machine. Two summer visitors, the German Max Delbrück of Vanderbilt University and the Italian Salvador Luria of Indiana University, were hardly at risk of being drafted. In 1938, while on a Rockefeller Foundation fellowship at Caltech to learn fruitfly genetics with Thomas Hunt Morgan, Delbrück grew interested in Emory Ellis's work with bacteriophage, a virus that attacks bacteria. Trained in theoretical physics and an alumnus of the philosophically inclined Copenhagen group of Niels Bohr, Delbrück became fascinated with the simplicity of bacteriophage. It seemed to him the hydrogen atom of biology, the simplest particle capable of genetic replication. Delbrück took a physicist's approach to genetics: phage would allow him, he believed, to exclude all confounding variables and focus on the rudiments of heredity, the basis common to all organisms.

Delbrück and Luria began meeting at Cold Spring Harbor in the summer of 1941 to study bacteriophage, or simply phage. They brought their students and postdoctoral fellows and soon amassed a small cadre of workers who returned each summer. In 1945, near the north end of the campus in what was then called Davenport Laboratory, Delbrück began a new summer course on phage genetics. The tradition of summer courses at Cold Spring Harbor dated back to the Bilab's founding in 1890. Delbrück's course, however, was more intense, mathematical, and sophisticated than any taught at Cold Spring Harbor previously. The "phage course" was a two-week boot camp cum Chautauqua designed for graduate students and Ph.D.'s moving into phage genetics. Enrollment was just six the first year, but it swelled to a dozen or fifteen by 1950. Many of the students were established scientists, drawn, like Delbrück, from other fields to the new genetics. The rest were bright and mathematically oriented graduate students. This group of biologists, physicists, and students became known as the phage group, credited with beginning the molecular genetic revolution in the 1950s.[19] Since then, with few exceptions, science at Cold Spring Harbor has meant molecular biology.

This contact put McClintock in a context unique among maize geneticists. At a big, midwestern land-grant university, her natural audience would have been other maize geneticists. But at Cold Spring Harbor, she talked with and presented her work to the rising stars of genetics, bright and sometimes cocksure young men, many of whom had emigrated from physics. They were science snobs. They ignored cytology and embryology in favor of clean graphs, calculations, and crisp, all-or-nothing experiments. Where she worked on big, slow, complex corn, they studied tiny, fast, simple phage. One of her experiments took a year; if one of them botched an afternoon's experiment, he knew it by dinner and could repeat it in the morning. McClintock studied the most complex and naturalistic organism she could conveniently get into the lab. She sought to complexify biology, to bring her experimental system as close to the chaotic "real" world of nature as her mind would allow. The phage group took the simple systems approach. Their goal was to simplify life to first principles. McClintock and the phage group shared, however, a near-total disregard for biochemistry.

Making Mutations

As McClintock told the story, she never really committed herself emotionally to Cold Spring Harbor; she just never ended up leaving. "I didn't know then whether I wanted a job," she recalled in 1980. "I never thought that I would have to continue it any time. I was just having a good time. And I stayed on during the war."[20] Her research plan at Cold

Spring Harbor was not revolutionary: she would extend her broken chromosome studies and seek to uncover new maize genes.

In 1942, she used the BFB cycle to explore the effects of loss of increasing amounts of a chromosome. In BFB, chromosome breakage might occur anywhere along the arm. Often, a small fragment of chromosome would be lost at cell division. In genetics argot, loss of the chromosome tip was a "terminal deficiency." Terminal deficiencies of different lengths produced different patterns. When only the terminal knob was lost, the plant looked normal and was healthy. A slightly longer deletion produced a seedling that was pale yellow in color. A longer deletion still yielded a white seedling. If both chromosomes lacked too large a segment, the deficiency was terminal in the other sense: the embryo died. But if the terminal deficiencies remained small, the result pointed toward an interesting implication: the possibility of generating specific mutations at will. This would be a powerful tool for mapping genes along a given chromosome arm.[21]

The next year, she dusted pollen that had a newly broken chromosome 9 onto silks of plants in which one chromosome 9 was normal and one lacked the terminal third of the short arm. Half the female gametes, then, had a normal chromosome 9, and half had a deficient one. After fertilization, the broken end from the pollen grain underwent BFB, resulting in various degrees of loss of a terminal chromosome segment. If the pollen fertilized a cell with a normal chromosome 9, the fertilized egg would contain one complete set of chromosome 9 genes and no mutations would appear. A geneticist would say the deficiency was "covered" by the intact chromosome. But if the sperm fertilized a cell with a deficient chromosome 9, the progeny would be homozygous deficient for that minute segment, lacking both copies of a given gene.

The result was a technique for producing directed mutations, mutations within restricted regions of a single chromosome or even at a single locus. McClintock noted that, by carefully selecting her stocks, she could "independently produce the same mutations time and again."[22] Geneticists recognized the power of directed mutation in exploring the maize genome. The ability to reproduce mutations at a given locus could be a powerful tool in exploring gene function. Demerec remarked on this aspect of her findings in his summary of her work: "The chromosome-breakage mechanism is, thus, a means of independently and repeatedly producing the same mutation. In this respect, this mutation process is a directed one."[23]

The terminal-deficiency method worked fine for producing mutations at the chromosome tip, but the loss of too large a segment killed the embryo. To produce directed mutations further up the chromosome arm, McClintock needed to produce "internal" deficiencies, in the middle of the arm. There were two ways to do this. One, like her means of producing terminal deficiencies, involved the chromatid form of the BFB cycle. Sometimes BFB broke a chromosome in two places. If two of the main broken ends rejoined, a small fragment between the breaks would be lost, creating an internal deficiency. She designed crosses that would pair such chromosomes with a homologue that lacked a large portion of the short arm. In such an embryo, the chromosome with the internal deletion would cover the deficiencies of the other chromosome everywhere except where it was itself deficient. This would result in a homozygous internal mutation that McClintock could detect when the plants began to grow up. Going back over the material that led to the numerous pale-yellow and white-seedling mutants, she indeed discovered plants in which double breaks and internal deficiencies had occurred, yielding other mutants.

The second method of generating internal mutations exploited the chromosome type of BFB cycle. To initiate this type of BFB, she had to introduce two newly broken ends, one from each parent, into a nucleus. In chromosome BFB, breakages affect both chromosomes 9 and therefore may produce homozygous deficiencies in the daughter cell. Although McClintock did find isolated cases in which chromatid BFB persisted into the sporophyte, the chromosome type was a reliable way to generate mutations in leaf genes.[24]

"In working this out on paper, I decided that the mechanism should produce a series of little deficiencies, very tiny ones," she said in an interview decades later. This assumption came from her study of the fruitfly literature. Hermann J. Muller, Milislav Demerec, and Richard Goldschmidt had all studied minute deficiencies that caused mutations in Drosophila. McClintock followed the Drosophila literature closely and believed the BFB cycle produced mutations in this way. The mutations thus produced could be a means of mapping the short arm of chromosome 9.[25]

The 1944 Experiment

Scientists and mathematicians share a concept, never explicitly defined but universally understood, they call elegance. It is an aesthetic, a value, a quality of an experiment or proof. Its opposite is the "brute-force" approach, a blind, mechanical exhaustion of all possibilities until a solution is found. An example is the problem of summing all the integers from

1 to 100. Most people solve it by brute force: $1 + 2 + 3 + 4 + \ldots$ The elegant solution is to recognize the underlying pattern: $0 + 100$, $1 + 99$, $2 + 98 \ldots 49 + 51$, and then 50, which has no mate. The solution becomes trivial and can be done in one's head in seconds: $(50 \times 100 + 50 = 5{,}050)$. Elegance connotes simplicity, resourcefulness, economy, and above all, intelligence.

McClintock's experiments were the embodiment of elegance, but her rhetorical style was quirky. Often, she would state a widely recognized problem or theory—so far, standard scientific style. But she would follow with a hypothetical solution, typically an intricate or farfetched one. Then she would reveal that she had in fact invented the technique, found an example, or done the experiment that demonstrated it.

For example, she began her 1944–45 report to the Carnegie Institution by stating the problems presented by the scattershot method of X-ray-induced mutation. "In the past, many methods have been used to induce mutations. The majority of these methods do not give rise to specific mutations or to mutations confined to specific regions of the chromosome complement. Instead, a random assortment and distribution of mutations are obtained."[26] Radiation is a brute-force approach. It generates random mutations all over the genome, most of which produce either no visible effect or no viable organism. Microbial geneticists solved that problem by using organisms for which it was easy to screen enormous numbers quickly for interesting mutations.

McClintock, using big, slow-growing maize, needed another way. She continued, "A better understanding of the factors involved in the mutation processes would be possible if specific mutations associated with specific regions of the chromosomal complement could be effected."[27] Until then, such a technique had been only a dream, but McClintock proceeded to demonstrate that she had effected precisely that. The BFB cycle could substitute for x rays. By constructing maize stocks with chromosomes that favored BFB, she could generate random mutations within a small region of one chromosome arm. This allowed her to map the genes in one region, which was both useful and interesting genetically. The technique also eliminated the need for expensive and dangerous X-ray equipment. She called it a "nonselective" method for producing mutations, though "semidirected" might have been more accurate. Five years earlier, the BFB cycle had been a genetic novelty. She had now modeled it into a powerful tool for uncovering new mutations and new genes. It was an elegant and original approach to a perfectly conventional genetic problem.

She set out to use both the chromatid and chromosome types of BFB to generate internal mutations.[28] Though at the time she reported having examined plants from both types of experiment, it was only the latter, the plants that received two broken ends, that made it into her private myth. As she reconstructed the experiment many years later: "I figured what we'd do next was to get a lot of plants that had undergone a chromosome type of breakage-fusion-bridge cycle. That cycle I found would be in many cases stuck, but before it stuck . . . [t]here were lots of deficiencies."[29] McClintock often spoke metaphorically of her material; by the BFB cycle getting "stuck," she probably meant that it stopped for unknown reasons, showing no further mutations in descendant cells. One reason for this might be that the broken ends healed. The single-broken-end, chromatid-BFB-type plants she examined may have been material she had collected in years past. The double-broken-end, chromosome-type plants constituted her summer 1944 crop.

The previous year, she had generated kernels that contained two broken chromosomes 9, one from each parent. On 11 May 1944, she planted approximately four hundred and fifty of these, starting them in the greenhouse and then moving them to the field when the plants were strong and the weather warm. Each seedling should have undergone the chromosome type of BFB cycle. From them she hoped to recover new mutations, which she could then map.[30]

The family of plants resulting from a specific genetic cross is called a culture. McClintock recorded most of her cultures by number, in ascending order.[31] When she had a special set of plants, however, she grouped them under a series. In her records, she classified these broken-X-broken cultures as the "B" series, for "broken." McClintock recorded over two hundred cultures in the B series.

Walking her field on 27 June, she noted that the eighty-seventh plant of the B series had a "wide sector of white on tiller."[32] A tiller, or side branch, produces its own leaves, shoot, and tassel. Usually, a tiller is genetically identical to the main stalk. The long white stripe, or sector, down the leaf in plant B-87 occurred only on the tiller. Because it was solid white and not variegated, its appearance suggested that the BFB cycle had stopped—but only in the tiller. She further noted that the plant did not show a "dicentric" pattern of pigmentation, the kind of pattern of stripes and spots that characterized the BFB cycle. Though most of the mutants carried the dicentric pattern she expected, this one was odd. And that the tiller differed from the main ear was especially puzzling.

"So my plan was to self-pollinate them," she said, "to see if the recessive changes would show up" as new mutants.[33] Self-pollination uncovers mutations for the same reason that it is unwise to marry your sibling. Because maize leaf cells, like human cells, have two copies of each chromosome, one donated by each parent, recessive mutations are compensated for by a normal copy of the gene; often, a single good copy is sufficient for normal or near-normal function. But if two bad copies are present, or both copies are absent, the mutation is expressed. Self-fertilizing, or selfing—the ultimate inbreeding—maximizes the likelihood of combining two bad copies of a gene and thus exposing it in the progeny. McClintock planned to self-fertilize plant B-87 and the others in the B series. Since her corn tended to have many tillers, selfing often involved several fertilizations per plant. Since some plants, like B-87, showed differences between main and tiller ears, she had to be extremely careful lest even a single pollen grain drop onto the wrong silks and contaminate the experiment.

McClintock made crosses using the same techniques and materials her teachers had used, and which are still used by maize geneticists today. The day of pollinating is determined by the maturation of both male and female parts of the plant. Silks must be emerging from the nascent ear and be ready to receive pollen. To avoid contamination from the pollen of nearby plants, each shoot is covered with a slender, pleated glassine bag when the shoot is small and hard, long before the silks began to emerge. "Shoot-bagging" is part of the ritual of the maize geneticist's daily field check. The shoot continues to grow inside the bag. The semitransparent shoot bag allowed McClintock to check whether a spray of silks had emerged from the ear, meaning it was ready to be pollinated, without exposing them to contaminating pollen floating in the air.

Early in the morning on the day of the fertilization, McClintock visited the plant and collected the pollen. First, she stripped off all the existing anthers, or pollen-producing organs, from the tassel capping the plant. Maize pollen is produced rapidly but is fertile for only four to five hours. Stripping off the old anthers stimulates the production of new anthers and pollen, ensuring fresh pollen later that morning. She then covered the tassel with a brown paper bag, which she pinched at the bottom with a paper clip. Since her field was just steps from her laboratory, it was easy to check shoots and bag tassels first thing in the morning, then go to her lab, have some breakfast, and plan her strategy for the rest of the day. Later in the morning she returned to her brown-headed plant, inverted the bag, shook the pollen from the tassel, and pulled the tassel out of the sack, again stripping off all the anthers. She pinched the end of the pollen bag and tapped the pollen toward the bag's mouth. She then removed the shoot bag, poured the powdery, saffron-colored pollen onto the silks, and rebagged the shoot. The fresh silks are sticky; a coating of pollen gives them the fuzzy look of a newborn chick.[34]

She recorded an identifier on a wooden paddle that she poked into the ground next to the plant. Detailed information about the cross went on a four-by-six-inch card that she stored in a catalog in her laboratory. She learned the system from Rollins Emerson at Cornell and modified it; it is still used by maize geneticists today. In the top right corner she wrote the culture number—this was not assigned until the kernels were planted in the field. A plant's parents were recorded in the upper left corner of the card, with the genotype and chromosomal configuration of each as well as the date of the cross. When the resulting ears had matured, McClintock would record on the back of the card the number of each type of kernel planted and, the following season, the appearance and fate of the plants they produced. Of those that did not survive, the most common fate was "pulled up by birds," specifically crows, the maize geneticist's bête noire. When the ear developed, late in the summer, McClintock collected it and plucked out and analyzed the kernels.

As expected, she turned up a number of new mutations in the B series. These included, she wrote in about 1950, "alterations in growth giving dwarf plants, plants with fused parts, abnormal ear or tassel development . . . [w]hite seedlings, yellow-green seedlings, virescents [those with an abnormality in chlorophyll accumulation], luteus [yellow] seedlings, various types of pale-green seedlings."[35] Several mutations were located on chromosome 9, such as small-kernel *(smk)*, spotted-leaf *(spl)*, and pale-green *(pg)*. All of these, she emphasized, were the sort of mutants she expected to find as a result of the BFB cycle.

But something else had happened as well. Among the new mutants, she discovered a large number of mosaics, or variegates. In all, she found seven or eight new mutable genes, several of which were located on the short arm of chromosome 9—in total, four times as many in one experiment as had been isolated in the previous thirty years of maize genetics. She soon realized that this was but the first crack in the dam; from this initial burst of variation would flow a torrent of new and increasingly complex variation. She studied the descendants of this crop for the rest of her life.

The main ear of plant B-87—the plant whose tiller had the wide, white-leaf sector—had four types of kernels. She designated them A, B, C, and

D. Plant B87-1 should have had two different chromosomes 9. One had a duplication of part of the short arm and carried the *C* gene, which produces the normal red kernel color, and recessive *wx*, for waxy endosperm. The other chromosome had a small deficiency at the tip. This chromosome carried the *I* gene, which inhibits red kernel color, and dominant *Wx*, nonwaxy. The A group comprised a single kernel. It was covered with small red spots and was nonwaxy throughout. Group B comprised four kernels, also with many small spots of red, but those spots were waxy, while the unpigmented areas were nonwaxy. Group C consisted of two spotted waxy kernels. Some of these kernels, she noted, resembled Rhoades's *Dotted*.[36] The A, B, and C kernels all showed evidence of break-age-fusion-bridge.

Most of the kernels fell into group D. The D kernels were plain white and nonwaxy *(I Wx)*, lacking the pigment spots that would have indicated a dicentric chromosome. McClintock wrote on the culture card "not dicentric." Not carrying a dicentric chromosome, they were not undergoing the BFB cycle. Plants grown from these kernels should not show mottling or streaking. The kernels went into packets by group, ready for the next planting. She did the same with the kernels from the tiller ear of B-87 and with the other ears produced by the self-fertilizations. As it turned out, the D kernels, those with the least interesting patterns, produced the most interesting plants.

In October, after the harvest, McClintock left Cold Spring Harbor for ten weeks at Stanford University in the lab of George Beadle, her old friend from Cornell. Beadle had left maize genetics and had begun to work on the fungus *Neurospora crassa*. Genetically, Neurospora was becoming well understood. Many traits had been identified and several genetic linkage groups had been assembled. But cytologically it remained enigmatic.

The situation was exactly analogous to that in maize in the mid-1920s, before McClintock solved it. Not even the number of chromosomes was known, let alone which linkage groups went with which chromosomes. At Cornell, McClintock had shown Beadle how to see maize chromosomes; Beadle knew that if anyone could solve the problem it was McClintock. She was passing familiar with Neurospora. In 1940, Carl Lindegren, the father of Neurospora genetics (and later a pioneer in yeast genetics), had spent a year in Stadler's laboratory. McClintock saw quite a lot of him, she said at the time. He brought her some slides of Neurospora cells. When she looked at them, she recalled, she "saw right away what the chromosome number was, and that it could be worked on . . . But I didn't try any more than that." So when Beadle asked her, as he had done in graduate school, to help him with the cytology, she was primed.[37]

McClintock went out to Stanford and pored over Neurospora chromosomes. After several days, she had gotten nowhere. "I wasn't seeing things, I wasn't integrating," she said later.[38] Then she took a walk and sat on a bench under Stanford's fragrant eucalyptus trees. "I sat out there, and I must have done this very intense, subconscious thinking. And suddenly I knew everything was going to be just fine, and I couldn't wait to get back. In five days I had everything major solved." The patterns crystallized suddenly and she worked out the peculiar cytology rapidly. She was able to follow the paths of the chromosomes through the cell division cycle. She discovered that after the chromosomes had synapsed, they elongated, swelling to fifty times their usual size. At this stage, she could distinguish banding patterns and other chromosomal details. This was the Neurospora equivalent of the pachytenes in maize or the Drosophila giant salivary chromosomes. McClintock's work made cytogenetics of Neurospora possible.

It is the sort of story that has nurtured McClintock's reputation as an intuitive mystic. But the eucalyptical epiphany was preceded by many hours of looking and thinking. If the solution fell suddenly into place, it was because she had spent time moving the pieces into the right positions. She then applied the "Feynman method" of thinking very hard and then finding the answer. Her intuition was a combination of experience, observation, and integration.

Twin Sectors

The Stanford trip provided a welcome respite from life at Cold Spring Harbor, about which she was still ambivalent. The following May, McClintock wrote to Beadle that she was considering leaving Cold Spring Harbor. "It is so dull and uninspired here," she wrote, "and the leadership"—Demerec—"shows no signs of improving or benefitting from past experiences." Demerec had high ambitions for attracting big-name scientists, but he was not getting them. Several recruitments had fallen through, McClintock continued: "We did not get [Tracy] Sonneborn and we did not get [Hermann J.] Muller." In her current mood, this made her pessimistic: "I am a little afraid to remain here." What kept her was the science. Her work of the winter and spring of 1945 had revealed a kaleidoscopic set of intriguing patterns. "I have just started an interesting problem in maize," she wrote, perhaps a little coyly.[39]

To get a jump on the summer growing season, in late winter she planted the forty D kernels from plant B-87, in a seedling bench in her greenhouse, and watched the seedlings come up. Thirty-two plants survived. Of these, eighteen were green and fourteen were white. "Now the little seedlings that came up were extraordinary," she recalled. "About half of them seemed quite normal. And the other half were full of streaks." These seedlings had streaks, she said, "because they had lost some necessary genic materials." As the plant grew, new cells would be produced, and some would have complete genomes. The seedlings, she said, "kept sending out little side branches, trying very hard to make new plants from the cells that had the full genome complement. Finally, as many as fourteen of these little side branches would come out of a seedling. And finally one would come out that was fully green, and make the plant."[40]

Because the D kernels were white and nonwaxy, the plants grown from them should not have contained the dicentric chromosome 9—in which case, they should not have undergone the breakage-fusion-bridge cycle. But they were producing a wide variety of variegation patterns, similar but not identical to BFB-type patterns. A researcher with less confidence would have thought the plants had been contaminated.

The seedlings had variegated chlorophyll patterns, stripes and streaks, as though the pigment gene was being turned off and on as the leaf grew. This variegation was occurring in several plants, resulting in a riot of new patterns: "One [plant] would give wide stripes of green and wide stripes of yellow. Others would be yellow with tiny little green streaks that were in a very special pattern. Others would have these spots of colorless in a green background. There were so many different ones."[41]

Mutations from wild-type, or normal, to recessive were easy to explain: a gene had been altered or eliminated and no longer produced its pigment. Several known cytological mechanisms could account for this, including breakages, rearrangements, deficiencies, even "transpositions," the term given to genes that changed position when a part of a chromosome arm broke and reattached to a different chromosome, by translocation or inversion. Such could account for streaks of yellow or no color on a healthy green background. But leaves with streaks of green on a yellow background were problematic. They were called revertants, because a mutant form was mutating "back" to wild-type. Revertants were difficult to account for under standard mutation theory, which held that mutations represented gene damage: a second lightning strike rarely restores vigor to a victim. The rarity of revertants had made them easy to ignore. When McClintock uncovered a great number of them, revertants became a significant problem.

McClintock wondered whether such revertant patterns occurred in other maize stocks as well but had been ignored for lack of explanation. "I am beginning to suspect that the phenomena of variegation or mutability from recessive to dominant is fairly common in maize," she wrote to Rhoades. "Many reverse mutations may have been observed but discarded for fear that contamination was the cause."[42] She knew contamination was not the cause in her plants, because she did all the fertilizations herself. With this suspicion, McClintock began to question the mutation theory, a central tenet of 1940s genetics.

"I am beginning to feel a little mental indigestion," she wrote Rhoades, "in trying to figure out just what to grow, how much and what to do with it after it is grown. Also, I am beginning to feel some amazement over the variegation stories. In this material I have uncovered 7 or possibly 8 variegates from recessive to 'wild-type.' I am beginning to wonder if there are not many of them in our ordinary stocks and that they just have been neglected because no particular thought has been given to the appearance of changed sectors."[43]

She grew up the plants derived from the self-fertilization of plant B-87 and planned the crosses she would make with them. She had sufficient skill in the field, and was stingy enough with her pollen, that she could fertilize many plants with the pollen from one tassel. Three of the resulting cultures were particularly curious. In her annual report to the Carnegie Institution, she wrote that in each of the three, "one of the broken chromosomes 9 is continually being lost from cells during development."[44] In certain cells, all the known traits on the short arm of chromosome 9 mutated. Moreover, she determined that the loss was not occurring by either of the known means: bridge formation or ring formation. Rather, she wrote, it "appears to be caused by the inability of the two halves [chromatids] of this chromosome to migrate to opposite poles" in some cells. Some mutation had occurred in the self-fertilization of B-87 that disrupted the machinery of cell division. The more often loss occurred—the greater the rate of loss—the smaller and more numerous the streaks.

Furthermore, of the kernels she planted from this cross, several were "individually provocative." Some showed reversion or back-mutation, from mutant to wild-type. "As the leaves grew," she said, "the background pattern was obvious in most of the leaf. That was the pattern that was started with. That might be small green streaks in a white background."[45] This was the revertant pattern—odd in itself. But superimposed on this was another pattern that was totally unfamiliar. To the Carnegie department that year, McClintock wrote in her characteristically flat professional style that it was "significant that twin sectors accompany many if not most of the alterations in rate" of chromosome loss.[46] Recalling the discovery years later, she said, "I could see sectors appear in the leaves in which the pattern of variegation was very different from that of the rest of the leaf or the pattern that was seen in the seedling when it was growing. And then I would see cases in which there were two sectors side by side, one of them having a pattern that was the reciprocal of the other." These side-by-side sectors, she said, "were twin sectors."[47]

McClintock's terminology is evocative but a bit misleading. The sectors were twins as yin and yang are twins. On a leaf with a background pattern of streaks, she saw a pair of broad adjacent stripes, one with more streaks than the background and the other with *proportionately* fewer.[48] It was as though mutation rate were controlled by a volume knob. In most of the leaf the level was set at, say, 5. But in the twin sectors, one sector was raised to 7 and the other was lowered to 3.

It was significant that the twin sectors were adjacent. The leaf develops by the law of geometric increase: one primordial cell divides in two, each of which divides again, and so on, such that a given longitudinal sector arises from a cell that was the sister of the progenitor cell of the adjacent sector. The twin sectors, then, must have arisen from a common ancestor cell. McClintock reasoned that at that first division, some factor must have been differentially distributed between the daughter cells. This factor must control the mutation rate and therefore the variegation pattern. Most of the cells in the leaf got a standard dose of the factor, but at this cell division, something went wrong, so that one daughter cell got a superdose and the other got a minidose. A greater mutation rate meant more white streaks; a slower rate meant fewer streaks. "One cell gained what the other cell lost," she said. "I couldn't get it out of my head."[49]

"I was astounded," McClintock later said of the seedlings produced from the forty white D kernels. The parent stocks had no genetic instabilities of the sort that turned up in the new cultures. She had expected mutations, but not this riot of different variegation patterns. Something had happened in the genes of these seedlings that had sent them out of control: "It had gone wild," she said. "The genome had gone wild."[50]

Wild indeed. The 1944 experiment had turned up many new mutations, as she had expected. Besides the conventional, stable mutations were seven or eight mutable genes, which spontaneously mutated and then reverted as the plant grew. This many mutables at one time was itself unusual—previously, only two mutable genes had been described in maize. Nevertheless, McClintock could explain most of the new mutables in terms of breakage-fusion-bridge phenomena. One of them, however, the chromosome-loss mutable in the descendants of B-87, was especially striking. In those plants, some unknown mechanism—not BFB, not ring behavior—was causing the loss of a large chunk of the chromosome. The difference in pattern between them and the other variegates was subtle, really just a matter of regularity. In her annual report to the Carnegie Institution, McClintock wrote that the patterns on the kernels were "distinguished by the number of i [colored] spots, their uniform distribution . . . and their relatively similar size."[51] The chromosome loss mutant indicated that there were other kinds of genes besides conventional trait-producing genes. Later, she would consider that these were not genes at all, but some other, nongenic chromosomal material.

Among the chromosome-loss plants, moreover, there were the twin sectors. These indicated that the rate and timing of mutation was controlled: "The frequent appearance of twin sectors showing inverse rates of mutation leads to the suggestion that the controlling factor for future mutational occurrences may be altered or segregated at a somatic mitosis," she wrote in her Carnegie report.[52]

With the chromosome-loss mutable and the twin sectors, McClintock had the germ of the hypothesis that formed the basis of her science for the next twenty or more years: Genes are controlled by other kinds of chromosomal elements, and this control is responsible for the patterns that compose biological form.

Pattern and Chromosome Behavior

McClintock's realization that one cell gained what the other cell lost derived from her recognition of layers of pattern. The twin sectors and the chromosome-loss mutations were second-order variegations, variegations of the variegation, in her already-mottled leaves and kernels. Because the twin sectors were sister sectors, she realized that the twin sector pattern is an outcome of the growth of the plant. It was therefore a developmental or embryological problem. But it was also a genetic one, because genes, she knew, determine traits such as pigmentation. In the years following her discovery of the twin sectors, her work on mutable genes led her to a new theory of the genetic control of development. This led her out of mainstream genetics and into new territory that other geneticists would not begin to explore seriously until the late 1950s. As she began to focus on development, pattern remained central to her thinking and in fact took on diverse new meanings.

Pattern was not a static arrangement of pigments or even pigment genes but rather a dynamic, interactive process, shaping and shaped by the growth of the plant through time. It was, in other words, both a reflection and an agent of control. Control reached from the superficial appearance of the kernels and leaves to the deepest levels of biological information that determined the action of genes and the growth of the plant. The key insight of McClintock's work on controlling elements was her realization that patterns in the leaves and kernels indicated underlying genetic control, which in turn was determined by patterns of gene action.

"Pattern" became a standard heading in her notes and files after 1946. She used the word in many different senses, which suggests that for her it was less a descriptor than an organizing principle. She wrote of "patterns of variegation" in kernels and leaves, of course, but also in the pollen.[53] Some of the variegation patterns became distinct enough to acquire names, such as "early loss" pattern and "late loss" pattern.[54] Patterns of variegation were evidence of "patterns of mutation." Although a straightforward step given the acknowledged genetic basis of the variegation patterns, the concept of patterns of mutation indicated a higher-order genetic process: something controlling the action of the pigment genes.

McClintock extended her sense of pattern to include processes. Patterns of mutation were related to "patterns of action" of genes and gene regulators, which further indicated that genes were not autonomous, independent entities. Moreover, she wrote that "patterns of action may vary in different parts of the plant." The patterns of gene action themselves had patterns, which were tied to the different tissues and growth stages of the plant.[55]

Pattern even came to represent for her a quality of a genetic locus. By the late 1940s she hypothesized "pattern alleles," which she distinguished from "quantitative alleles."[56] Because the patterns of action were expressed differently in different parts of the plant, they gave rise to "patterns of differentiation"; differential gene expression not only reflecting but determining the growth of the plant and the formation of the various plant tissues.[57] Finally, she wrote of "patterns of gene organization and regulation," the highest-order level of control.[58] Genes were "organized," not randomly assorted, and their order was linked to the regulation of gene action. Although patterns might control gene action, there could also be "conditions controlling pattern." These included genetic controls, but might also include environmental controls—the response of the genome to change. By the mid-1960s, in her discussion of pattern alleles, she seems to have thought that a controlling element could convert a pigment gene into a pattern gene, a gene that encodes, instead of a biochemical product, a pattern.[59]

McClintock relied on pattern in communicating her science to colleagues. Such a reliance, Carla Keirns has pointed out, was characteristic of cytogenetics. But increasingly, McClintock addressed the new group of molecular biologists, not cytogeneticists. The reasons were both geographical and intellectual. At Cold Spring Harbor, microbial geneticists and molecular biologists constituted the bulk of the year-round staff and summer visitors. McClintock grew increasingly reluctant to attend professional meetings, so she spoke mainly before audiences that came to her. But she also sought out the molecular biologists. She enjoyed their intelligence and, by the 1960s, she saw them recognize the importance of taking development into account. Molecular biologists and cytogeneticists had different visual styles; the former relied (and continue to rely) on "cartoons," schematic representations of invisible molecular mechanisms, both for presenting data and for building models. Late in her career, McClintock's elements were absorbed by molecular biology, were represented in cartoons. McClintock, however, came to lean ever more heavily on pattern to describe her findings. Her late papers and reports are rich with photographs of kernel patterns but shy on molecular models and even cytogenetic tables and statistics.[60]

She became so attuned to pattern, in fact, that she sometimes saw it where only randomness existed. Fortunately, she had the humility and humor to laugh at herself. Bacterial geneticist Evelyn Witkin, who came to Cold Spring Harbor in 1945 and soon became McClintock's closest friend there, recalled a day probably in the late spring or early summer of 1948:

> One day, very soon after her first insight that transposition was responsible for her variegations, she called me very excitedly, asking me to come down to see something new. As I always did when I got that kind of call, I dropped everything and ran. Barbara showed me a ker-

nel and asked me what I saw. I told her that I saw purple spots of various sizes. She then showed me that the spots were size-paired: small with small, medium with medium, large with large.[61]

McClintock could be very persuasive, and her enthusiasm was infectious. "As soon as she pointed this out," Witkin said, "I immediately saw the pairing." Part of McClintock's genius was to instantly produce a plausible biological explanation for a new pattern: "She was as excited as I ever remember her, and she thought that it meant twin transposition events were taking place simultaneously in different cells, perhaps predetermined by an earlier event in a single ancestral cell." Witkin returned to her lab in the mental state that "usually followed these visits: with my head buzzing with wonder and puzzlement." But the excitement was short-lived:

> I think it was the next morning that she called me, with one of her loudest laughs (the kind she reserved for laughing at herself), telling me to FORGET the paired spots! She had decided that the pairing was an illusion, and that random spots could easily seem paired as long as the distance between them was unrestricted.[62]

In a random array, one can find any pattern one wishes. The heavens were not sprinkled with stars so they might be viewed from our watery speck in a remote corner of a nondescript galaxy, yet gazing at the night sky our Sumerian ancestors saw wheels, animals, human figures—meaning. Correct interpretation of pattern means distinguishing correlation from cause. In the case of dots and specks in corn, is there a mechanism and is there a function for the patterns? Because the changes in gene action were linked to the growth of the plant—because of the twin sectors—McClintock believed these patterns had meaning. By stimulating in her a new sense of pattern, the discovery of twin sectors in 1945 set a new direction to McClintock's thinking and her research program. It changed the pattern of her work. To her, pattern meant control. As her studies of mutable genes carried on into the late 1940s, she argued from pattern to develop a theory of developmental genetic control. "That," she said, "was the basis of continuing."[63]

5

CONTROL

The real secret of all of this is control. It is *not* transposition.
—BARBARA MCCLINTOCK

In the standard interpretation, mutation represented gene damage. It was as though geneticists stood outside on the street at night, looking at well-lit houses. The lights, they assumed, were always on. If a window suddenly went dark, they figured a sharpshooter with a pellet gun had burst the bulb. Embryologists accepted the theory that the lights were always on, but they noticed that over time different windows were lighted at different times. They invoked elaborate systems of shades, blinds, and curtains to explain the bright and dark windows.

First with X rays and then with the breakage-fusion-bridge cycle, McClintock had created in her maize a seedy genetic neighborhood, full of flickering bulbs, frayed wiring, and snapping shutters. Wandering this district, she recognized flicker patterns that were not random. Her conceptual leap was to see that the patterns indicated not just process but control.

She discovered a house in which the lights flickered in a complex but recognizable pattern. Rather than ignoring this variability or invoking shades and blinds to explain it, she imagined the lights had switches. Mutable or mosaic genes, in which the light alternated on-again, off-again, would then have nothing to do with the bulb; rather, the switch had gone haywire. The switches, she conjectured, were normally undetectable—no wiring diagrams were available in those premolecular days. But they were revealed in this one house, and she saw no reason not to assume the same electrician had wired all houses.

The genetic factor or element controlling mutation rates in the twin sectors seemed to be such a switch. Between 1946 and 1948, McClintock's primary goals were to identify and isolate this element. By the spring of 1948, she had accomplished this, but she was surprised by two wrinkles: there were two elements, and they moved. She soon saw that they provided her with an explanation of mutable genes. And mutable genes were simply ordinary genes to which something funny had happened. They became her key to the genetics of development.

98

Self-Control

Barbara McClintock was hardly a pure observer, averse to intervention, as the McClintock myth characterizes her. In both her personal life and in her science, she was active, dominating, and sometimes manipulative.

Each morning during the growing season, May through October, McClintock walked her field. It is a quiet, meditative process, requiring both focussed concentration and peripheral alertness. One must attend both to each plant individually and to the field as a whole. Was the plant all right, straight, bug-free, moist? Was it producing shoots yet, or tasseling? Was anything strange happening? McClintock knew each plant, because she kept her crops relatively small. She would cross an interesting plant to ten or twenty others, each slightly different in genetic constitution, each able to give a hint about the unknown processes stirring the chromosomes. That plant may itself have come from a one-of-a-kind kernel. She could not afford to lose even one plant, though of course she sometimes did.

Unable to tolerate sloppiness or uncertainty, she maintained total control over her material. She did not trust staff gardeners even to water her plants. Her occasional trips often ended with agonized homecomings when she returned to find plants underwatered, pulled up, or blown over. The Department of Genetics had a "greenhouse man but he is not too bright," she wrote to George Beadle in 1951—though she might have made the same complaint any time in her career. "Whenever I have been away for a few days during this part of the year—Feb. and March,—one of the assistants here has taken charge but I can't ask anyone to do it for more than a few days."[1] Evelyn Witkin recalled that McClintock's only permanent field hand was a human scarecrow: she armed him with a shotgun and told him to stand in the field and make sure no birds pulled up her plants.[2]

In 1953, she wrote a similar harangue to Beadle about her frustrations in trying to get the most basic help with scientific agriculture. She began by telling him what he already knew: "I do my best work with a small project, thought out carefully so that I can handle it myself." Nevertheless, there was always the problem of caring for the plants. "The question of vigilance comes in," she told him, "for a variety of things can happen during the growing season, either in the field or in the greenhouse, that must be controlled."

At Cold Spring Harbor there was no support community of maize geneticists and students to take care of each others' material, and the maintenance of the plants was in general "conducted in the most primitive manner," she continued. A few men did all the odd jobs. These were, McClintock said, "older men who are unable to obtain jobs paying adequate salaries"—by implication, weak, incompetent, and half-senile. Turnover was high; the labor pool was unreliable and unstable. "Often, labor has been either unavailable to me or has been so incompetent" that she had been forced to do the hard labor herself. "This past summer, it was even necessary for me to do some of the tractor work." She would have cut quite a figure, clenching in her teeth one of the long FDR-style cigarette holders she then favored.

Any emergency, such as flood or windstorm, meant she had to make repairs. Yet she was reluctant to try new people, to hire local high school students, for example. She told Beadle, "I had experience with one and he proved to be quite a failure. The plants just went unwatered." As a group, therefore, high school students could not be trusted. Director Demerec, she said, believed she was "too fussy about maintenance," and, she added incredulously, he was "honestly convinced that my attitude is an unreasonable one and that I expend too much of my energies on the maintenance foolishly."[3] Demerec may simply have been concerned for her health—he often admonished her to take it easy. But McClintock could hardly relax her standards. In science, as in art, the virtuosi tend to be control freaks.

Yet for most biologists, science is also a gregarious, cooperative activity. Not so McClintock. Occasionally she would take on an apprentice, only to be pained by his or her sloppiness. In 1942, she wrote to Marcus Rhoades, "If I only would not insist on doing my own work things would go swifter and easier. I had a girl working for me this summer but she was so careless, it grieved me to have her touch my material. I am going to stick it out alone for a while even if the pressure here is strong for my getting an assistant."[4] She took pride in her fierce independence and meticulousness— better to complain about the pace of work than to sacrifice certainty.

Perhaps the only assistant who met McClintock's standards was Sophie Dobzhansky, daughter of Drosophila geneticist Theodosius Dobzhansky, a regular summer visitor to Cold Spring Harbor in the 1940s and 1950s. Sophie seemed to have the needed gentleness and care to be trusted with McClintock's plants. But even Sophie suffered criticism. In 1949, McClintock, complaining about hot weather, a poor crop, and general fatigue, clucked to Rhoades: "Sophie did not arrive until the 26th of July— after the worst was over; her help now is welcome as I am tired, cross, and somewhat disinterested."[5]

In early 1950, Zlata Demerec, the director's daughter, won a brief, intense tutorial with the master cytogeneticist. The following fall, Zlata planned to go to graduate school at Cornell to study with Adrian Srb, a distinguished cytogeneticist. At the beginning of February, a quiet month in the maize geneticist's calendar, she asked McClintock to train her in the basics of Neurospora cytogenetics—or, as McClintock told Beadle, "I am elected to get her started." She threw herself into the tutorial, taking off a full week from her research, working late each evening to prepare examples and materials for the next day. Zlata, she wrote Rhoades, "seemed to progress well with the work although her background needs much filling out. Actually, I had a very good time as the N[eurospora] chromosomes are fun to work with. It was just the feeling of pressure and responsibility. That session is over but the feeling of squeeze continues. 'Taint fun!"[6]

Her need to control extended to other aspects of her life as well. She was always especially interested in younger scientists, who had a fresh outlook on science and did not question her authority. She herself said that Stadler had often talked about her "mother interest," in regard to students.[7] Sometimes people thought she was overbearing. In early 1942, her erstwhile advisor, Fitz Randolph at Cornell, exploded at her for what he thought was an overzealous editing job on his star student's paper, probably submitted to the journal *Genetics*. To McClintock herself, Randolph raged about her "garrulous, grandmotherly advice." The same day, he fired off a note to Rhoades, editor of *Genetics*, calling McClintock a "Prima Donna, as Emerson so often says," who wanted to be the "goddess of science" and "godmother to aspiring young scientists everywhere." Apart from the sexist tone that jangles modern ears, Randolph was surely wrong about McClintock's criticism when he told Rhoades that "women"—implying McClintock—"very often personalize their scientific thinking."[8] McClintock could be scathing, but she was rarely personal to someone's face.

In April 1947, McClintock was honored with the Achievement Award of the American Association of University Women (AAUW). She traveled to Dallas to receive the award at the association's annual meeting. The award carried a prize of $2,500. Though McClintock could be prickly, she knew how to be gracious. In her acceptance speech, she attributed the extraordinarily fertile period of her career in the early 1930s to her fellowships. She framed her paean to institutional support in terms of freedom. The National Research Council fellowship, she said, gave you the chance "to do something where you wanted to, when you wanted to, no restrictions, no obligations, you could work all night if you wanted. You did not have to get up and teach the next morning. To do what you wanted and really have a wonderful time doing it." Such freedom was a benefit to anyone doing creative work: "Anybody interested in art or any field where you need time and freedom should have fellowships and I am very strong for them."[9] The AAUW women were "exceptionally fine," she wrote Beadle afterward, "good looking and gracious. I labeled them a very handsome lot." Nevertheless, as she told Gertrude Zilske, Beadle's assistant, receiving the award was an ordeal, because she dreaded "submitting to Stuffed-shirt policies."[10]

The Source of Control

None of these distractions kept McClintock from her work for long. The mutable-gene problem consumed her, and control was the essence of the problem. Control had been on her mind even before she saw the twin sectors. In her 1944 Carnegie report, she had described plants in which the chromatid type of breakage-fusion-bridge cycle, normally confined to kernels and gametes, had persisted into the sporophyte. The fact that this had happened repeatedly, but in only 4 of the 188 families of plants she had examined, suggested "that some controlling genetic factors may be responsible for the continuation of the chromatid bridge cycle into the sporophytic tissues."[11]

Most of the mutants arising from the 1944 experiment were consistent with the BFB cycle. But in several of her mutable-gene cultures, part of one chromosome 9 seemed to be lost repeatedly as the plant grew. The pattern of this chromosome-loss mutant was different from that of the BFB losses, in two ways. First, the chromosome-loss mutant lost the same set of genes each time. Thus the dissociation of the chromosome always occurred at the same locus. She confirmed this finding both genetically, by showing that the same traits were affected each time a break occurred, and visually, under the microscope. Second, the spots were uniform in size and distribution. The BFB cycle produced spots of various sizes distributed randomly over the kernel. The uniformity of the chromosome-loss mutations meant that the dissociation was occurring at the same point in the development of the kernel. But what was causing this dissociation and loss? She designed her experiments of 1945 and 1946 to home in on the site of breakage.

The most striking plant she had by the end of 1945 was 3503D-17.

Culture 3503 was the family of plants resulting from the self-fertilization of plant B87-1. Plant D-17 of this culture was the seventeenth plant of the D kernels. The seedling was green, though the leaves had many fine, white streaks. McClintock selfed 3503D-17 to further reveal any new misbehaving genes, and she also crossed it to other "tester" stocks of known genetic constitutions. One of these crosses, made during the winter of 1945–46, resulted in an ear bearing what she called a special kernel. Most of the kernel had a colorless (white) background speckled with red. The red spots were clusters of cells with the dominant *C* gene active as well as the dominant *Bz (bronze)* gene. Over most of the kernel, then, *C* and *Bz* had been mutable, off in most cells but repeatedly turned on briefly to form the pigmented spots. The kernel, however, had one large spot, or sector, of *C Bz*. Whatever had been flipping the switch during the development of the rest of the kernel had stopped in that sector, yielding a simple dominant pattern with no variegation.[12]

McClintock saw this pattern as striking new evidence of the control of mutation. The twin sectors in the leaves had revealed changes—and exchanges—in mutation rate. This new kernel showed that the rate could go to zero—no mutation at all. She plucked out the kernel and grew it up, and it produced a plant in the summer of 1946 to which she gave the culture number 3745. Culture 3745 was the progenitor of a new series she called the chromosome-loss or Cl series. The lineage deriving from culture 3745 came to play a central role in her experiments.[13]

The descendants of the plants in the 1944 experiment continued to throw off new mutations. She carried out further crosses and analysis with the Cl plants and their great-uncles and second cousins. In early October 1946, she wrote to Beadle, "I have what I think is an extremely interesting, quite new phenomenon." She had twenty-five to thirty new mutable genes, all following the same "laws of behavior," she said. These "laws" were actually McClintock's generalizations about maize mutable genes. They amounted to the as yet dimly understood mechanisms of reduplication, or copying, of chromosomes and its disruption by chromosome breakage. She continued, "I believe that all of the mutable genes may be involved in some form of irregular chromosomal reduplication."[14]

In October, after the harvest, McClintock reaped generalizations from her data. It was a good time to present her current work to the other scientists at Cold Spring Harbor in a "staff meeting," or labwide research seminar. Preparation for this staff meeting seems to have spurred her into a period of intense intellectual activity. In the weeks before the late October meeting, she wrote out many pages of notes, tables, and diagrams. These notes, intended for no eyes but her own, offer a fine-grained image of her thought process as she first named what would become controlling elements.

If breakage was occurring consistently at the same location, then there was likely a genetic locus predisposed to breakage. A "locus" is literally merely a site, but geneticists had long used the term interchangeably with "gene." On 20 October, she wrote about a new locus, *"D,"* related to the dissociation events.[15] The *D* locus lay where the breaks were occurring, one third of the way down the short arm of chromosome 9. It "must be present to show breakage-detachment phenomena," she wrote, and then laid out the crosses that proved this, from 3503D-17 and its descendants. The *D* locus was a new kind of gene: rather than produce a pigment or other physical trait, it caused the breakage of the chromosome. McClintock's notes give no sign of surprise at this.

The "presence of *D* locus does not ensure that detachments will occur," she continued. There must be a second locus. When this second locus was present, "variegation by *D* may occur." Four days later, the second locus had a name: *"V,"* undoubtedly for "variegation"—a nod to Emerson and his nomenclature for the *P* locus.[16] She then asked rhetorically, "Does another locus, other than *D*, affect the behavior of the *D* locus?" By reanalyzing her crosses from the past three summers, she determined which plants carried no *D* locus, which carried a single *D* locus, and which carried two *D* loci. In plants where both chromosomes 9 should have carried *D*, she found that only about half the kernels showed variegation. Only when *V* was present would breakage occur at *D*.

Further, she believed the *"V* locus may itself be a variegating locus."[17] Although *V* behaved for the most part in a conventional Mendelian manner, its presence resulted in "new action of loci appearing in all variegating cultures." Nevertheless, its presence should be detectable. On 25 October, she worked out for several of her plants which ones should have *V* and laid out several tests and predicted outcomes.[18] She thus invested the *V* locus with a measure of autonomy. That autonomy made it capable of controlling *D*.

The next question was, Where is *V*? Classical linkage analysis suggested that *V* and *D* were inherited independently. She reasoned that "rates of test crosses involving Cl cultures and 3882B do not fit unless *V* is 50 [crossover] units or more" from the *C* gene. Fifty crossover units or more mean the two loci assorted randomly, or were "unlinked." This would place *V* either on the long arm of chromosome 9 or on another chromosome.[19]

McClintock spoke at the staff meeting on 28 October. It was her first public talk on her new mutable genes. She titled it blandly, "Behavior of a locus on one chromosome of the complement." The chromosome in question was of course chromosome 9, and the locus's behavior was quite strange. In preparation, she digested her notes from the previous days into an outline for her talk.

The outline begins by sketching her problem and showing the variegations and twin sectors. It sets out the cytological and genetic evidence for the existence and location of the *D* locus. Its presence caused breakage, resulting in loss of the distal two thirds of the short arm. The outline continues with a discussion of the genetic consequences of breakage and diagrams of the appearance of the chromosomes during breakage. Presence of *D*, she explained, was necessary but not sufficient for breakage at this site. This, she wrote in her outline, "brings up question of why?" The answer lay in the pattern: "Involvement of 1 factor is certain: this is the pattern of the variegation: Patterns very different" (underscores in original). She did not mention the *V* locus. "Would one expect all *D* carrying chromosomes to show variegation? Answer is NO." This note suggests she argued that *D* not being sufficient for chromosome breakage did not necessarily imply a second, *V* locus. The *D* locus might simply operate parttime. This thinking was precisely the opposite of her private reasoning of the preceding days.[20]

It is striking that her notes do not mention *V* at this point, and that she seems to have covered for it with a subjective argument based on pattern. The other geneticists in the audience would have recognized the 50 percent expression rate of the *D* locus as indicating the involvement of a second dominant locus; certainly one could imagine alternative explanations, but a second locus is the simplest. Why did she not mention the *V* locus? She may have conceived of its existence only a few days earlier. Perhaps it was too new and she did not yet have enough confidence in that result to broach it publicly. The *D* locus was the one for which she had solid genetic evidence and the one that was more consistent with conventional genetics. But it was *V* that drew her interest more and more.

She stressed that both the timing and the frequency of chromosome breaks were controlled. Breakage was not a random phenomenon. Further, two variables were controlled independently: timing of mutations and frequency of mutations. She then described four variegation patterns in terms of this control: (1) time of loss, which affected the size of spots (the later the loss, the smaller the spots); (2) frequency of loss, which affected the number of spots; (3) time and frequency of loss through the generations (for example, evolutionary change); and (4) time and frequency of loss within a tissue in a single plant, or developmental change. Under this last, she asked: "If change occurs, what is the nature of this change? How are sister nuclei affected?"[21] In other words, what is it that one cell gains and the other cell loses? No record survives of the reaction to the talk among the Carnegie scientists, but from her notes it is clear she intended to present an argument about mutable genes and control.

An Integrative Interlude

Two weeks after the staff meeting, on 10 November, McClintock, still unable to tie herself down to one institution, set off again by car for California. Beadle had moved to Caltech that year to take over the directorship of the biology division, the position held since 1928 by Thomas Hunt Morgan, who was retiring. Beadle himself was thus leaving the lab bench for a desk job, but he maintained his Neurospora laboratory and continued to take students. One was Jesse Singleton, who had started out in maize genetics at Missouri during McClintock's last year there. He had come to McClintock "in desperation" for help and guidance, which apparently brought out her mother interest. He asked McClintock for some suggestions of graduate programs, and one of them was Beadle's group at Caltech. He chose Caltech and Neurospora genetics for his doctoral project. McClintock told Beadle that Singleton was "not a fool by any means but he sometimes does foolish things because of excessive shyness." He sometimes acted peculiarly and Beadle might "have to do a little excusing" for him. She added, "he will need encouragement to get over his fears. Discouragement with himself may cause him to fail utterly."[22] She agreed to serve on his dissertation committee, which greatly pleased both Singleton and Beadle. Beadle arranged for a nonsalaried visiting professorship for McClintock, which gave her access to Caltech's laboratories and libraries. She roughed it for the trip out, bivouacking in the back seat of her car in a sleeping bag. She remained in Pasadena for two months, staying in a house in town. After the intellectual binge of the fall, her spirits seem to have wilted. Beadle wrote to her the day after Christmas, in response to some domestic complaint, "I hope your situation at the house improves and that you don't get too discouraged about it. Have a happy new year and don't let them get you down."[23]

Her spirits were back up by April, when even a seemingly interminable Long Island winter did not dampen them. She wrote to Gertrude Zilske, Beadle's secretary, "Don't worry too much about our weather—it ain't so bad, after all. I really quite like to beat at the weather when it beats at me. It is fun to wrap up in a fur coat and let the cold bring on the rosey [*sic*] cheeks. You could be envious if you only just wanted to remember! Spring is about to appear and where is it nicer than in the east?"[24]

Puzzles

By early 1947, McClintock fully accepted the reality of her new loci, *D* and *V*. She had been using the *"D"* and *"V"* notation in her culture cards since late fall 1946, and had even returned to culture cards of some plants grown earlier and modified the genotype, adding in *D* and *V* where appropriate, suggesting she was tracing the lineage of these loci, trying to figure out when, where, and how they had appeared.

She planted her summer crop on 13 May, about as early as the Long Island climate permits. She did most or all of it by herself, and it was tough work. The next day, she wrote to Beadle, "Got my planting done yesterday and the muscles are screaming today."[25] Her goals for the summer included further studies of the *V* locus, particularly to determine whether the *V* locus in each of several new mutants was the same *V* in all cases.

She also had a new *V* that acted on the *A2* locus on chromosome 5. Was this the same *V* that acted on the *D* locus on chromosome 9? Could *V*, in other words, affect other loci, or did the *D* and *A* loci each have their own modifier?[26] This was an important question in regard to developmental control. If *V* could affect multiple loci, perhaps it was a master switch that regulated large numbers of genes, determining when and to what degree they were active. McClintock did little with the new controller at the *A* locus until the mid-1950s. At that time she returned to it as the first instance in her plants of her second major control system, *Suppressor-mutator.*

The highlight of McClintock's summer was the discovery of a new mutant that she named *c-m1.* It was a mutable *c* allele, the recessive form of the dominant *C* gene. The *C* locus is peculiar. The dominant allele, *C*, produces red pigment in the aleurone, or outer layer of the kernel. The recessive form, *c*, produces no color. But there is also another gene at the same locus. It is designated *I*, for inhibitor of color. Like *c*, its phenotype is no color, but *I* behaves as a dominant. Aleurone cells that are *Cc* will have color, but *IC* cells will be colorless. *C* and *I*, then, refer to the same physical site on the chromosome. The *"m1"* in *c-m1* indicated optimistically the first mutable *c* allele.

McClintock had crossed ten female plants that lacked both *D* and *V* and were recessive for the *C* gene with a male that had two copies of *D*, one of *V*, and a double dose of *C*. When these crosses produced ears, she expected a number of kernels to show variegations in the *C* gene, when *D* breaks caused loss of *C*. These should appear as colorless spots on a red background, and they did. But on one of the ears there also appeared a single kernel with red spots on a colorless background, a revertant from *c* to *C*. McClintock designated these *"c→C"* mutations. These were unexpected. She grew a plant from this kernel in the greenhouse during the winter. The plant had yellow-green streaks on the leaves, indicating that *D* and *V* were present and had been contributed by the male parent. This meant the kernel was not the result of contaminating pollen from another plant. She gave this plant the number 4204.[27]

McClintock had seen *c→C* mutations before, but they had always been accompanied by mutations at the other loci on the chromosome arm. Plant 4204 was unusual in that only the *C* locus had become mutable. Also, *c-m1* did not produce chromosome breaks as *D* mutations did. It was clearly a different type of mutable gene. McClintock wondered whether the *c-m1* plant carried *V*; whether *V* was needed for *c→C* mutations; if so, whether the same *V* activated *c-m1* as activated *D*; and finally, whether *c-m1* activity was related to *D* activity. Culture 4204 and the *c-m1* mutant would occupy much of her thought for the next year.

By the end of July she had nearly finished her pollinations. Planting and harvest are hard work, but pollination season is the toughest. One's schedule is set by the plants. When they begin tasseling and producing shoots, the scientist has to be ready; the window for collecting pollen from a given plant is only about five hours. That plant can be crossed only to plants bearing shoots that are ready to be pollinated. The work is demanding, both physically and mentally. The geneticist must both plan carefully and improvise. First, she designs her experiments, often days or months in advance. During pollination season, the morning inspection of the cornfield reveals which plants are ready to be crossed. She then must quickly solve a puzzle, matching up males and females that are ready to be crossed within the constraints of what she wants to know. On 30 July, McClintock wrote to Beadle, "Am just on the ebb of the pollination season and feel somewhat in the ebb state myself! It will be good to have it all over."[28]

In late summer or early fall, McClintock altered the nomenclature for her controlling loci. On 28 September 1947 she referred to *"Ds,"* her new designation for the *D* locus. In those same notes, she also wrote that "pattern depends on condition of *Ac* locus,"[29] changing her *V* designation to *Ac,* for "activator." Since she had first hypothesized *V*'s existence, she had thought of it as making *D* "go into action"; the new name expressed this more accurately.[30] A risk of employing such functionally-descriptive nomenclature is that a name may turn out to be too restrictive. McClintock soon found that *Ds* did other things besides break chromosomes; by 1949 she had begun to regret naming *Ds* "dissociator." This misnomer still ate at her in 1972.[31] She never regretted naming *Ac* the "activator"; to her, it was always the element that controlled *Ds.*

As she planned for the winter greenhouse crop, McClintock pondered the relationship between *Ds* and *Ac* and their physical nature on the chromosome. She was beginning to think *Ds* and *Ac* were complex or compound loci, composed of several, probably identical, units. Their action seemed to depend on their state or composition. *Ds* seemed to "give something" to *Ac* when dissociation occurred, doubling *Ac* and reducing *Ds.* "Very probably, the state of any one *Ds* locus affects the type and frequency of mutations," she wrote. "This, however, is only one of the factors involved. With any one particular *Ac* locus, the pattern will depend on the differences in the *Ds* locus."[32]

An important piece of the puzzle was dosage, the number of copies of *Ds* or *Ac* present in the cell. McClintock exploited the fact that the endosperm, the fleshy portion of the maize kernel, is triploid—it contains three copies of each chromosome. She designed experiments in which the endosperm got zero, one, two, or three doses of *Ac,* and the same with *Ds.*[33]

The third dose of *Ac* proved puzzling. In December, McClintock analyzed crosses in which *Ds* dosage was kept constant at one, while *Ac* dosage was varied from one to three. One dose of *Ac* produced moderate speckling. Two doses produced heavy speckling. But with three doses the speckling was lighter even than with one dose—the kernels were almost colorless. A triple dose of *Ac,* she wrote in her research notes, "either has no losses or very delayed losses that usually show as a lightly speckled pattern."[34]

She solved this puzzle with a flash of insight, equal in its originality to any in her career. She surmised that in addition to determining the frequency of *Ds* mutation, *Ac* must influence the timing of the onset of mutations. She saw that the appearance of the triple-*Ac* kernels "suggests that dosage of *Ac* results in a much delayed appearance of dissociation."[35] The more *Ac* present in a cell, the later and more frequent the mutations. With one or two doses, more and later mutations meant more and smaller speckles. But with three doses, the onset of mutation was so late that the kernel was almost fully developed before any mutations occurred. The development of the kernel was the limiting factor.

"This Thing Was Moving Around"

In late 1947 and early 1948, anomalies began to mount in several quarters of McClintock's investigations. The *c-m1* mutant, for example, interacted in some way with *Ds*, and both responded to *Ac*. When *c-m1* and *Ds* were both present in a nucleus, fewer *Ds* mutations occurred than with *Ds* alone. The *c-m1* mutant, she wrote, "acts as a modifier of *Ds* but both are dependent on *Ac* for primary action. If c-m1 could *not* be seen as a visible mutation, its presence would be manifest by a *reduction* in the frequency of *Ds* mutations."[36] There seemed to be a zero-sum relationship between *c-m1* and *Ds*, with *c-m1* taking away some of the *Ds* mutations.

McClintock had planned to go to Caltech again in December, but she had to cancel the trip. She apologized to Beadle that her work was going too well to leave it, and she was not happy about the fact. She referred to Beadle's visit to Cold Spring Harbor the previous October, writing, "When you were here I stressed the frustration I felt over the problem I was working. For the past month I have been working frantically day and night, to get my ideas organized and make plans for the winter planting." The pressure she put on herself at last overcame her, and "the night before last," she wrote, "I just broke up." Ernst Caspari, another confidant and colleague at Cold Spring Harbor, "stated if he had my problem, breaking the way it is now, one couldn't move him from it. I don't feel quite that strongly, however!" This moment of apathy was fatigue-induced. As a Carnegie scientist, McClintock's only formal duties were to do science at her own pace. Yet she placed such high demands on herself that she paid a physical price: "Because of the frantic work, I am terribly tired and nervous . . . the muscles in my face feel as if they haven't relaxed in days."[37] Beadle was encouraging and sympathetic. He understood that she should stay at Cold Spring Harbor but wished she were "running down the hot lead" in Pasadena, "so we could watch from the sidelines." He signed off, "Don't let the mutable genes get you down."[38]

After the new year, McClintock continued her analysis of *Ds, Ac,* and *c-m1*. The winter was a hard one, "the worst for me in years," she wrote to Beadle. "My muscles are bulging from the exercise of shoveling snow—no kidding! It snows about every four days and no thaws."[39] By late January she had confirmed that c-$m1$ was close to C, twenty crossover units from the end of the arm, and that Ds was toward the centromere from Wx.

She hypothesized that c-$m1$ and Ds competed for Ac, and that the $c{\rightarrow}C$ mutations somehow "protected" the Ds locus from breakages. Perhaps a given Ac could act only on a single locus; if an Ac caused $c{\rightarrow}C$ mutations, it could not cause breakages at Ds. Also, or in addition, the C locus might harbor a "modifier"—perhaps c-$m1$ itself—that affected Ac action. In describing this hypothesis McClintock seemed to have come to a firm conviction and then softened it by mentioning, parenthetically, its contingency: "modifier may be at C locus itself resulting from competition of changed c and Ds for Ac action. This shown by c-m1 activity. Change to C gives protection from mutations at Ds (To be proven)."[40] In mid-February, she still needed "proof of need for Ac to activate c-m1 dosage of Ac and rate & time of mutations[;] effect of c-m1 as modifier of Ds action."[41] Her faith in her analyses grew so great that she sometimes conflated assumptions, hypotheses, and conclusions.

Another puzzle concerned the location of Ac. Normally, the process of localizing a new element to a specific locus was straightforward: just perform the right crosses and do the math. But mapping Ac was proving frustrating. At first, Ac and Ds had seemed to be unlinked, implying that Ac was located far from Ds, either on the other side of the centromere or on another chromosome altogether. But by January 1948, she had found plants in which Ac and Ds appeared to be linked.[42] In other words, in some plants it seemed to be located near Ds and in others not.

In March, McClintock analyzed her greenhouse material from the previous winter and began to plan out her crosses for the summer. She was still mulling over the location of Ac. The c-$m1$ locus and its apparent competition with Ds also absorbed her. She began to consider c-$m1$ behavior as a modified type of Ds action, to link it to dosage effects—the "few-late loss" pattern, which resulted from a triple dose of Ac—and, tentatively, to the twin sectors. In her research notes, she followed the standard convention of "right" and "left" when referring to gene locations. The image is of the chromosome laid out before one, with the short arm to the left and the long arm to the right. "To the right" in reference to the short arm of chromosome 9 thus means "toward the centromere," and "to the left" means toward the telomere, or tip. She wrote,

If in the modified type of Ds action where modifier is located 20 units from end of arm [for example, c-$m1$], it is suggestive that some factor at this locus is required in order that this particular Ds locus functions. In other words, there may be 2 segments of the Ds locus, one located to left of I [C] and the [other at the] regular Ds locus. Possibly, only when 2 are present will Ds behavior occur. Breaks can occur at *either* Ds locus. This would account for the twin sectors and the relatively frequent appearance of . . . chromosome patterns that are sectional in the kernel. It would also account for the few-late losses appearance because of interference with the regular Ds locus—such as occurs with c-ml.[43]

In other words, *Ds* seemed to be in two places at once. To explain this, McClintock seems to have had in mind an established genetic mechanism, such as duplication of the *Ds* element followed by a chromosomal rearrangement that resulted in the insertion of a new *Ds* near the *C* locus.[44] (See Figure 5.1.)

Yet another anomaly appeared in regard to *Ds*. In the summer of 1947, McClintock crossed a plant carrying *Ds* and *Ac* to a standard tester stock recessive for all genes on the short arm of chromosome 9 except for the *C* locus. One of the resulting kernels showed an unexpected type of variegation. "In this kernel, many chromosome breaks of the Ds mutational type had occurred, but the position of the breaks was just to the right of the I locus, instead of to the right of the Wx locus," McClintock wrote, summarizing these results in April 1949. The kernel showed typical *Ds*-induced mutations at the *I* locus, but *Sh, Bz,* and *Wx* were unaffected.[45]

She treated the kernel gingerly. She did not slice away any tissue to examine the chromosomes. She planted it in the greenhouse in the winter of

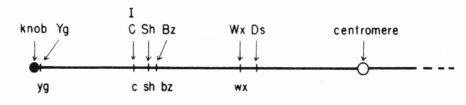

Figure 5.1. Diagram of chromosome 9, with most of the long arm removed. The *V* or *Ac* locus (not shown) seemed very far from *D (Ds)*, either on the long arm of chromosome 9, or on another chromosome altogether. Reprinted with permission from Barbara McClintock, "Chromosome organization and gene expression," *Cold Spring Harbor Symposia on Quantitative Biology* 16 (1951): 22, fig. 9.

1947–48. It matured early; by the time it tasseled, no other plants were ready to which it would be useful to pollinate it. All she could do was self it and see what mutations were exposed. The plant received number 4306. Analysis of the resulting kernels confirmed the presence of *Ds* activity—but between *I* and *Sh,* not to the right of *Wx.*[46]

Evelyn Witkin was witness to all of this. While training in Drosophila genetics with Theodosius Dobzhansky at Columbia, Witkin had taken a maize genetics course with Marcus Rhoades. As a postdoctoral fellow, Witkin worked on bacterial genetics in Demerec's lab at Cold Spring Harbor, and in 1945 Demerec appointed her a staff investigator at the Carnegie Department of Genetics. Her eyes are bright, her smile quick and inviting. She and McClintock fast became friends. In an interview, McClintock said that Witkin "was somebody I would go to whenever a new finding came along. And Evelyn was the only one that really understood what I was doing and why I was doing it."[47]

When Witkin entered McClintock's laboratory, McClintock "was usually at the microscope, and there were stacks of reprints and journals on her desk, all with little slips of paper marking things." One of Witkin's strongest impressions was of McClintock's kinetics. "She moved around very quickly. She hardly ever stayed in one place, even at the microscope she would jump up and do something else and walk around. She always had some things to show me."[48]

One day, in March or early April, McClintock called Witkin in an unusually agitated state. "She was just beside herself with excitement, and was almost incoherent, she was talking so fast. She had drawn the conclusion that this thing was moving around. And she was very very excited about it. I was not at the moment able to follow the experimental evidence for this, not right away, because she was in a very interesting state of excitement. I mean more than usual. Because I think she saw something of the outcome of this."[49]

It is unclear whether "this thing" that was moving around was *Ac* or *Ds.* On 18 April 1948, McClintock wrote out an outline of some planned crosses under the heading: "For change in position of *Ac* locus."[50] By this time she had clearly hypothesized that the reason *Ac* and *Ds* sometimes seemed linked and sometimes "free" or unlinked was that *Ac* had moved, from a site far from *Ds* to one near it.

Ten days later, on 28 April, she headed a notebook sheet "4306—Transposed *Ds.*" On it she wrote out the first explicit description of transposition. The explanation for *Ds*-like activity at the *I* locus in that plant was that *Ds* had been "inserted into short arm of chromosome 9 to left of *Sh*." When this happened, it left a "deficiency where *Ds* moved."[51] A few months later, in a letter to cotton geneticist S. G. Stephens, she explained that transposition of *Ac* was supported by "some evidence that has recently turned up. It involves the insertion of the Ds locus from its position immediately to the right of Wx into a new position between I and Sh." This case further suggested that *Ds* was small in size, because while *I* and *Sh* lay near one another, the inserted *Ds* element did not disrupt the normal crossover relations between them. "This particular case is very interesting and I am pushing it hard this summer."[52]

In a summary of these results, probably written about two years later, McClintock observed, "The known transposition of Ds serves as a basis for assuming that the different locations of Ac are the consequences of transpositions of a single Ac locus possibly by a method comparable to that observed for Ds."[53] Thus, although the first mention of transposition in her research notes concerns *Ac*, soon after the discovery she referred to the transposition of *Ds* as indicating that *Ac* also moved. Combined with the fact that she had been puzzling over *Ac* free versus *Ac* linked since the previous winter, this evidence suggests—but only suggests—it was *Ds* first, *Ac* second.

Nothing in McClintock's research notes supports Witkin's memory of McClintock's great state of excitement at recognizing transposition. In 1946, McClintock used exclamation points and underscores when she realized that a subgene explanation for *Ac* and *Ds* could account for an exchange between chromosomes. But her research notes in the spring and summer of 1948 are free of hyperbolic punctuation. If McClintock in fact realized *Ds* transposed before she realized *Ac* did, there is even less to support a transposition revelation, because she saved no notes on the transposition of *Ds* until ten days after her first notes on transposition of *Ac*. McClintock was meticulous in saving her notes from this period; it seems unlikely that she would have not taken notes or thrown them away—indeed, if either were the case, it would further weaken the case for transposition being the major discovery of controlling elements. On the other hand, given transposition's later significance, it would be natural for Witkin's memory to exaggerate the "Eureka!" moment, or to associate a remembered day of excitement with the discovery of transposition.

This is not to say that transposition was insignificant to McClintock. Immediately, it solved two problems that had been vexing her. But it was not the discovery of a new mechanism of chromosomal behavior that thrilled her. She does not appear to have even considered the mechanism of transposition for at least several months. When she did so, it was with the language of conventional classical genetics and cytology.

Rather, McClintock's own comments suggest it was the *role* of transposition, more than transposition itself, that excited her. This emphasis emerged gradually over the following months. Transposition was the process "responsible for the change in pattern," she said in an interview.[54] The change in pattern reflected developmental control. In 1980, after transposition was discovered in bacteria, yeast, and fruitflies, McClintock recollected that "people have said how pleased I must be that all of this transposition is going on and I said, 'Don't take that for granted. The real point is control. The real secret of all this is control. It is *not* transposition.'"[55]

A Developmental Genetics of Her Own

When McClintock had seen the twin sectors in 1945, she "decided that what had occurred was that a control of the action of genes was involved. That this control of the action of genes was similar to the determination event of the embryologist," she said in 1980.[56]

She had learned some embryology in college. She invoked this training retrospectively, when reflecting on her work of the late 1940s, as her "zoology." With the exception of the nucleolar organizer, her work of the 1930s shows little influence of embryology. Crossing-over, ring chromosomes, and the breakage-fusion-bridge cycle were straight classical cytogenetics. But the twin sectors reawakened her interest in development. On first sight, the twin sectors suggested to her a regulation of mutation that changed through development.

As a cytogeneticist whose early background included zoology and embryology, she was attuned to the questions of growth and form. But as one with twenty years in classical genetics, her outlook was rigorously nucleocentric. Others interested in developmental genetics were concerned with how the cell could control or modify the genes. McClintock worried about how the genes controlled the cell.

Her scientific problem became one of using nuclear genetics to solve a problem of development. Her data seemed to answer the paradox of nuclear equivalence, the question of how, if all cells have the same set of genes, and all genes are turned on all the time, one organism can produce different tissues. Embryologists had used this paradox as an argument that the Mendelians, in ignoring the cytoplasm, had failed to link genetics to the rest of biology. McClintock, however, answered the paradox without invoking the cytoplasm. The chromosomes controlled themselves, she said, turning on the correct set of genes at the right time for each cell type. The twin sectors were "a case of a determining event occurring, as a consequence of a mitotic cell division event, in which the two daughter cells differed by a determining event that would control what happened *later.* And if that was related to development, I should try to find what it was that one cell gained that the other cell lost (emphasis hers)."[57]

The language of "determining events" traced back through the embryological literature to August Weismann's idioplasm theory, in which determinants segregated in each cell division, with each cell ending up with a single determinant, or cluster of biophors, that determined what sort of cell it would become. Though the idioplasm was long gone, embryologists maintained and modified the concept of determining events, to refer to an unseen event whose effects were expressed later in development. A determining event was a decision point in development, at which a cell proceeded down one path or another; past that point, there were certain kinds of tissue that a cell could no longer become. Determining events gave embryologists access to the domain of time, which geneticists focussing on the abstractions of hereditary transmission ignored. McClintock needed access to the time domain as well. All of her most important insights and discoveries had come from applying her cytology skills to genetic problems. Classical cytology had strong ties to embryology. So here again, it was from her cytological background that the apparent solution came.

By the 1940s, the German tradition of cytoplasmic inheritance studies had begun to influence American genetics. The pioneer and leading researcher in American cytoplasmic inheritance was Tracy Sonneborn, of Indiana University. Sonneborn had trained in protozoology at Johns Hopkins under Herbert Spencer Jennings, who had carried out classic research on heredity and selection in the single-celled ciliate Paramecium. Sonneborn adopted both Jennings's research organism and something of his holistic, organicist view of nature. A warm, gentle man, Sonneborn spoke of Paramecium in anthropomorphic terms and described the need for "empathy" and "understanding" of one's experimental material. He was a naturalist at heart and a sympathetic observer of his organism who had, as Judy Johns Schloegel cites, an "intimate knowledge" gained through "intimate acquaintance" with his protozoans.[58]

Sonneborn investigated the relationship among genes, cytoplasm, and environment. His classic study began in 1943, with the published description of a trait he called "killer." Paramecium that carry the killer trait kill animals sensitive to it. Other strains are immune to the killer trait. Sonneborn isolated a genetic factor, *kappa,* that was responsible for the trait. He found that *kappa* was inherited cytoplasmically and that its effects could be modulated by altering environmental variables such as nutrition and temperature.[59]

Sonneborn's work sparked the development of a new line of research into what C. D. Darlington called plasmagenes, or heritable cytoplasmic particles.[60] In 1950, Darlington gave a paper entitled "Mendel and the determinants," for "The Golden Jubilee of Genetics," a symposium celebrating the fiftieth anniversary of the rediscovery of Mendel's principles. In it Darlington ranked cytoplasmic inheritance with genetic crossing-over as one of the two great exceptions to Mendel's law of segregation. Its study, he concluded, "will in future enable us to see the relations of heredity, development and infection and thus be the means of establishing genetic principles as the central framework of biology."[61] In other words, cytoplasmic inheritance would be crucial to the reintegration of development and heredity—with the bonus of potential medical applications.

For a time in the mid-1940s, plasmagenes and cytoplasmic inheritance captured the fancy of the genetics community; non-chromosomal genetics appeared to be the new frontier in heredity studies. But by the turn of the decade, *kappa* was shown to be a virus, and other examples of cytoplasmic or non-Mendelian inheritance proved to be Mendelian after all. This work did, however, have a lasting impact on genetic thought. It challenged once again the dominance of the string-of-pearls gene. Geneticists were forced to reconsider the gene theory; the effects would reverberate through the 1951 Cold Spring Harbor Symposium, where McClintock made her famous presentation of controlling elements.

McClintock shared the philosophy of the cytoplasmic inheritance crowd, but she never aligned herself with them. She knew the principal figures and counted some of them among her friends. In 1944, for example, Sonneborn sent her congratulations on her election to the National Academy of Sciences, and they continued to exchange amicable notes—he congratulating her, she thanking him warmly—until the 1970s. In none of their surviving correspondence, however, is there any scientific discussion.[62] Darlington, in his essay on Mendel and the determinants, admired

McClintock's work: he dilated on the nucleolar organizer as an example of evidence of the structural integration and complexity of the chromosomes. But he did not mention McClintock's new findings on transposable controlling units.[63]

Ruth Sager was another leading figure in American cytoplasmic inheritance research. A student of Marcus Rhoades at Columbia in the late 1940s (Rhoades's dissertation had been on cytoplasmic inheritance of pollen sterility), she went on to a long career at Hunter College in New York. At Hunter, she worked on the genetics of chloroplasts and pioneered the use of the single-celled organism *Chlamydomonas* as a model system. By the 1960s, she had established herself as a leader in the field of cytoplasmic inheritance, a phase of her career that culminated in the authorship of a textbook entitled *Cytoplasmic Genes and Organelles,* published in 1972. In 1975, she moved to the Dana Farber Institute in Boston, where she made important contributions to our understanding of breast cancer and eventually became chief of cancer genetics. But before she did all this, she had to run the McClintock gauntlet.[64]

In 1949, Sager asked McClintock to serve on her Ph.D. committee. Her thesis concerned starch production in maize kernels, which is influenced by the *Waxy* locus, and she used some of McClintock's stocks. McClintock agreed to serve on Sager's committee, but immediately there was friction between them. On 6 May, she complained to Rhoades, "A copy of her thesis came yesterday morning. I have only just glanced at it but was appalled at a statement she made about the rearranged chromosome 9." Long explanation. Then: "Marcus, I just do not trust her. I do not trust the accuracy of the stated observations. She has been under great emotional strain. This, I suspect, has been due to the realization on her part of the sloppiness of her whole research efforts [*sic*]." She continued, "Sager's reaction to me is hostile. Because of the position she finds herself in, this attitude is understandable. For this reason, it is unfortunate that I must be on the committee to examine the thesis."[65] Sager later developed a reputation as a determined, competitive researcher[66]—little wonder, with this introduction to scientific repartee.

Other well-known geneticists who were interested in development, such as Salome Glueckson-Schoenheimer, Conrad Waddington, and Leslie C. Dunn, do not appear in her surviving correspondence or private writing during this period. In a brief conversation in 1996, Glueckson-Schoenheimer told me that in the 1940s and 1950s, she did not recognize McClintock as having done developmental genetics. Nor does

McClintock appear in recent historical accounts of developmental genetics and cytoplasmic inheritance.[67] The plasmagene people simply did not see her as one of them.

There is no reason they should have. McClintock stayed within the milieu of nuclear genetics. She used the tools and methods of classical cytogenetics. She remained steadfastly nuclear in her orientation, rarely considering the cytoplasm as a significant influence on heredity, as other developmental geneticists did. When she began presenting this research at scientific meetings in the 1950s, she did so mainly before gatherings of classical geneticists and the bacterial and viral geneticists who were developing the new molecular biology.

A few biologists investigated questions of gene action and differentiation from a nucleocentric point of view. In 1941, George Beadle and Edward Tatum first put forward their hypothesis, based on the genetics of Neurospora, that a gene was a particle that determined an enzymatic reaction. Their paper, "Genetic control of biochemical reactions in Neurospora," began thus: "From the standpoint of physiological genetics the development and functioning of an organism consist essentially of an integrated system of chemical reactions controlled in some manner by genes." In their model, each gene controlled a single enzymatic reaction in a biochemical pathway, such as the synthesis of vitamin B6. Their theory, a landmark in the history of genetics, became known as the one gene–one enzyme hypothesis.[68]

It was a theory of genetic control, but it differed in two important respects from that which McClintock was developing in the late 1940s. First, Beadle and Tatum's approach was strongly chemical. This reflected the background of Tatum, who was trained as a chemist. Beadle, trained in genetics alongside McClintock, recognized the power of chemistry in explaining gene action and had the mental and technical flexibility to adopt a chemical approach to genetics. McClintock never evinced any real interest in chemistry. Second, the control in their system was by the genes on the physiology of the cell. They did not address what controlled the genes themselves.

McClintock made regular visits to Beadle's laboratory beginning in 1944, when she helped him work out the cytology of Neurospora. No correspondence survives to tell us how McClintock reacted to this work at the time, but by 1951 Beadle had become the figurehead of a group of scientists more committed than ever to the string-of-pearls vision of the gene. By transforming the paradox of nuclear inheritance into the more tractable questions of physiological genetics, Beadle and Tatum steered the problems of the genetics of development away from the question of gene action. Though McClintock and Beadle remained friends, they came to represent opposite ends of the spectrum on the theory of the gene.

Boris Ephrussi was another leading scientist in genetic investigations into development. Ephrussi, who was Russian-born but had lived in France since the 1920s, introduced Beadle and, later, Jacques Monod to questions of gene action.[69] His experiments with Beadle on eye development in Drosophila were a technical and analytical tour de force.[70] In the 1920s, he helped initiate investigations into gene action. He was, in short, one of the brightest of the developmental geneticists of the time and one of the best experimentalists. Jan Sapp has described Ephrussi as "polarized on the gap between embryology and Mendelism."[71]

Ephrussi therefore ought to have been a prime ally of McClintock's. Yet in a 1950 letter to Rhoades, McClintock mentioned a recent visit to Cold Spring Harbor by Ephrussi and his new wife, the noted crystallographer Harriet Taylor. Ephrussi was by this time working in France on gene action in yeast. Taylor's research concerned virally induced transformations in Pneumococcus—fine work, though less developmental than Ephrussi's. Ephrussi was apparently eager to speak with McClintock, but the feeling was less than mutual: "Boris was plenty anxious to tell me of what he was doing but I had to make an engagement with Harriet to get her to talk. In my opinion, her work is far more important than his."[72]

According to bacterial geneticist Rollin Hotchkiss, Ephrussi was aware of McClintock's work, got the argument, but did not see how the interpretation followed from the data. In Hotchkiss's telling, when presented with McClintock's findings, the distinguished and brilliant Ephrussi seems reduced to head-scratching gibber: "I remember Boris Ephrussi, certainly one of the more sophisticated geneticists of higher organisms, coming away from talking with her and saying, 'It . . . what . . . Ds . . . it's very troubling."[73]

New Suggestions from Strange Quarters

The atmosphere at Cold Spring Harbor was changing for the better. The end of the war brought a general social warming. The summers, particularly, regained their prewar liveliness—a new experience for McClintock, who had come to Cold Spring Harbor just as the United States was entering the war. The annual symposium was reinstated in the summer of 1946, with a big, vigorous meeting that reflected Cold Spring Harbor's increasing orientation toward bacterial genetics: it was titled "Heredity and Variation in Microorganisms."

The phage group was an increasingly important part of the Cold Spring Harbor environment, especially in the summer. It was bigger than ever, and growing fast. The phage students brought both great intellectual talent and a boisterous fraternity sensibility to the placid seaside laboratory. In 1948, graduate student James Watson did not take the course but did spend the first of many summers at Cold Spring Harbor among the phage group. He and his cohort would often "go into the village to drink beer at Neptune's Cave. On other evenings, we played baseball next to Barbara McClintock's cornfield, into which the ball all too often went."[74] In early 2000, Watson told me, entirely plausibly, that the remark reflects the younger scientists' view of McClintock as an intimidating and distant figure. "We were like children," he said. "We knew Barbara would get pretty mad if the ball went into her cornfield. It was like your mother," who would yell at you for breaking a window. He added that the games included nonphage people as well, among them Sophie Dobzhansky, Theodosius Dobzhansky's fifteen-year-old daughter who helped McClintock in that cornfield.[75]

Under Delbrück's charismatic leadership, the phage group tackled the fundamental questions of life. Delbrück, a romantic theoretical physicist, hoped bacterial viruses would reveal new physical laws that would explain the self-reproduction, self-organization, and hereditary transmission that made living things unique.

McClintock followed the phage research, but for her, it raised different questions. While Delbrück was trying to discover new laws of physics, McClintock was interested in the developmental and evolutionary implications of the phage work. Surprisingly, it was phage genetics, not plasmagenes, that led her to a rare meditation on cytoplasmic inheritance.

She had considered the cytoplasm in terms of regulation of development at least once previously, in 1942. Strikingly, she seems to have considered nucleic acid, not protein, as the hereditary material—two years before Oswald Avery, Colin McLeod, and Maclyn McCarty's discovery of the "transforming principle" in bacteria. In notes dated 13 October 1942 and filed under "Problems general," she jotted down what she thought were crucial experiments: "To determine whether nucleic acid stored in cytoplasm of developing egg during long growth period (see Painter and

Taylor—Proc. Nat. Acad. Sci. Aug. 1942) is related to utilization for chrommeres in subsequent egg cleavages." And further, "Determine by photometry the gradual decrease of nucleic acid in cytoplasm during cleavages and accumulation in nuclei. Would suggest nucleic acid stored in cytoplasm to take care of rapid divisions of chromosomes [?] where no time allowed for new synthesis of nucleic acid."[76]

In the late 1940s, she kept a file of notes on her reflections concerning phage and bacteria. Bacteria and viruses, she wrote, had taught us that hereditary units may lie scattered throughout the cytoplasm, not tied up in chromosomes within a membrane-bound nucleus. (The genes of a bacterium or virus, we now know, are in fact collected into a chromosome, but she was right that they are not in a nucleus.)[77] She suggested that the scatter of hereditary particles in these organisms might have evolved from the more organized system of higher organisms that string their genes together onto chromosomes. This conjecture turned the mainstream evolutionary view on its head. Most scientists believed (and still do) that most evolution is a process of increasing order: that more complex organisms evolved from less complex ones, and that genes coalesced onto chromosomes and became bounded by nuclear membrane. McClintock's conjecture also contradicted the mainstream genetic view that the chromosomes were the only significant hereditary particles. She suggested that "many organisms possess both types, a chromosome component and a cytoplasm component of nucleoprotein having *genic* action" (emphasis hers).[78]

This brief consideration of cytoplasmic inheritance, written about the time she was unraveling mutable genes and transposition, may indicate an effort to place her new results along a spectrum of genic action. That it emerged in a written reflection on phage suggests she was integrating the two poles of contemporary genetic research. At one end lay stationary, chromosomal genes, as described by geneticists. At the other end lay cytoplasmic genes, the interest of embryologists and a handful of developmental geneticists. In the middle, perhaps, she placed her new movable elements, genelike particles that integrated into the chromosomes but which might at any moment be found floating about the nucleus. Her contact with the phage group, then, loosened up her thinking about genes and chromosomes. She was beginning to see hereditary material as fluid, acting both inside and outside the nucleus, and possibly having unforeseen properties.

"The future in genetics lies in an analysis of genic action," she wrote, again some time in the late 1940s. The notes got filed with her other miscellaneous thoughts on phage and bacterial genetics. The analysis of genic action, of course, she was coming to see as *her* future. Given the way the problem was consuming her, it was natural enough to see it as the most important problem in genetics. "We know," she continued,

1. That self-reproducing elements in cells are nucleoproteins
2. That the chromosome is composed of nucleoprotein
3. That the chromosome carries many of the factors, or genes, responsible for differentiation and growth
4. That genic action is related to control of enzymatic action
5. That the stream of enzymatic reactions controls the process of differentiation and thus development.[79]

In these assumptions, one can see the influence not only of the phage group but also of Beadle and Tatum. McClintock clearly recognized that the action of the genes was related to enzyme action. Her fourth and fifth assumptions show she had accepted the one gene–one enzyme hypothesis at least by the late 1940s, well before it was widely adopted by the genetics community. But although the two theories shared the language of control, they operated on different logical levels. Beadle and Tatum were interested in control *by* the genes; McClintock was pursuing the control *of* the genes.

This thinking, spanning the living world from corn to viruses, led her momentarily to contemplate evolutionary change. Evolution has created little that is new, she wrote. What is new has come about in one of three ways. The first was "loss of certain actions in the cell," or mutation. Second was "a change in timing of particular action, and third was "a change in the special relations in the cell of a particular reaction."[80] These last described her work leading up to and immediately following her discovery of transposition. Changes in timing were caused by dosage and perhaps by other means. Changes in "special relations" referred to what she soon called "changes in state" of a mutable locus. In order for timing and "special relations" to be changed, they had to be set in the first place.

The focus of her work during this period, then, was on what set that timing and those relations. By the summer of 1948, she suspected that it was *Ac,* and possibly other elements like it, that controlled gene action. She concluded: "This conception brings about a[n] element of unity in theories about a number of diverse reactions in cells and organisms."[81]

Control, Pattern, and Integration

Transposition, McClintock's most famous discovery and that for which she won the Nobel Prize, must be understood in terms of her search for the source of control of the genes. Her interest in control of mutable genes preceded the discovery of transposition by more than three years. Within ten years, by the late 1950s, she largely ceased her in-depth exploration of transposition, though control remained central to all of her science for the rest of her life.

Proximately, transposition was McClintock's solution to the twin puzzles of how *Ds* could seem to be in two places at once and how *Ac* and *Ds* could be unlinked in some cultures and linked in others. It was never, however, what interested McClintock most about either locus. *Ds* was interesting because it was a genetic locus that, instead of producing a trait, broke the chromosome. It was a genetic site, perhaps a gene, that initiated chromosome breakage. It was also unusual in that it only acted when another locus, *Ac*, was present. *Ac* engaged her because it, instead of producing a trait, controlled the action of another locus, namely *Ds* (and possibly others). Transposition soon provided McClintock with an explanation of mutability. It gave her the vehicle by which *Ds* and *Ac* could affect multiple genes.

As she continued to investigate *Ds* and *Ac*, she found new inceptions of control at other loci. The patterns they produced became more various and more intricate. In seeking to explain the patterns, she held fast to one axiom: that a single basic process lay at the heart of all of them. Control was a response to a mounting tension between her integrative intellectual style and the increasing complexity of the phenomena she studied. Unless she posited an underlying system of control, the patterns she observed would have dissolved into chaos—something she could not accept.

There are two kinds of people in the world: those who group people into lumpers and splitters and those who don't. Lumpers tend to see the connections among organisms or phenomena. They group their data, synthesize them into theories. Splitters see distinctions. They are analysts, observers. McClintock had made her reputation as a splitter. Her eye for detail was unmatched, and she was extraordinarily skilled at designing experiments that teased apart distinct genetic phenomena. But she was becoming a lumper. Perhaps it was her natural bent, certainly it was enhanced by her exposure in college to the classical traditions of natural history and developmental mechanics. From her discovery of twin sectors until the end of her career, she sought unifying theories that would explain her increasingly complex data.

In her mind, mutable genes were transforming into a problem concerning biology's "big" questions: growth, form, and evolution. The question of control lies at the heart of all of these—what controls, drives, or guides changes in organisms?

6

COMPLEXITY

As Mrs. Witkin states, the problem is "fantastically complex."
—BARBARA MCCLINTOCK

Walk into the cornfield of most maize geneticists and you will soon be strolling through long, straight rows of tall, straight plants. A single stalk rises six to eight feet. Long leaves radiate from the stalk. By midsummer, one or two or three shoots emerge and a single tassel crowns the plant. From a distance, a research cornfield is indistinguishable from an agricultural cornfield. The connection between corn genetics and agriculture has always been tight. Much funding for maize genetics is based on the promise of agricultural benefit—higher yields, healthier crops. Only on closer inspection would the mutants in the geneticist's plants become obvious: odd pigmentation in the kernels, strange leaf formation or color, puckering or other abnormalities in the plant's tissues, or more dramatic distortions of growth.

Freed from the stringent selection pressure of practical benefit, McClintock's cornfield harbored a motley crowd: short, scruffy, bushy plants, utterly useless to agriculture. In fact, her plants were highly tailored to her needs. At barely over five feet tall, she could not reach the tassels of normal maize plants, so she bred her corn to be short. The ears on her plants were unusually large, with comparatively enormous numbers of kernels. Perhaps most strikingly, her corn had many tillers, or side branches.[1]

Tillers seem to be a throwback to teosinte, a stubby, highly branched wild relative of maize, with many tassels and ears on each plant. (The evolution of maize will be taken up in Chapter 8.) Most maize geneticists, like corn farmers, find it easier simply to grow more plants than to deal with tillers; besides, most maize stocks derive from tillerless agricultural varieties.

Since 1944, every significant step in McClintock's work had come with the appearance of a single "special kernel" out of perhaps thousands, an anomaly from the expected pattern. Explaining the anomaly required as many crosses as possible to tester stocks of known genetic constitution. Tillers gave her several times more pollen and eggs to work with. Occasionally, as with plant B-87, some developmental event occurred that made the tiller different from the main stalk. The large ears and multiple tillers allowed her to get more experimental material from each plant. This was essential, since she kept her crops small, in order to maintain tight control over her material. Complexity stemmed from control.

The Pattern Generator

Imagine a large field planted in quadrants, with each sector containing characteristic plants from the stocks of four of the most important maize geneticists: Marcus Rhoades, Rollins Emerson, Lewis Stadler, and Barbara McClintock. I asked Robert Martienssen, a maize geneticist at Cold Spring Harbor Laboratory who knew McClintock and her corn, whether, led out into such a field, he would be able to pick out McClintock's plants on sight. He did not hesitate: "Absolutely." The other three quadrants, however, would be indistinguishable from one another, and from commercial corn.

Standardization of a model organism is a basic process in genetics and other experimental biology.[2] To have confidence that she has identified the cause of a phenomenon, the researcher must hold constant all variables except the one of interest. In genetics, this is done by repeatedly crossing individuals of known genetic background until the background traits are constant and breed true. Most maize geneticists find the highly inbred and genetically uniform commercial strains of corn useful for this process. Standardizing a strain gives it a transparent background that makes mutations stand out.

McClintock mystified some of her colleagues by her apparent lack of need to standardize her strains. One of those colleagues is Jerry Kermicle, a maize geneticist from the University of Wisconsin, who was a postdoctoral fellow with Royal A. Brink in the late 1950s and early 1960s. From then through the 1970s, Kermicle worked with mobile elements, interacted with McClintock extensively, received from her many samples of kernels. In 1996, he recalled having been amazed at how unstandardized her cultures seemed. Before the Wisconsin group could use the stocks they received from her, they had to standardize the genetic background.

McClintock seemed to do this mentally. "She had the capacity to somehow, in her mind, even out the effects of these other genes." His praise has a backhanded quality. "Relatively few people have this gift of focusing when there are so many distractions," he said. "She was a wizard in being able to focus on what was significant when there were an awful lot of spurious things going on."[3]

To her eye, the plants *were* standardized. By the end of the 1940s, she was looking at suites of genes that all interacted with one another and were set in motion by other, nongenic elements. The phenomena she was studying were merely complex, as were her interpretations of them.

Back in the 1910s and 1920s, T. H. Morgan's fruitflies had begun to "throw" new mutations at an accelerating rate. Within three years of the appearance of the white-eyed fly, Morgan's boys could no longer keep up with the mutations and began sending flies to colleagues. Robert Kohler called this chain reaction of mutations the breeder reactor. The result of many factors—Morgan's skill as an administrator, the relative simplicity of the experiments and analysis, the social and material culture of Drosophila genetics that soon developed—the breeder reactor was essential to Drosophila's becoming the premier model system in genetics.[4]

McClintock's maize exploded in a similar way but with important differences. We may call it the pattern generator. As the mutations in McClintock's corn came ever faster, their complexity increased dramatically. The techniques of fly genetics became standardized, and therefore were easily adopted by other labs. Because McClintock's material grew increasingly complex, her techniques and analyses became customized. To be sure, many maize geneticists used her seed. But as Kermicle points out, the material was hard to use. Simply put, no one else could do the things she could. The breeder reactor created the community of drosophilists, but the pattern generator isolated McClintock from other maize geneticists.

Unstringing the Pearls

In April 1948, when McClintock realized *Ac* and *Ds* transposed, she began to integrate her research and reading into a new model of gene structure. In the 1930s, the simple string-of-pearls model of the gene had begun to erode under the pressure of more flexible models. Position-effect variegation, in which a gene's location on the chromosome relative to heterochromatin affected its action, indicated interactions between different parts of the chromosome. The subgene hypothesis held that genes were not atoms of heredity; they comprised series of identical subunits whose actions were additive. With the development of radiation genetics beginning in the late 1920s, however, this fertile period of alternative theories of the gene gave way to a newly rigid conception. The use of X rays and other mutagens strengthened the concept of the gene as a molecule—unitary, independent, and of fixed action—which was damaged by mutation. This model was powerful because it required that genetics and cellular physiology be explained in chemical terms. To most geneticists in this early period, the model's oversimplification of the gene and gene action was a useful constraint that aided analysis.

Mutable genes were a problem for the otherwise robust chemical-change-in-the-gene model of mutation. Rollins Emerson had made his reputation in the 1910s by reconciling what had seemed the anti-Mendelian qualities of mutable genes in maize with Mendelian genetics. In the 1940s, the model of the gene as a unit of mutation had been strengthened by the model of a mutation as a chemical change. Conventional mutagens knocked out genes, rendering them permanently nonfunctional. Under such a model it was difficult to account for genes that oscillated between two states.

Increasingly, McClintock came to think of the view of the gene as a unit of mutation as an unacceptable distortion. She revived the older, more flexible concepts of the gene to explain her mutable genes. The result was a model of the chromosome as a dynamic structure comprising many interacting parts. Transposition cemented the model's various components. The model was complex, integrative, theoretical, speculative. With it, McClintock left the shore of rigid empiricism and set sail in the open water of speculation.

Her foray into theoretical genetics began with connections to Richard Goldschmidt. Since his studies on sex determination in the moth Lymantria in the 1910s, Goldschmidt had been interested in the dynamics of gene action, in contrast to the transmission of static genes, as studied by Morgan's students and intellectual descendants—that is to say, the majority of American geneticists. Goldschmidt considered genes as quantitative, meaning their effects were additive, either by the addition of entire alleles, as in the early work on the Drosophila *bar* gene, or by the addition of gene segments. He developed what he called a dynamic theory of the gene, in which the action of genes through time was taken into account. Goldschmidt believed that genes either made or were enzymes. Therefore, they catalyzed chemical reactions. Quantitative genes had quantitative effects: the more subunits a gene had, the greater the velocity of the chemical reaction it catalyzed. His *Physiological Genetics*, published in 1938, established him as a major if controversial figure in this alternative branch of heredity studies. *The Material Basis of Evolution*, published in 1940, was a remarkable effort to synthesize his developmental genetics with evolution—to integrate, in other words, the three major problems of growth and form.[5]

About the time McClintock discovered transposition, she adopted and extended the terminology of quantitative genes. In research notes dated 14 April 1948 she thought this through. Quantitative genes show greater and greater expression as the number of subunits increases. Recalling Sturtevant's studies of position effects at the Drosophila *Bar* locus, McClintock wrote that new alleles "will arise from old by crossing over (like Bar)." Other loci, however, she said were qualitative. The *R* locus in maize, for example, either produces or does not produce pigment. Her own mutable *c-m1* was qualitative.[6]

For McClintock, quantity and quality were properties not only of genes but of alleles. Her notes list quantitative types, qualitative types, and even "quantitative and qualitative together." The *C* locus, she discovered, could throw mutations resulting in quantitative or qualitative alleles. She had discovered a new mutable quantitative allele of the *C* gene. She called it *c-m2*. It arose in plant 4000B-2, and she recognized immediately that its pattern was distinct from that of *c-m1*. The *c-m2* allele produced various shades of pink color in the outer layer of the kernel.

Invoking a variant of the subgene hypothesis, she supposed that *c-m1* and *c-m2* represented mutations in different "blocks" of the *C* locus. The *c-m1* allele represented a mutation in the qualitative block, *c-m2* a mutation in the quantitative block. The qualitative block—which she called the "color producer block"—controlled whether the *C* pigment was produced at all, while the quantitative block controlled the "depth of color," or amount of *C* pigment produced. Mutations in one block did not affect the other.[7]

Qualitative alleles, she believed, evolved from quantitative ones. First, unequal crossing-over duplicated the gene. Then, through mutation, one copy of the gene was converted from quantitative to qualitative. Qualitative genes were, she wrote later that summer, "duplicate genes that have mutation by single step to give same general reaction but on slightly changed substrate = 'new' product recognized."[8] At the *C* gene, various alleles produced "various grades of color depending on # of C genes present." But the mutable allele *c-m1* was qualitative: as it oscillated, pigment production was switched on and off. "Evolutionary relations clearer," she wrote. "Quantitative genes multiply at locus and then diverge step-wise— either way, qualitative genes arise from quantitative but remain related to similar processes."[9]

Qualitative and quantitative genes consolidated McClintock's thinking on mutable genes. She sometimes referred to qualitative alleles as pattern alleles, because, by controlling the expression of traits, they determined the pattern of the tissue. Quantitative genes, by their mechanism of addition and subtraction of subunits, at last explained the twin sectors. Most important, she related them to *Ac* and *Ds* through a remarkable process of analogical reasoning.

Ac and *Ds* exhibited different "states." She first mentioned states in late June 1947, in reference to the *D* (now *Ds*) locus.[10] The concept seems nebulous and indeed superfluous in this first instance. It seems that a plant appeared to undergo a change in the action of *D* during the growth of the plant: an early loss pattern (large spots), for example, might change to a few-late loss pattern (small number of tiny dots). The pattern was similar to a change in dosage of *V* (*Ac*). But since it occurred spontaneously, and subsequent crosses showed that *V* dosage did not change, she assumed that *D* must have undergone an alteration. Changes in state during development of a kernel resulted in changes in pattern. It is not clear precisely how she thought state related to the other properties of the *D* locus, timing and frequency of mutations. In the same notes she referred to it both as determining timing and frequency and as a third property, distinct from timing and frequency.

By April 1948, she had concluded that both *Ac* and *Ds* could be found in two states, high and low. The high state of an *Ac* locus resembled the dosage effect: one high-state *Ac* was like several low-state *Ac* loci. Both produced kernels with few small speckles. A kernel with multiple high-state *Ac* loci might have its action so far delayed that the kernel finished developing before the mutations occurred and no speckles would form; it would appear as a uniform recessive. *Ds* occurred in similar states, which she referred to by their speckling patterns: *Ds*-early and *Ds*-few-late. By holding *Ac* dosage constant, she determined that these states of *Ds* were independent of *Ac*.[11]

She soon saw the parallels between the states of *Ac* and *Ds* and the qualitative and quantitative genes. The states of *Ac* and *Ds* could be caused by a common mechanism: exchange of subunits that influence the degree of expression. Instead of altering pigment intensity, for example, an increase in the number of subunits could move *Ac* from low to high state, delaying the onset of mutations at *Ds*. Similarly, the patterns of expression of *Ds* in what she now called its standard position matched, with a little imagination, the patterns of the *c-m1* allele. Integrated into conventional genetic theory, states and subunits seemed to solve the puzzle she'd been worrying since 1945. The well-known process of unequal crossing-over could account for changes in state. When she diagrammed this process in her research notes, she concluded, with rare double exclamation points, "One chromatid gains what the other chromatid loses!!"[12] (See Figure 6.1.)

In June, McClintock wrote two essays on qualitative and quantitative genes, gene states, and gene blocks. One she kept to herself; the other she wrote as a letter to cotton geneticist S. G. Stephens. Both were probably warm-ups for her annual report to the Carnegie Institution, which she wrote later that summer. Stephens had spent a sabbatical year as a research associate at the Department of Genetics in 1945–46, after which he went to the Texas Agricultural Experiment Station.[13] Stephens worked on what he called pseudo-alleles of cotton, which McClintock likened to her qualitative genes. She befriended Stephens while he was at Cold Spring Harbor, and after he left they maintained a correspondence. By June 1948, she had fallen behind in their correspondence for a few months, owing no doubt in part to the intense thinking of that spring, and she wrote him a twenty-seven-page letter updating him on her current work and theories.

She discussed all the major ideas described above and explicitly related the states of Ac to gene blocks. "I am viewing the mechanism of building up the units of a block to be similar to the mechanism of building-up of blocks in changes of the Ac locus. With this mechanism, one builds up blocks and reduces them in the same mitosis—one chromatid gains what the other loses."[14] The "building-up" was the accretion of subunits by unequal crossing-over. In her personal essay, she related this to the idea of genetic control. A locus may become "susceptible" to breakage as stimulated by Ac: "We may view Ac, then as a locus that controls the frequency of somatic breakages at loci that have become susceptible to such breakages because of some specific alteration."[15]

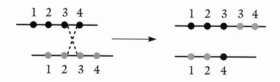

Figure 6.1. Unequal crossing-over. The paired chromosome strands on the left both have four units of Ac. By unequal crossing-over, the top strand ends up with five units and the bottom with three. Having more subunits means that the copy of Ac is in a higher state.

Although McClintock's experiments and techniques remained firmly within the tradition of classical genetics, her ideas about quantitative and qualitative genes and gene blocks undermined the string-of-pearls model. In McClintock's model, genes were neither unitary nor independent. Quantitative genes implied that genes were divisible, while qualitative genes and gene blocks implied there were higher orders of organization among genes. The gene had become for her a middle level of chromosome organization, with both higher and lower levels important in the expression of traits.

In private notes that July, McClintock directly questioned the theory that mutations represented chemical changes in the gene. "Assumption that mutations occurring frequently are *chromosomal changes*," changes in chromosome structure, or in chromosome elements other than genes, "is more reasonable than an assumption of a *change in a gene* in certain direction" (emphasis hers).[16] This was perhaps her first explicit statement of what would become her primary argument on behalf of controlling elements.

She had begun to consider *Ac* and *Ds* as physically or chemically related. "A locus," she wrote, meaning *Ds, c-m1*, or *c-m2*, "becomes activated to respond to *Ac*." This activation could not be understood unless "part of *Ac* locus is inserted into another locus making it susceptible to action of *Ac* when *Ac* present. *Ac* locus probably breaks frequently and part or all of *Ac* locus removed to another locus when it also has broken in same nucleus."[17] Since 1947, she had thought that *Ds* might "give something" to *Ac;* now she wrote explicitly that they might exchange subunits. Strikingly, this implies that *Ds* is in fact a partial *Ac* locus, as it is understood today. *Ac*, then, is in a new way the controller of the mutation process. *Ds,* missing part of the *Ac* structure, can only be controlled by *Ac*. *Ac* and *Ds* are made of the same stuff.

In September 1948, probably just after the harvest, she got a viral infection and by late October she had still "not thrown off the secondary infections." Explaining her situation to Beadle, she apologized that she would be unable to visit Caltech that year. "That tale-of-woe will explain why I am not able to make any moves of any kind for a while! Would like to sleep for weeks."[18]

Meanwhile, Indiana University phage geneticist Salvador Luria had somehow gotten hold of McClintock's twenty-seven-page letter to Stephens of the previous June. McClintock had sent a copy of the letter to Rhoades, and perhaps to others, as a concise statement of her thinking on mutable genes. Luria was no maize geneticist, although he may have had contact with some of the midwestern corn community and he knew McClintock from his summers at Cold Spring Harbor. Luria thought her work so important that he gave a seminar on it to the geneticists at Indiana, based on that letter.

In the audience was none other than H. J. Muller, discoverer of radiation-induced mutation and champion of the chemical-change-in-the-gene theory of mutation. On hearing Luria's talk, Muller apparently bolted out of the seminar room and headed straight for Western Union. "I tried to telegraph you immediately afterwards," he wrote on 27 October, "but the telegraph office was apparently closed so I am sending you this now." He congratulated her on her "magnificent achievement" in solving the old problem of mutable genes—"at least in maize." McClintock responded, characteristically, with an annotated copy of her forthcoming Carnegie report, written in August and containing what would become the first publication documenting transposition, as well as photographs illustrating *Ac* dosage effects and kernel sectors.[19]

Muller's comments show that he was aware of and interested in McClintock's work early on. But they also show that he either did not understand or did not accept her full interpretation. "What a relief it is," his congratulatory note concluded, "for us mutation workers to know that the mutable genes are in a different class from . . . 'ordinary gene mutations,' after all."[20] He was delighted to find that the gene-damage model did not have to account for mutable genes; they resulted from a different process. His theory was saved.

McClintock, however, saw her theory as an *alternative* to the chemical-change-in-the-gene model. Muller's model posited autonomous, particulate genes, in which mutations were "true"—that is, chemical. McClintock's postulated an integrative genome in which genes acted in suites, controlled by regulatory elements. In her Carnegie report of that year, she wrote that her investigations "cast doubt on interpretations that postulate a 'true gene mutation,' that is, a chemical change in a gene molecule," and suggested that phenotypic change was rather the result of reversible inhibition and modulation of genes.[21]

Coincident Breaks and Repulsive Forces

By November 1948, McClintock developed a model for the mechanism of transposition of *Ds*, which I will call her "coincident-break" hypothesis. She reasoned that when *Ds* caused a breakage, it was torn out of the chromosome at both ends and floated free in the nucleus. If, at the same time, another break occurred elsewhere, on the same or another chromosome, two more broken ends would form. McClintock knew from her years of study that broken ends tend to fuse. The *Ds* element could find—might even be attracted to—the new spontaneous breakage site and would fuse to and reunite the ends.[22]

The coincident-break hypothesis is a revealing example of McClintock's research style. It demonstrates how her mutable-gene work was rooted in her earlier studies of the breakage-fusion-bridge cycle and broken chromosome ends. It also shows there is no magic to the transposition phenomenon. She explained the mechanism of movement by well-documented, uncontroversial means. The chromosome breakage cycles McClintock observed in maize had not been detected in other organisms, but there could be no arguing that they occurred in corn.

The coincident-break hypothesis also illustrates McClintock's willingness to postulate unseen events. She had not seen a coincident break; rather, she reasoned that it must occur. Hypothetical mechanisms were of course characteristic of classical genetics, and she had a remarkable track record for farfetched hypotheses that turned out to be correct. But in the past there had usually been a simple test, such as the presence of a ring chromosome, that could confirm or refute her speculations. Increasingly as she moved into more theoretical territory, she was forced to trust her reasoning, without the security of cytological corroboration.

Finally, and most important, the coincident-break hypothesis undermines her argument that transposition mediates the genetics of development. Because under it transposition depends on a stochastic event, the hypothesis precludes transposition as the source of control of gene action. Within a few years, McClintock would abandon coincident breaks and all other efforts to explain the mechanism of transposition, concentrating instead on the effects of controlling elements.

Also in November, she realized what had been happening with the *c-m1* mutable. She had been puzzling over this allele, which produced revertant red spots on a colorless kernel, for more than a year. The *c-m1* mutable appeared in a plant known to have both *Ds* and *Ac*. When *c-m1* appeared, there were rarely any further *Ds* mutations. She now recognized that *c-m1* was a case of *Ds* having transposed into or just to the left of the *C* locus. She now had examples of *Ds* in three locations along the short arm of chromosome 9: in "standard" position, to the right of *Wx;* to the right of *C* and to the left of *Sh* (case 4306); and at or just to the left of *C*.[23]

With *c-m1*, McClintock had a case of *Ds* residing in a known gene. This case complicated her picture of *Ds* action. When *Ds* lay between known genes, its observable action was limited to chromosome breakage, followed by dicentric formation and cycles of breakage-fusion-bridge. She began to recognize the *c*→*C* mutations, reversions of the recessive to dominant form first seen in plant 4204, as in fact a new class of *Ds* behavior. The result of *c*→*C* mutations, patches of red on a colorless background, were solid red, not variegated for *C* or any other gene. This meant that once a mutation from *c* to *C* occurred in a cell, *Ds* activity stopped. When she looked at the chromosomes of such cells, she confirmed that no further breakages occurred.

She outlined a long memorandum to Marcus Rhoades on "c-m1—a transposed Ds locus."[24] Dated January 1949 and clocking in at fifty-seven pages plus tables and figures, the final memo contains all the elements of her writings of the previous summer—qualitative and quantitative genes, gene blocks, coincident breaks—and restates her contradiction of the standard theory of mutation. Further, it draws her onto unfamiliar ground, speculating on biochemical and physical mechanisms to explain transposition, gene expression, and mutation.

The first section of the memorandum deals mainly with a comparison of *c-m1* and *Ds*. McClintock had written to Stephens of the striking similarities in the action of the *Ds* and *c-m1* mutants. A primary purpose of the memorandum is to analyze *Ds* action at these two loci and determine whether one set of phenomena could explain both sets of data. The second section, much briefer and containing far less analysis, is mainly a description of the behavior and an outline of the genetic origins of *c-m2*.

McClintock again postulated invisible substances and undetectable phenomena to explain events she could not see. She diagrammed a mechanism for transposition of *Ds*, in which, prior to the separation of chromatids, *Ds* elements on each chromatid bound to one another. She hypothesized a "repulsive force" that drove the chromatids apart. (See Figure 6.2.) This repulsive force was reminiscent of the forces hypothesized early in the century by colloid chemists. Colloids are complex, gelatinous suspensions of molecules; in the days before the centrifuge and electron microscope, the cytoplasm was seen as a colloid. Colloid chemists sought new laws of chemistry in biological colloids; as a consequence, the colloid chemists came to be seen as mystical or vitalistic.[25] McClintock's mutual admirer Richard Goldschmidt worked in colloid chemistry before moving into genetics.

Her explanation of transposition, with the repulsive force here black-boxed as "some process," illustrates how clear her writing could be when she was confident of the data and on solid empirical ground ("unsaturated" broken ends are ends that have not healed and are thus capable of fusion):

> Ds mutations are associated with some process that often results in tearing out of the Ds locus from the chromosome. This tearing-out process produces broken ends capable of fusing with other broken ends, not only in the torn chromosome but also in the torn-out Ds locus. The torn-out Ds locus, with broken ends capable of fusion with other broken ends, may be inserted into a new location if a coincident break occurs elsewhere in the chromosome complement. Fusion of unsaturated broken ends, a well established phenomenon, is all that is required to complete the process of change in location of Ds.[26]

Turning next to a comparison of c-m1 and Ds, she hypothesized another mysterious process: hidden mutations. The c-m1 allele showed two distinct classes of phenomena: breakage, followed by formation of dicentric chromosomes, and c→C mutations. In its standard position, Ds

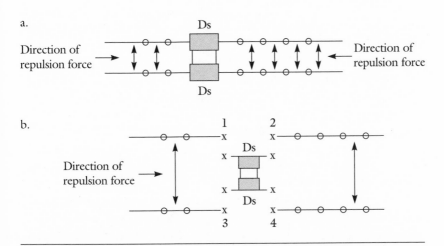

Figure 6.2. McClintock's diagram illustrating the "repulsion force" she hypothesized in 1949. The repulsion force tended to push the paired chromosome halves apart *(a)*; the doubled Ds element, bound to itself, would resist the force and be ripped out of the chromosome, initiating transposition *(b)*. Redrawn and used with permission from the American Philosophical Society Library.

showed only breakage. It required no great leap of imagination to recognize that chromosome breakage in *c-m1* corresponded to breakage at *Ds*-standard position. But for a true correspondence between *c-m1* and *Ds*, McClintock needed a parallel to the *c→C* mutations. She therefore postulated "hidden" mutations at *Ds*-standard. These would correspond to *c→C* mutations, but, since the absence of *Ds*, denoted *ds*, has no phenotype, there is no visible reversion of character as in the case of the *C* locus.

Repeated insertion and removal of *Ds* into and out of the *C* locus explained the *c-m1* pattern. The initial mutation, early in the formation of the kernel, was caused by insertion of a *Ds* at the *C* locus. *Ds* caused chromosome breakage, with subsequent loss of the *C* locus, resulting in a mutation from dominant *C* to recessive *c* in every cell where breakage occurred. "Conclusion. C to c arose as consequence of insertion of Ds into C locus." Occasional removal of *Ds* during kernel development restored the *C* locus in those kernels, resulting in a spot of *c→C* reversion. This reversion apparently happened many times, because there were many spots of *C* on a *c-m1* kernel. To create a new mutable gene, she reasoned, "It is only necessary, it would appear, to insert a Ds locus into (or adjacent to) a normal dominant locus. Inhibition of the dominant locus must be associated with this insertion, if a changed phenotypic expression is to be recognized following insertion."[27] Later, she would try to create new mutable alleles by setting up crosses to favor insertion of *Ds* into specific loci; those efforts led to an important paper in 1953 and jokes with Rhoades about going into the mutable-gene business (see Chapter 7).

Point 4b under "General considerations" in an outline for the memorandum reads: "Inhibition of C action by presence of Ds at locus = a position effect."[28] Here, McClintock adopted Jack Schultz's sense of position effect, as gene inhibition caused by proximity to heterochromatin. But there was a twist: the *C* locus is not near a knob or centromere. For the presence of *Ds* to cause a position effect, *Ds* itself must be a tiny—and transposable—piece of heterochromatin.

Thinking of *Ds* as made of heterochromatin makes her hypotheses about strange repulsive forces and the like easier to swallow. Heterochromatin had many strange properties—who knew what other bizarre things it might do? Why couldn't there be a repulsive force?

Her speculations on the patterns of *c-m2* action were less successful. The *c-m2* mutable produced a graded series of shades of pink, which she interpreted as the effects of increasing numbers of subunits of the quantitative block of the *C* gene. The *c-m2* locus itself, then, would be compound, producing "two distinctly different types of visible mutation." To explain the deficits she observed at the *C* locus, she hypothesized two "diffusible, colorless substances" that would be produced by the *C* gene. Both substance 1 and substance 2 would be needed for full *C* expression. She may have had in mind Sewall Wright's classic paper of 1917, in which he proposed a two-enzyme model to explain the inheritance of coat color in mice.[29] Spemann's organizer was also a "diffusible substance," invoked to explain lens induction—it was just this that opened Spemann to charges of mystical vitalism. On this subject, her writing grew vague and elliptical:

> That the weakened color intensity of pink mutations appears to be associated with some specific deficiency of a needed substance is suggested by the dosage responses of the pink mutations. This substance cannot be substance 1 for substance 1 is often produced in excess even in the mutations giving very faint pink. If so, then all mutations giving pink must likewise produce some substance 2 . . . If the intensity of the pink color is an expression of the level of some limiting substance this substance must be one of the products of mutations of *c-m2* . . . This may be substance 2 . . . or it may be a third substance associated with genic action of a normal *C* locus.[30]

McClintock, as usual, maintained some skepticism—and humor—about the scheme, and she recognized its overweening complexity. "Although this substance-producer interpretation fits the observations so far made, I believe that one should be canny about accepting it as it has been presented. I have that uncomfortable feeling of having mentally over-looked the missing link that could simplify the whole interpretation. This is by way of confession and not of retraction."[31]

In the eight months preceding the January 1949 memorandum, she had already gotten enormous mileage out of transposition. Though the mechanism she proposed was conventional, the phenomenon of transposition unlocked the problem of the twin sectors and gave her the basis of a solution to the problem of mutable genes. This much was solid enough to convince Rhoades, Luria, Muller, and possibly others including George Beadle that she had solved the problem of mutable genes—at least in maize. In October 1948, she wrote to Beadle that she had "finally got Marcus convinced of my logic although it was hard going with him for quite a while."[32] At the same time, the memorandum shows that as her corn yielded increasingly complex behavior, she suggested increasingly speculative mechanisms to explain it. To Rhoades, who had followed McClintock's work since the early days at Cornell, the memorandum must have seemed a dramatic departure from her usual scientific style.

The memorandum was the first installment of a flash flood of written material she sent to Rhoades in 1949 and 1950. She next went back and assembled all her evidence on *Ac* and *Ds* into a series of "reports." These were private communications to Rhoades, though written in formal scientific style. The first, on *Ac*, was completed in March 1949, with the *Ds* report written in April and given to Rhoades in May.[33] These reports cover much of the same territory as the Stephens letter and the 1949 memorandum, but in far greater detail and with all the evidentiary support she had available. Gene blocks, coincident breaks, qualitative and quantitative genes all receive thorough treatment. The reports are exhaustive catalogs of her findings of the previous five years, plant by plant. All the culture numbers are given, as are the kernel ratios and statistical analyses. In short, in these reports are all the data the wider genetics community never saw.

To conclude the *Ds* report, McClintock offered her best reconstruction of the initial appearance of the transposed *Ds* element in 1947 (recognized in 1948). It arose on a chromosome 9 with a duplication on the short arm that led to BFB when crossing-over occurred with another chromosome 9 containing a broken end.

It is possible, now, to reconstruct the events that gave rise to this Duplication chromosome 9 with a transposed Ds locus. Three assumptions regarding these events are required. (1) A Ds mutation occurred at i[t]s usual time—late in the development of the sporophytic tissues [that is, in the maize plant, not the kernel] in a cell of plant 4108C-1. The chromosome in which this Ds mutation occurred was normal in morphology and carried [the genes] I Sh Bz Wx and Ds in its standard location. The Ds mutation resulted in breakage of two sister chromatids at the position of the Ds locus in each chromatid. Evidence that at Ds mutation brings about breaks in sister chromatids at the locus of Ds is well established. This assumption is therefore legitimate. (2) The Ds mutation not only caused breaks to occur at the position of the Ds locus but resulted in the release of a submicroscopic chromatin segment that carries a Ds locus. This released segment carrying Ds has unsaturated broken ends. It could be lost from the chromosome complement if fusion with some other broken locus did not occur. Loss of the Ds locus as a consequence of Ds mutations has been considered in detail elsewhere (see report on c-m1 mutations, January, 1949). The manner of this loss may be suggested from cases such as the one being described. (3) At the same time that the events in (1) and (2) occurred, a spontaneous *chromosome* break occurred just to the right of the I locus in this chromosome. (Both sister chromatids were broken at the same locus.) Evidence for frequent spontaneous breaks in maize is good (McClintock, unpublished). This assumption is therefore legitimately taken. These three events would give a series of broken ends . . . Fusion of broken ends could readily occur to give rise to the configuration shown . . . A Duplication chromosome 9, with an inverted order of genes in the proximal duplicated segment and having a transposed Ds locus just to the right of the I locus is now formed.[34]

McClintock's explanation makes clear that she did not think of the transposition mechanism as out of the ordinary. It occurred by the confluence of two well-established events and one highly plausible one: a mutation at *Ds* (established), removal of a segment containing *Ds* (plausible), and a coincident break elsewhere in the chromosome complement (established, though not as to frequency). In assumption 2, McClintock was careful not to assume that only the *Ds* element detached; she could say with confidence only that a segment containing *Ds* had been released. Step 4 was left to implication. She had established in the late 1930s that fusion of unsaturated broken ends occurred; simple probability governed the likelihood of insertion of the released *Ds* element at the point of the coincident break.

She also proposed, in the *Ds* report, an "organizer" theory of developmental control. This idea harked back to her own work on the nucleolar organizer and ultimately to Spemann. The organizer theory was based on her observations of chromosomal knobs. The knobs are composed of heterochromatin. McClintock had seen that, at certain phases of the cell cycle, knobs tended to associate with the nuclear membrane. She reasoned that they were somehow involved with the transfer of enzymes or other substances out into the cytoplasm. The nucleolar organizer was a region of heterochromatin on chromosome 6 that orchestrated the condensation of the nucleolus. Analogously, she suggested the knobs organized the genes. Her theory brimmed with integration and interaction: between *Ac* and *Ds*, between chromatin and heterochromatin, between nuclear and cytoplasmic elements.

Much of the complexity of her theory derived from the need to accommodate the rapidly growing bank of data she was collecting. She now had several examples, or cases, from which to draw. "Case I" was plant 4306, in which *Ds* had moved to a new position between the *I* gene and the *Sh* (shrunken) gene. She had plant 4204, in which *Ds* had transposed to the *C* locus to give *c-m1*. The *c-m1* allele was a qualitative or "pattern" allele, *c-m2* a quantitative allele. She had another pattern allele, a mutable form of the *Waxy* gene, which produces starch in the kernel. The *wx-m1* mutable had turned up in the same plant as *c-m2*. With two mutable loci near each other that affected different components of the kernel, McClintock used clever genetics, dexterous technique, and sharp observation to see whether *wx* and *c* mutated in the same cells.

In 1949, McClintock wanted to zero in on the *waxy* mutable. In so doing, she dipped her toe into the waters of quantitative biochemical analysis—and quickly withdrew it. She sent samples of kernels with different *Waxy* constitutions to Ruth Sager, on whose dissertation committee she was serving, for biochemical analysis. Production of the sugar amylose was taken as a measure of *Waxy* production. If the *waxy* mutable was in fact a quantitative allele, amylose production should increase linearly with *Ac* dosage. With Charles Otto Beckmann, a starch chemist at Columbia, Sager analyzed the amylose content of kernels with various doses of one of McClintock's mutable *waxy* alleles. They confirmed McClintock's hypotheses, but McClintock was hardly bowled over by the power of biochemical analysis: "I rather suspect," she wrote to Rhoades, "that my visual analyses are better indices of the amount of amylose present than the chemical analyses."[35] She placed her faith in the low-tech, old-fashioned, qualitative measures of classical cytogenetics.

She continued to write through the fall, pulling together all that she had found in the past two years. Beadle again invited her out to Caltech, but in January 1950 she declined. "Although I know it would be wise to stop working morning, noon, and night, as I have been doing for two years," she wrote, "just the same, I don't seem to see how to do it now. The story on the mutable genes is over some of the major hurdles. It is necessary to write up much that has accumulated. That is now being done. It will take all the time I can give to it between now and summer. The job is immense. Don't worry—I won't publish all of it!"[36]

Indeed, she published none of it. During 1949 and 1950, her initial reports to Rhoades expanded into at least eight long chapters. Apparently, she initially intended to write a book-length monograph on *Ac* and *Ds*.[37] The text was supported with nearly as many pages of tables and provided culture numbers, kernel ratios, and cytological descriptions. All or nearly all was sent to Rhoades, who dutifully annotated it and returned comments to her. Publication of this material, either as a book or as a series of journal articles, would certainly have been a major event for geneticists. But it remained in Rhoades's files, with carbons in McClintock's own file cabinets. With one exception, a paper in 1953, all of her publications on mutable genes and controlling elements were summaries, reviews, and conference papers, mostly in non-peer-reviewed publications.

According to maize geneticist Jerry Kermicle, the lack of published data contributed to her subsequent reputation as someone who simply intuited the bizarre relationships she postulated, without going through the necessary logical steps to substantiate them.[38] The reports were likely a casualty of McClintock's response to her colleagues' reception of her work, which I will examine in Chapter 8.

She was interrupted at least a couple of times during the fall and winter by requests for talks and lectures. In late 1949, Vannevar Bush, president of the Carnegie Institution of Washington, asked her down to Washington to present a lecture to the trustees. Director Demerec was apparently nervous about it, owing no doubt to his concurrent effort to fund two new buildings on the Cold Spring Harbor campus. He offended McClintock by asking her if she wished to rehearse her talk before the Carnegie staff.[39] Nevertheless, since Bush asked, she felt obliged to go. She wrote to Rhoades on 7 December, complaining good-naturedly that preparing for the trip was a "chore with the addition of a great clothes burden for decorating the body[, which] is one of the necessities requiring taking three suits and a dinner dress with all that goes with it—and that ain't little! Can't wait to get back!"[40]

Also that fall, John Moore, chairman of the zoology department at Columbia University, asked McClintock and Rhoades to give the prestigious Jesup Lectures. The lecture series was established in 1905, on the retirement of Morris Ketchum Jesup, a New York banking magnate, from the directorship of the American Museum of Natural History. Until 1930, the lectures were a collaboration between the American Museum and Columbia; since then, they have been administered by Columbia alone. Invited speakers give a series of three lectures on biological topics of general interest, with some expectation though no requirement of producing a book from them.[41]

McClintock and Rhoades corresponded about the invitation. Though she initially thought it sounded feasible, McClintock seems to have demurred on the grounds that their work did not fit within the confines of conventional genetics.[42] "I wish you would be kinder to your friends at Columbia and stop getting us upset over the prospect of your not giving the Jesup lectures next year!" Moore responded. "The Jesup lectures are not restricted to any one field of biology and certainly not to a particular phase of current genetic thought. We asked you and Marcus to give the lectures because it was felt that your work is highly significant in biology.[43] Despite Moore's pleas, they apparently never gave the lecture.

Despite these distractions, she did manage quite a lot of analysis during this period. She had by late fall thirteen new cases of transposition of *Ds* and had completed her analysis of the first case to her own satisfaction. "The general picture," she wrote to Rhoades, "seems to have become stable and I now feel the required confidence that justifies the efforts." The analysis of the summer's crop and the actions of *Ds*, *Ac*, and *c-m1*, she continued, "has given me enough confidence to feel free to discuss the general aspects publicly." Lectures at the University of Pennsylvania and Harvard were coming up, and she anticipated a skeptical reaction: "Although I expect criticism of the views expressed, I now feel that the views can be defended with quite precise experimental data."[44]

The crowds at the two universities heard the most mature story McClintock had on mutable genes, but it was essentially the story of the previous year, bolstered by much more data. Two months later, the story mutated once again; McClintock pulled together the various elements of the story she had been chewing on for several months into her most cohesive theory yet on mutable genes, transposition, and genetic control. The new theory was speculative, integrative, and visionary. It was also, at least in the universal sense she intended it, wrong.

The March Burst

In March 1950, McClintock had a burst of intellectual activity that may have been greater than any she had experienced before. It followed a low period in February, during which she wrote to Rhoades that things "have been too strenuous." The strain of constant work took its toll: "I have just bitten off more than I can handle and lately have been very tired. Twice during the past week I have felt really exhausted. This is a rare reaction. I don't like it because it interferes with my plans and activities."[45]

Within two weeks, however, she was back up again. On 4 March, she sent Rhoades three packed, typewritten pages that she said amounted to her third try at a letter to him. The letter and one that followed the next day integrated her thinking of the past year on mutable genes, gene control, heterochromatin, and transposition. From them emerges a sweeping theory of genetic control of development by what she came to call controlling elements. Developing, substantiating, and revising this theory guided most of her research for the rest of her career. The letters include numerous addenda, emendations, and marginalia. (In the quotes that follow, braces indicate text McClintock inserted above the line or in the margin.)

In the past two weeks, she said, she had been doing a lot of thinking that pulled together many of the lines of reasoning she had been following. She believed her thoughts to be revolutionary, "but I doubt if they would surprise some of the *Drosophila* geneticists. In other words, I have become *completely convinced* that we are working with the same kinds of events in our mutable loci that the *Drosophila* people have been investigating for years, that is, position-effects, and that the primary cause of our mutable behavior is also *heterochromatin*" (emphasis hers).[46]

In fact, many drosophilists do seem to have accepted her ideas on heterochromatin. However, it was also drosophilists who had built what she would soon call the "good old chemical change in the gene" model she was attacking. This was the sticking point.

Recall that cytologists had distinguished between euchromatin, which contained the genes, and heterochromatin, the dense, geneless material that comprised knobs and centromeres. McClintock posited a second type of heterochromatin, "gene heterochromatin," which occurred in lengths too short to be seen but which acted as spacers between the genes.

She stated her model as a list of "assumptions," which were actually her conclusions. These included, first, that "the genes are separated, one from another, by {gene} heterochromatin." Second, "it is this heterochromatin that is capable of the breakage-fusion phenomena." In BFB, it is not the genes but the heterochromatic spacers that break. Third, "the knobs and the regions about the centromeres have the {main mass of} heterochromatin that controls the action of the heterochromatin between the blocks of genes—in a very particular way at a very specific time in development."[47] Genic heterochromatin is controlled by knob heterochromatin.

The fourth assumption was essentially the organizer model of genic control. She assumed that chromosome knobs and centromeres were in contact with the nuclear membrane and were the "receivers and givers of substance" between the nucleus and cytoplasm. Logistics determines the position of knobs and centromeres on the chromosomes. Interactions between these heterochromatic regions "are a reflection of this organizational {working} phase of the heterochromatin."[48] Heterochromatin was not junk at all; it served a vital purpose in negotiating the complex and completely unknown interactions between the nucleus and cytoplasm.

Fifth and most speculative, the two types of heterochromatin orchestrated the action of the genes. For this assumption, she merged the concepts of subgenes, heterochromatin, multilevel genic control, and position effects. "The heterochromatin of the controlling group (knobs, etc.)," she continued, "has sub-units (heterochromatin genes) and these control when certain euchromatin genes in the complement will function." Knob chromatin held the highest position of control. It regulated gene action "by controlling the action of the heterochromatin between the genes of the euchromatin." The nature of the control was determined by position effects *within* the heterochromatin: "The positions of the heterochromatin genes in the controlling group, with respect to one another, are important. If they are transposed by translocation, their action changes."[49] (Her phrase "transposed by translocation" shows the original sense of transposition, dating to the early days of classical cytogenetics, as any movement of a locus by conventional means, such as translocation or inversion.)

In short, she proposed a two-tiered system of control, in which knob heterochromatin controlled the intergenic heterochromatin, which in turn controlled the genes.

McClintock's thinking was informed by deep reading of the Drosophila and other genetics literature. McClintock cited Demerec and others in her letters and memoranda to Rhoades. Other scientists' research on mutable genes, however, resembled McClintock's theory only superficially. The drosophilists never proposed that either mutable genes or heterochromatin exerted any coordinated controlling effect on normal, stable, euchromatic genes. To them, mutable genes were a valuable phenomenon for studying the nature of mutation, not a system for understanding growth and development.

She also borrowed from her long-time friend George Beadle. In 1932, Beadle had published a description of a mutant allele he called *sticky*, the presence of which resulted in the failure of paired maize chromosomes to separate at cell division.[50] McClintock took this idea of stickiness and applied it to the heterochromatin. Noting that raising plants in low temperature increased the mutability of her mutable genes, she wrote,

> it looks like the cold affects the heterochromatin, changes its action and in doing so makes it sticky. This stickiness may be the answer to the events occurring at Ds and Ac and thus to the events occurring at all mutable loci when breaks and fusions occur. It may be responsible for the transpositions of Ds and Ac; in fact this stickiness may result in the breakage in a position other than Ds and Ac because of the sticking of them to the heterochromatin of other chromosomes. Mutability is the expression of the changes occurring at the heterochromatin. Sometimes it is such that the heterochromatin becomes sticky. When this occurs, then comes the transpositions, dicentrics, deficiencies, etc.[51]

The next night, McClintock took another stab at explaining herself in an even longer letter to Rhoades. "After writing to you last night," she began, "my brain kept on going and this morning I sat down to fill out some of the thoughts in sequence. The result was startling but along the lines that I mentioned to you last night. It was just a lot clearer." The letter is a tangle of marginal notes, lines crossed out, and sentences struck through and rewritten above, obviously written in a great intellectual heat. "It looks a mess," she confessed below the signature, "but it is nearly 12 P.M. I should go to bed (but have had a hard time going to bed for several weeks!) so will not retype."[52]

McClintock wrote that she suspected both *Ac* and *Ds* were transposed segments of heterochromatin from the knob at the end of chromosome 9. The BFB cycle had splintered the knob chromatin and sprayed fragments of it around the chromosomes. The broken-X-broken experiment of 1944 had violently perturbed the delicate system controlling gene action, creating the "wild" genome she had first seen in 1944 and 1945. The heterochromatin elements now behaved unpredictably. They had gone haywire and were causing breakages and other mutation phenomena in her strains. This was what had enabled her to detect them. In normal plants, the elements exerted the control necessary to guide the growth of the plant. "By them," she wrote in the letter of 4 March, "by transposition, may come all of the mutable loci." The knob and centric heterochromatin "was being broken up and its sequence of action along the chromosomes consequently disturbed."

The most startling idea was that elements such as *Ds* and *Ac* might control *all* genes: "I decided," she wrote in the 5 March letter, "that all genes act in the system by being differentially inhibited and released from inhibition {by the inter-gene heterochromatin}." Further, "this was the reason for the development of the organism in an orderly fashion."

The problem she believed she had solved was not the geneticists' problem of the nature of the gene, but the embryologists' (and Goldschmidt's) problem of how to make different tissues out of one set of genes. The answer was that dancing bits of heterochromatin switched the genes on and off, executing the developmental program.

The crux of the theory was that transposition was nonrandom. The whole system was controlled by some as yet undetermined force that initiated coordinated mass-transpositions. At any given moment in any given cell, most genes would be silenced. Those genes needed at that moment would be disinhibited by transposition of nearby *Ds*-like elements away to other sites, perhaps inhibiting genes that had been needed only a cell division before.

With this concept, McClintock's mutable genes were no longer merely a model of mutation. They were the core of a comprehensive theory of organismal development. Once again, however, the concept represented an extrapolation from her data. The evidence for it was correlative, not inductive. How she justified the leap from control of mutation to control of development continues to baffle many geneticists. It was an integration, a synthesis of all she knew about genes and development into an interlocking whole.

"From the two letters I wrote you over the week-end you must consider that I have gone completely screw-ball," she wrote to Rhoades on 9 March.[53] The mania seemed to have passed, her humor returned. But there followed another twenty-two pages of explanation. The following Sunday she wrote to Rhoades yet again, saying, "It was with relief that I read your letter which came last night. I did not know what kind of a reaction you would have to my two hyper-thyroid letters of last week-end." McClintock told Rhoades that the "rather sudden 'click-over' to a comprehension of the primary events that interrelated all of these seemingly unrelated phenomenon [*sic*] rather shocked me. You got the letters written in a state of shock."[54]

This "click-over" sounds like the sort of intuitive leap for which McClintock has since become famous. Yet it came at the end of years of experiments, months of intense thinking on this problem, and at least two weeks of intense mental work, represented in more than thirty pages of tightly packed, narrow-margined, heavily worked letters. The sudden moment of understanding resulted from gathering all the pieces of this intricate puzzle and working over them, examining each one from all angles, trying out different combinations, discarding systematically all those that were logically excluded, until at last everything fell into place. The letters of 4 and 5 March are typewritten artifacts of McClintock's process of integration.

Three more weeks of reworking her ideas and writing resulted in a paper she hoped to send out shortly. Her earlier published work was distinguished by its rigorous empiricism. Her interpretations were conservative and well-substantiated. She had avoided the theoretical debates of the 1930s. Her reputation was built on the absolute trustworthiness of what she set down in print; she wrote nothing that was not backed up by reams of observations and calculations. This paper was a dramatic departure from that style.

This paper would have no data. It would discuss only her new theory of mutation and developmental control; the data would come later. She wrote Rhoades, "I should very much like to get this information distributed about as soon as it is in the state to do so without going through a detailed discussion with all the data. If the ideas could get flooting [*sic*] about, the data could follow shortly." Like Darwin, who considered the *Origin of Species* an "abstract" of his full argument, McClintock wanted to publish a "preliminary note" to "set the stage."[55] Science has a long tradition of "review" articles, data-free papers in which a researcher surveys recent work and points to future directions. The article McClintock planned, however, was a true research report, coupled with a speculative interpretation, in which she did not present data.

Rhoades received, read, and returned comments on her paper in four days. In her reply, McClintock articulated the problems with the (classical) gene theory as it stood, and how she hoped to address them. A revolution was occurring in genetics, she believed, and she wanted to aid the cause:

It has become obvious to most active geneticists that the good old days of mapping genes . . . without the main clarifying objective are over . . . From now on, it seems to me, there must be a phase of integration where the various isolated phenomenon [*sic*] are drawn together and where the biochemical, histochemical, chromosomal, cytological, developmental etc. phases are more clearly integrated.

Some of the geneticists see this very clearly but a great many still cling to the good old chemical change in the gene to explain the change in phenotypic expression. The dilemma is becoming more obvious each year. Just now we seem to be going through a distinct change and re-evaluation. My material may help in making this reevaluation.[56]

McClintock's contribution to the reevaluation was in fact to broaden the scope of genetics. Geneticists should not restrict themselves to studying only mutational elements, she argued. She envisioned the genome as a complex, integrated hierarchy of different types of units. Some of these units acted as Beadle-and-Tatum genes. Others, however, seemed to be made of different stuff and to control the action of the genes. Assuming that nuclear genes direct development, they must be regulated in some fashion. McClintock believed they were regulated by controlling units such as *Ds* and *Ac*.

She knew the editors of *Genetics,* the premier American genetics journal. She published seldom enough and had sufficient reputation that they almost certainly would have accepted her paper. Yet she wanted to send it to the *Proceedings of the National Academy of Sciences (PNAS)*. As a National Academy member, McClintock could publish essentially anything she wanted in the journal. It was not a fully peer-reviewed publication. Nor was it the most widely read or prestigious journal among geneticists. National Academy members often used *PNAS* as a forum for speculative work perhaps not quite ready for rigorous peer review. The paper she had written was in fact longer than the standard for *PNAS,* but McClintock told Rhoades she was "quite willing to pay for the extra charges," though only if Rhoades thought the paper was worth sending in. She knew the arguments were speculative and likely to be controversial. She seems to have wanted to publish her theory as she conceived it, without going through the usual channels of peer review and critique.[57]

The 1950 *PNAS* article marked McClintock's "dive into the deep,"[58] her immersion in fundamental theoretical problems of genetics. She was uneasy about it. In January 1951 she wrote to Beadle that she had "always enjoyed working on nice, circumscribed little problems that could be neatly tied up and sealed." This problem, however, took her to the theoretical roots of biology. "I don't like 'hitting the fundamentals,'" she continued, "but, unfortunately, the problem now rests in that area even if it was an off-shoot of a neat, little, sealed-in affair."[59]

Ghost in the Organism

McClintock began to grow increasingly philosophical about her science. While all good scientists seek to identify the assumptions underlying their experiments and observations, McClintock began to question the most fundamental assumptions of genetics. When she was preparing her 1950 manuscript, she wrote to Rhoades,

> Are we letting a philosophy of the gene control [our] reasoning? What, then, is the philosophy of the gene? Is it a valid philosophy? It is the historical understanding of the evolution of this philosophy that is of prime importance in understanding the state that genetics has gotten into. There has been too much acceptance of one philosophy without questioning the origin of this philosophy. When one starts to question the reasoning behind the origin of the present notion of the gene (held by most geneticists) the opportunity for questioning its validity becomes apparent.[60]

These are not the words of a loner, alienated from her scientific community. She enjoyed the role of the heretic, but she challenged dogmatic assumptions because she was *part* of the genetics community, not because she was outside it.

At about this time, McClintock began to follow mystical-sounding ideas outside her science. In 1950, Immanuel Velikovsky published a book called *Worlds in Collision,* which created a controversy large enough to be called an affair.[61] Velikovsky, a psychiatrist by training, pureed myths and legends, historical data, and scientific evidence, and baked the mixture into a cosmic pudding. He argued that the planet Venus was in fact a comet that had spun off from Jupiter and had struck the earth in biblical times. This collision theory led him to many conclusions, including that the theory of gravity was wrong and gravitational phenomena were actually electromagnetic, that our petroleum deposits were impregnations of Venusian detritus, and that lunar craters were the result of the moon's passing through the tail of the comet. He cited passages from the Bible and other ancient writings to support his assertion about the Venus-comet collision. Velikovsky has been called a crank, a pseudoscientist, a charlatan. Most astronomers denied his assertions out of hand. Carl Sagan staged a public debate to debunk him. Some persisted in criticizing him even when the evidence for their criticism was shown to be wrong; clearly he touched a nerve in many scientists and provoked highly emotional reactions in them.[62] McClintock is said to have told friends not to dismiss Velikovsky so quickly, that there might be a grain of truth in his work.[63]

In 1952, Carnegie scientist Richard Roberts gave a series of lectures on extrasensory perception (ESP). Predictably, the lectures did not go over well among the other scientists, but McClintock was fascinated. She also became interested in unidentified flying objects (UFOs). At some point, she became interested in Buddhism and biofeedback techniques. She experimented with controlling "involuntary" body processes such as body temperature and pulse.[64] Her bookshelf included volumes on acupuncture and poltergeists, and handbooks of parapsychology and unusual phenomena.[65]

McClintock's interest in Velikovsky, UFOs, and Buddhism seems to support her current reputation as an intuitive, mystical figure. Indeed, late in life, she enjoyed calling herself a mystic. It became part of her private myth, a means by which she could set herself apart from those she felt ignored her. McClintock's biographers have generally interpreted this bent as part of an intuitive, holistic, feminine style of science, a "corrective" to the limitations of (masculine) Western science.[66]

Some scientists, too, take McClintock literally and assume she believed in all these phenomena. Norton Zinder, a bacterial geneticist famous for his discovery of bacterial transduction, swears she believed in poltergeists.[67] Waclaw Szybalski left his native Poland and joined the staff at Cold Spring Harbor in 1950 to pursue bacterial genetics. He became a close personal friend of McClintock's. It is easy to see why; he is instantly warm, eternally sunny. According to Szybalski, "She believed very strongly in extrasensory perception. Although I was giving her examples that this was fraud and bologna"—he pronounced it "bologna"—"she was saying, 'No no no, really we have to look more into it.' I think she believed in it. I think she wanted to believe in it." To him, this attitude carried over into her science and her philosophy of science: "Even when she was talking about genes, she started to say that there was something more to it, and it might have something to do with things we don't understand. And using all these words and sentences that were expressing doubt: 'What's the truth?' et cetera."[68]

Szybalski had the impression that this doubting, philosophical, mystical mood was related to her periods of depression. "She was varying between being mystic and being scientific," he said. "And that to me was her two moods . . . Maybe when she was depressed that was the time when she was disappearing and just before or after that she talked about extrasensory perception and she had a lot of books on it, and mystical things."[69] It is tempting to speculate that McClintock's feelings of being ignored and rejected might be connected to this mood; when one is depressed, it is easy to feel misunderstood.

Several of McClintock's intimates, however, help us understand McClintock's "mysticism." James Shapiro, a bacterial geneticist at the University of Chicago and one of the principal players in the so-called rediscovery of transposition, does believe McClintock was a mystic. But Shapiro strips away the occult connotations of mysticism and focusses on the unknown, the incomprehensible. To him, a mystic is someone immersed in the mysteries of the natural world. A mystic may be logical, rigorous, skeptical. To Shapiro, then, McClintock's mysticism was consistent with good science and central to her creative genius.[70]

Most scientists I spoke with, however, linked mysticism with superstition and pseudoscience. I asked Rollin Hotchkiss, a bacterial geneticist at the Rockefeller Institution, if he thought that she was a mystic. "That she believed in little extra things?" he asked. "Gee, I don't think so. I always had the feeling that she was hard-headed, and looking for the actual cytological, physical, chemical, something mechanism that was involved with everything."[71]

McClintock was enough of an empiricist to recognize that, in the absence of conclusive evidence, it requires as much faith to disbelieve in UFOs as it does to believe in them. She had no arrogance about the depth of human understanding. "I never thought that the word mystical was the right word for this aspect of her," Evelyn Witkin said. "It was more a willingness to admit that there was a lot that we didn't understand."[72] Nina Fedoroff, the scientist who has done the most to explain controlling elements in molecular terms, has a similar view. To her, mysticism is a belief in the unknowable. McClintock was interested in the unknown. "She wasn't willing to disbelieve things," Fedoroff said. Unlike most people, "she really loved things she didn't understand." McClintock a mystic? "Hogwash."[73]

It seems naive to assume that a scientist with McClintock's record was somehow duped by spoonbenders, mountebanks, and frauds. For all her mystical talk, McClintock did not literally believe in ESP and UFOs; she merely believed that no conclusive disproof had been produced. In 1980, she told William Provine and Paul Sisco, "There was just so much that we didn't know" about ESP. "I realized there were a lot of frauds, but there were some things that weren't fraudulent."[74] This version of mysticism is simply an appreciation of the complexity of nature, combined with an unwillingness to claim certainty about that which is necessarily uncertain.

While some scientists link mysticism with UFOs and ESP, to others it explains McClintock's celebrated problem-solving ability, her vaunted intuition. Oliver Nelson, a sober, meticulous maize geneticist several years younger than McClintock and now emeritus at the University of Wisconsin, explained it to me patiently. "I'm not talking about flying saucers," he said, the phrase leaving his mouth like a lemon seed. "But I often came away from talking to Barbara with the idea that her ideas had been derived in a somewhat mystical sense. That is, that she didn't have to go through what I have to go through, that is a logical progression to arrive at a particular point. And that's what I mean by being a mystic."[75]

For David Botstein, a hard-nosed bacterial geneticist of a younger generation, mysticism was simply a word for what we do not understand about McClintock. "She believed in the great inner light bulb. That is, there is a certain time when the light bulb goes on in your head and you understand." Was she a mystic? "I think her communication was sufficiently ineffective, and yet her work was so good, that a lot of people just used that word."[76]

McClintock's mysticism, then, can best be understood as a result of the interplay among themes of her intellectual style: integration, pattern, and complexity. Pattern recognition is a form of integration; it is not far from the way many of us think of intuition. It is a sudden understanding of many elements and their relationships to one another. When the elements she was integrating grew extremely complex—both individually and as a set—it became impossible for her to know consciously how she put them together.

Further, the complexity of her science led her to complexity in the rest of nature. Scientists often take plausibility as a quick and dirty test of a hypothesis. If they cannot think of a mechanism by which something might occur, they are unlikely to take it seriously. This of course is a weak test; it neither supports nor refutes a hypothesis with any strength. But it often works, and it relies on the faith that scientists have discovered the basic elements of the way nature really works. McClintock's private myth of individualism and freedom gradually drained away that faith. She stopped believing that extrasensory perception could not exist simply because we knew of no mechanism for it, or that UFOs could not exist because we had never confirmed a sighting. Her corn showed her that nature was filled with new forces and unexplained behavior.

An Aesthetic of Complexity

At a time when simplicity was becoming increasingly fashionable in science, McClintock was drawn ever more toward complexity. Eukaryotic organisms in general were rapidly going out of style in genetics, with even Drosophila, the tiny workhorse of classical genetics, diminishingly represented in journal articles and conference papers. Maize is at least as complex as Drosophila: on the one hand, maize has more than twice as many chromosomes as Drosophila, shows alternation of generations, and naturally produces haploid, diploid, and triploid tissues; on the other, Drosophila undergoes metamorphosis and has a nervous system. As a consequence, while many scientists were familiar with the broad outlines of maize genetics, ever fewer learned the special characteristics and nomenclature of maize and maize genetics.

McClintock's interest in complexity, however, was unusual even among maize geneticists. Her turn of mind enabled her to grasp, embrace, and even exploit the complexity of her organism. While other geneticists sought to reduce genetic systems to their simplest, most fundamental components, she sought to incorporate new variables, to synthesize all the evidence she could muster into a grand, holistic theory.

The theory that emerged was a direct and explicit attack on mainstream genetics, first on mutation theory and then, increasingly, on the nature of the gene. The heresy in her argument was not transposition—there were several ways to account for gene movement with uncontroversial mechanisms, and McClintock's own explanation was conventional. It was, rather, gene regulation. Genes were neither independent nor autonomous, she argued. They were part of a network, a complex system that could not be detected until geneticists stopped concentrating on mutation.

Articulating this theory from Cold Spring Harbor, McClintock became a sort of scientific fifth column. Among the phage and bacterial geneticists who made up most of the staff and summer visitors, the scientific aesthetic was that simplicity was elegance, and elegance was truth. McClintock's aesthetic embraced complexity as nature's expression of beauty and truth. Her work on mutable genes and controlling elements led her down an in-

tellectual path that diverged sharply from that followed by most other geneticists. The complexity of her evidence, compounded by the complexity
of her theory, gave some of her colleagues the impression that there was
something preternatural about how she reached her conclusions. This was
exacerbated by the growing speculation in her arguments, her failure to
publish her data, and her increasingly cosmic conversations and interests.

It was on this stage, richly set with competing ideologies, changing
styles, accidents of geography, and quirks of personality, that McClintock's
controlling units made their public debut.

7

RECEPTION

The issue really wasn't transposability, it was the question of whether these transposable elements were part of the normal developmental process.

—JAMES F. CROW

By 1950, Barbara McClintock was one of the stars of Cold Spring Harbor. As a staff scientist, she was a fixture at the annual symposium, regardless of topic. Scientifically, her reputation was unshakable: she had two decades of pioneering cytogenetics to her credit. The fact that she worked on corn set her apart from most of the other senior scientists and nearly all of the young ones. Like Drosophila, maize had become old-fashioned; bacteria and viruses were the systems of choice. Moreover, she had a strange new story about movable elements and development that seemed partly astonishing, partly mysterious.

McClintock, though, also had a personal mystique. As a woman, she was already part of a small minority. But she had an exotic flamboyance as she lounged on the patio between sessions, dressed in her white shirt and khaki slacks, smoking cigarettes with a long holder. She was a favorite subject of graduate students snapping pictures at their first symposium. They caught her deep in conversation with another legend or alone, stretched out in an Adirondack chair, smiling at the sky. Her intelligence was of that startling sort that when you spoke to her, regardless of the topic, she seemed to rise above your train of thought and point out features of the landscape beyond your horizon. Many were intimidated by her. Some she sliced to ribbons. But those who were bright and unpretentious found this terrifying person to be intellectually passionate and intensely interested in ideas. They might end up talking for hours, or going back to her lab to look at corn.

It was thus, at the height of her reputation, that McClintock went public with controlling elements. Since 1948 she had been talking about them and reporting on them piecemeal in the Carnegie yearbooks. But in 1950, she compiled three years of experiments and hard thinking into a summary article that for the first time laid out her argument of control. The following year, she presented her findings at the Cold Spring Harbor Symposium. She produced more papers and gave lectures through the mid-1950s.

McClintock was nervous about going public with a theoretical argument. Her previous work had been resolutely empirical. She told Marcus Rhoades she was afraid it might not be understood. Her fears were partly realized. Transposition was accepted immediately. Any skeptics in the audience in 1951 were silenced by 1953, by which time transposition in maize had been confirmed twice independently. But the argument that the elements that produced the spots and blotches on her kernels also guided the development of the plant left other scientists scratching their heads.

Going Public

With her first article on mutable genes outside the Carnegie yearbooks, McClintock's published work shifted suddenly from data to theory, from evidence to interpretation. The paper, published in the *Proceedings of the National Academy of Sciences* in June 1950, lacked the empirical emphasis, the tables and careful documentation, that had characterized her work up to then. McClintock sensed that the field of genetics needed shaking up and a new theoretical framework, a "main clarifying objective." She believed that objective ought to be an understanding of the processes that guide the formation and growth of the entire organism. Her mutable-gene work, she believed, revealed those processes.[1]

Her paper would address, she wrote in the introduction, a phenomenon well-known to Drosophila geneticists and, though less prevalent in maize, one that had been analyzed by Rollins Emerson in the 1920s. It had been called variously, "mutable genes, unstable genes, variegation, mosaicism, mutable loci or 'position-effect.'"[2] No one to date, however, had suggested that the cause of mutable genes might be the same in flies, corn, and other organisms. McClintock suggested exactly that.

She was forthcoming about the lack of data in the paper: "The accumulated observations and data" from her studies, she wrote, "are so extensive that no short account would give sufficient information to prepare the reader for an independent judgment of the nature of the phenomenon. It is realized that this is unfortunate." She then alluded to the large reports she had been sending to Rhoades—the closest these documents have come to publication: "Manuscripts giving full accounts of some of this phenomenon are in preparation. Since this task will require much time to fulfill, the author has decided to present this short account of the general nature of the study, and the conclusions and interpretations that have been drawn."[3]

One of her primary conclusions was that genes and their controllers were made of different stuff. The variegated phenotypes in her cultures were "related to changes in a chromatin element other than that composing the genes themselves," she wrote. "Mutable loci arise when such chromatin is inserted adjacent to the genes that are showing the variegated expression." The *Ds* locus was "composed of this kind of material." She continued with an abbreviated discussion of "stickiness," the concept borrowed from George Beadle's 1932 paper, which she applied to controlling elements. The stickiness of *Ds* and *Ac* could change and did so "only at precise times in the development of a tissue."[4] Transpositions of this nongenic material into or next to a normal gene locus caused that locus to become mutable. The *Ac* element altered the stickiness of *Ds* and thereby initiated the transpositions. She suspected that "*Ds* and *Ac* are composed of the same or similar types of material." She believed "that this material is probably heterochromatin."[5]

To McClintock, this model explained all cases of variegation, in all organisms. The patterns on her mutable kernels and leaves were similar to the patterns in evening primrose, which had been attributed to position effect. Also, they were similar to the patterns in Drosophila that were believed to be caused by heterochromatin position effects. Although the conclusion was correlative, McClintock could not believe the parallels were coincidental: "The similarities are too great to be dismissed as being due to causally unrelated phenomena."[6]

More than that, she concluded that they *must* be related. A slippage thus crept into her reasoning. Similar patterns must have similar causes. Similar causes mean similar mechanisms: a primrose is a fly is a corn plant. Reasoning by pattern is reasoning by analogy, and in science only experiments can determine whether an analogy is a shortcut to understanding or a misleading dead end. Pattern and analogy are powerful tools for generating scientific questions. But, by assuming that the possible was certain, she allowed her sense of pattern to determine her conclusions.

The remainder of the paper addressed the new issue of control. "There can be little question," she wrote, "that transpositions of both *Ds* and *Ac* occur and that the time of their occurrence in the development of a tissue is under precise control." She offered neither an explanation for this statement nor any suggestion of the cellular materials that might instantiate that control. McClintock knew how such arguments were likely to be received, and she anticipated criticism by admitting, "It may be considered that these speculations with regard to heterochromatin behavior and function have been carried further than the evidence warrants." She conceded, "This may be true, but it cannot be denied that one basic kind of phenomenon appears to underlie the expression of variegation in maize."[7]

The paper is one of remarkable boldness; only a scientist with supreme confidence in her data and reputation could consider publishing such an article. As indicated in her letters to Rhoades during the months of its preparation, it was not a paper for which she expected to receive a warm response. It was intended to jar loose the assumptions and preconceived notions of her peers, calculated to stir up controversy.

Richard Goldschmidt enjoyed controversy himself, and he loved McClintock's paper. The brilliant but uneven German Jew who had harbored McClintock in Berlin in 1933–34 had fled Germany in 1936 and settled cantankerously at Berkeley, where he redoubled the fervor of his diatribes against mainstream genetics.[8] His *Physiological Genetics,* published in 1938, had clearly articulated the problem of the genetics of development. Goldschmidt distinguished between "static genetics," classical transmission genetics, and "dynamic genetics," the study of gene action and the genetics of development. "Development is, of course, the orderly production of pattern, and therefore, after all, genes must control pattern." At root, *Physiological Genetics* is an attempt to answer the paradox of nuclear equivalence with an idiosyncratic, quasi-biochemical theory of gene "reaction velocities" and the timing of gene action.[9]

Within months of the publication of McClintock's *PNAS* article, Goldschmidt published an encomium to it in *The American Naturalist.* The article was well-informed; he had studied her Carnegie reports. "In a series of reports since 1946," he wrote in the introduction, "Barbara McClintock has presented the results of her most remarkable work on so-called mutable loci in maize." Although her results were preliminary, Goldschmidt nevertheless felt that "the factual data as well as the interpretations are of such importance that a discussion at this moment seems to be in order."[10]

To Goldschmidt, one of the most interesting aspects of McClintock's paper was that it supported his theory of the gene. After reviewing McClintock's findings and tracing the logic of her analysis, he dwelled for a time on her notion of "states" of the locus. Goldschmidt saw she had unified his concept of quantitative gene effects with the subgene hypothesis, which postulated genes as composed of repeating functional units. Not only were entire genes additive in their effects, but genes could be subdivided into parts that were themselves additive. Goldschmidt also argued that her theory explained several "facts which have been known for a long time without ever having found their proper place within the Mendelian concepts and which now fall in line with the new data under discussion."[11] As a random example, he offered his own work from the 1910s and 1920s on sex determination in the moth Lymantria. Goldschmidt had shown complex dosage relationships and what at the time he had called different potencies of the sex chromosomes. He compared this research to McClintock's analysis of dosage effect and different states of *Ds* and *Ac*. McClintock took her elements to be heterochromatic; he had argued that sex-determining factors were heterochromatic. Most important, her results supported his long-held position that the classical gene, if it existed at all, was but one organizational unit in a hierarchy. Genes, if real, were no more important than subgenes, gene families, or whole chromosomes. The genome was a complex, interacting whole, and narrow-minded geneticists who focussed exclusively on mutations were missing the genome's fundamental structure.

Although apparently no correspondence survives to document McClintock's reaction to Goldschmidt's article, she must have been gratified by it. She had admired Goldschmidt at least since she got to know him in Germany. By 1950, she too seemed prepared to abandon the term "gene." In January 1951, she wrote Beadle about her theoretical leap. She echoed Goldschmidt's words when she told Beadle, "It seems to me that we might well stop dealing with genes in the old sense and attempt to view the nucleus as an organized system with various types of controls of the action of the determiners."[12]

The view of the nucleus as an "organized system" was embryological holism applied to the nucleus. McClintock substituted the word "determiner" for the word "gene," she explained to Beadle, "because so much of the emphasis on the gene has come from phenotypes observed." A gene was simply a symbol for a trait; a determiner took a more active role. This shift seems to have made her more comfortable about abandoning the chemical-change-in-the-gene theory of mutation. Determiners just kept on determining, but their action could be modulated by controlling elements. "The determiner may not have changed but when it acts and how much may well have been altered," she wrote.[13]

"All of the above may seem like the ravings of a lunatic," she told Beadle. "Sometimes I think so myself and then the evidence again stares me in the face and I am hard at work again." She was waist-deep in a major theoretical argument over the very nature of heredity and development. How far she had come in six years, from her earlier work mapping genes and studying chromosome structure! It was exciting but nerve-wracking and exhausting. "Nevertheless," she concluded. "I don't really like working on 'fundamental problems' and long for the old days."[14]

"Was There a Gene, or Was There Not?"

As the 1950s advanced, McClintock did work some data into her arguments. Nevertheless, the manuscripts that were to give a full account of the mutable-gene phenomena never saw the light of day. Rhoades, his students, and perhaps a few other close friends and colleagues saw them, but they were never released to the wider scientific community. Scientists' understanding of McClintock's data and theories derived mostly from her presentations at lectures and conferences and the publication of papers she delivered at symposia.

The most notorious of McClintock's public presentations was that at the 1951 Cold Spring Harbor Symposium. Under Demerec's direction, the once-shambling, month-long affair had slimmed down to a week and become almost exclusively dedicated to genetics, increasingly to phage and bacterial genetics. Through the 1940s and 1950s, Demerec consistently selected the most important and controversial topics of the day.

For McClintock, having this meeting literally next door to her laboratory was simultaneously a boon and a bane. It meant, on the one hand, that most of the world's most famous geneticists came to her each year. She had a guaranteed audience, dominated by the most progressive front in genetics. She was attending fewer and fewer professional meetings; yet the symposium was one she could hardly miss. On the other hand, the symposium audience comprised arguably the least sympathetic group of geneticists that could have been amassed. She was a star at the symposium, but part of the awe she inspired stemmed from her incomprehensibility. Although a few members of the Drosophila and maize communities did attend, by the late 1940s they were greatly outnumbered by microbial geneticists. Had she worked in the Midwest and gone to the annual maize meeting, the reception to her work might well have been different than it was.

The 1951 Cold Spring Harbor meeting was almost existential. Three hundred people, from distinguished senior scientists to hotshot junior faculty and graduate students, jammed the dining and lecture halls and swarmed Bungtown Road, the tennis courts, and the Sand Spit. Norton Zinder was an impressionable observer, twenty-two years old, just finishing his first year in graduate school at Wisconsin under Joshua Lederberg. He and Lederberg had recently gathered results suggesting that bacteria exchanged parts of their chromosomes when they mated; they called the phenomenon transduction. Lederberg was to discuss transduction in his talk at Cold Spring Harbor that year, while Zinder listened nervously. Zinder commuted out to Cold Spring Harbor from his parents' home in the Bronx. Sitting in his office at the Rockefeller University in 1996, his raspy voice still rose as he remembered that meeting. Daunted by all the stars, Zinder could do little but eavesdrop. "There were little knots of people standing around talking. I was certainly not one of the 'in' people, so I sort of hung on the periphery of the major players' discussions . . . And they were probably wondering who the hell I was. I may have been twenty-two, but I looked like I was twelve." His heroes were discussing the weightiest question of their science. "It was a very serious meeting as I recall. Everybody was serious. These were *real* issues. Was there a gene, or was there not? The question of whether there is a gene in genetics is not exactly a minor issue! People were really taking it seriously."[15]

Demerec wrote in the foreword to the symposium volume that "the original problem of defining the unit of heredity . . . has not yet been solved. In fact," he continued,

the large body of information accumulated since 1941 has made geneticists less certain than ever about the physical properties of genes. Ten years ago they were visualized as fixed units with precise boundaries, strung along chromosomes like beads on a thread, very stable, and almost immune to external influences. Now, however, they are regarded as much more loosely defined parts of an aggregate, the chromosome, which in itself is a unit and reacts readily to certain changes in the environment.[16]

Several factors had contributed to this change, in Demerec's view. One was a shift from flies (and corn) to bacteria and viruses. Ten years earlier, nearly a third of the papers at the Cold Spring Harbor Symposium dealt with Drosophila, and only 6 percent with microorganisms. In 1951, microorganisms were the subject of fully 70 percent of the papers, with only

9 percent devoted to fruitflies. Also, recent studies involving X rays and mutation, given a huge boost by military-sponsored research during and after the Second World War, as well as studies of chemical mutagens, had led to new views of the gene and the nature of mutation.

Norton Zinder, however, believed it was they, the bacterial geneticists, who had more confidence in the gene. He cast the 1951 meeting into two camps. "There was on the one hand a bunch of old geneticists, who had decided the gene didn't exist anymore. And there was no such thing as a gene, and they had some vague picture of chromosome structure and organization and whatnot, as to what was the genetic element." Demerec had invited distinguished classical geneticists such as Lewis J. Stadler, Royal Alexander Brink, and the always-entertaining Richard Goldschmidt, whose controversial theories had gained new credence in this uncertain atmosphere. In recent investigations they had pushed this concept and found that it broke down; many systems could be found in which genes did not act like simple, autonomous units. McClintock's recent findings fit squarely into this camp.

"And on the other hand," Zinder continued, were "the microbial people, who were coming out of [the] Beadle and Tatum one gene–one enzyme business, and were much more secure in the notion of a gene as a unit of function."[17] George Beadle and Edward Tatum's 1941 one gene–one enzyme hypothesis, recall, postulated that each step in the synthesis of a complex molecule such as vitamin B6 was controlled by a single enzyme, and that each enzyme was made by a single gene. Obviously, the idea depends on the existence of discrete genes; Beadle and Tatum's methods effectively defined a gene as a hereditary unit that when mutated disrupted a step in the biosynthetic pathway.

Though like most bold hypotheses it was oversimple, one gene–one enzyme contained a grain of profound biological truth: today, a gene is understood as the instructions for making a protein, such as an enzyme or part of an enzyme. Beadle and Tatum's model had not found easy acceptance. At the 1946 Cold Spring Harbor Symposium, for example, Max Delbrück had slammed the hypothesis as unfalsifiable, since Beadle and Tatum's experiments were designed to show the role of genes in enzyme synthesis and were therefore unlikely to uncover countervailing evidence. The 1951 symposium became a debating platform for the one gene–one enzyme model.[18]

The meeting's opening session dealt with the theory of the gene, and led off with a paper by Goldschmidt. He revisited the arguments many had heard before, but he had new evidence to support them. Early in his paper, he challenged Beadle, accusing him of "extrapolation from the mutant action to the existence of the original gene." McClintock's mutable genes provided Goldschmidt with a shining example of position effects and a dynamic genome. He referred to McClintock's elements as "invisible" (that is, submicroscopic) rearrangements that acted by position effects. To his full satisfaction, McClintock had proved that in corn, "mutable loci are actually position effects produced by genetically controlled and repeating transpositions and translocations." Goldschmidt waxed poetic in his analogy for how genetic function derived from position: "If the A-string on a violin is stopped an inch from the end the tone C is produced. Something has been done to a locus in the string, it has been changed in regard to its function. But nobody would conclude that there is a C-body at that point." In other words, what geneticists had been calling a gene was a transient property of the whole chromosome, not a physical object.[19]

McClintock followed Goldschmidt. Her talk was reportedly over two hours in length, and in printed form it filled thirty-five oversized pages of small type.[20] (Though long, it was not the longest presentation of the meeting: Joshua Lederberg's reportedly lasted six hours!) Twenty-three pages into the published version of the paper, she wrote, "It will be noted that use of the term gene has been avoided in the foregoing discussion of instability."[21] She continued, this "does not imply a denial of the existence within chromosomes of units or elements having specific functions. The evidence for such units seems clear." But the gene concept, she said, stemmed from studies of mutation, and so by definition excluded other kinds of chromosomal elements or metagenetic activity. She presented what she now called controlling units as a new sort of genetic material, not genes but regulators of genes. "The author agrees with Goldschmidt," she continued, "that it is not possible to arrive at any clear understanding of the nature of a gene, or the nature of a change in a gene, from mutational evidence alone."[22] She invoked Goldschmidt again in her conclusion: "Evidence, derived from *Drosophila* experimentation, of the influences of various known modifiers on expression of phenotypic characters has led Goldschmidt (1949, 1951) to conclusions that are essentially similar to those given here."[23]

Being a review, the paper contains little data beyond several stunning micrographs of chromosomes and photographs of kernels. What data she did present demonstrate transposition, not control. She provided one table: it shows that *Ac* was in some plants unlinked to *Ds* and therefore probably physically distant, yet in those plants' descendants it was tightly linked and therefore close to *Ds*. These linkage data are solid evidence that *Ac* transposed. The mechanism of the movement was conventional, "one that can give rise to translocations, deficiencies, inversions, ring-chromosomes, etc., as well as the more frequently occurring dicentric chromatid formations"—all standard and uncontroversial phenomena.[24] The second half of the paper is mainly a theoretical argument for transposition as a means of controlling gene action in development. In 1980, she said that her 1951 symposium paper was a "very important paper in the way that I tried to bring out that genes had to be controlled."[25]

Both Goldschmidt and McClintock were arguing for a dynamic genome, a self-regulating system defined by the interactions among its parts. Yet her model was more sophisticated than his. Unlike Goldschmidt, and despite more extreme statements in her private correspondence, McClintock believed in genes. She assumed the existence of discrete chromosomal units that produced traits. But in her model these were regulated by other units not detectable by mutation analysis. While Goldschmidt rejected the Beadle and Tatum model, McClintock was working outside it.

The 1951 symposium is the crux of the myth that McClintock was ignored. According to McClintock, the response to her paper was "puzzlement, even hostility." She told biologist Neville Symonds that in "1951 I gave a Symposium paper at the Cold Spring Harbor Symposium that nobody understood." To Evelyn Fox Keller she said, "the symposium in 1951 really knocked me." On the basis of such statements Keller wrote that the audience's reaction was "stony silence," punctuated only with muttering and muffled laughter. The talk, "dense with statistics and proofs," was not understood.[26]

McClintock had not expected it to be understood. She wrote several times to Rhoades that she thought her talks and her *PNAS* paper of 1950 were a speculative "dive into the deep."[27] She was uncomfortable about making that dive, but felt strongly enough that genetics needed a shaking up that she went ahead with it. While she clearly looked back on the 1951 meeting as a severe disappointment, it is difficult to believe that she suffered a sudden attack of naïveté between the fall of 1950 and the spring of 1951. More likely, the disappointment she expressed late in life over her 1951 presentation developed well after the fact. It became part of her private myth.[28]

It seems to be true, as the public myth maintains, that McClintock received no questions following her talk. No one who was present at the meeting seems to recall any, not even Waclaw Szybalski, who ran the slide projector during the talks and gathered commentary for publication in the symposium volume.[29] This does not mean, however, that no one spoke with McClintock. Photos from the meeting show her in animated conversation with many scientists. Joshua Lederberg recalled that there was "a lot of discussion—I can't understand this remark about 'stony silence' that others have remarked on."[30] There may have been a meeting after her talk in which McClintock spoke with several leading geneticists about her work.[31] Norman Horowitz, another symposium attendee, insists she repeated her talk later that evening—a remarkable claim, though it remains uncorroborated.[32]

Such memories may be exaggerated, but the core recollection, that her paper made an impression, is plausible. Despite the fact that many attendees rejected maize genetics as old-fashioned, in 1951 it was still hard for a geneticist to avoid completely either maize or McClintock. The field was sufficiently small that a sample of three hundred geneticists—even the three hundred least likely to understand her presentation—would necessarily include friends, close colleagues, and sympathetic outsiders.

Many attendees had already heard her present versions of her story, had talked with her extensively and seen her slides and kernels. Lewis Stadler followed McClintock in the session on the theory of the gene and cited her paper. Neurospora geneticist David Bonner cited her paper as well. Others attendees who had received, directly or indirectly, extended accounts included S. G. Stephens, Hermann J. Muller, and Salvador Luria. McClintock's longtime friend Harriet Creighton attended, as did Margaret Emmerling, who had worked with Rhoades and was now with Stadler. Ruth Sager and Neurospora geneticist Patricia St. Lawrence, two young scientists whose dissertations McClintock had read, were also there, as were many Cold Spring Harbor staff scientists who had heard McClintock's presentations at staff meetings. In all, at least thirty maize geneticists, drosophilists, and microbial geneticists in the audience were familiar with McClintock's work.[33]

Following the meeting, McClintock's friend and colleague Ernst Caspari summarized the symposium in the pages of *Science* magazine. Caspari noted the debate over the nature of the gene and placed McClintock's paper squarely within it: "Among the papers dealing with this aspect of genetics," he wrote, McClintock's work "aroused particular interest."[34] Even if Caspari were inflating the impact of her paper—his long acquaintance with her might well have made him more receptive than most to her ideas—such a statement in a widely read journal surely influenced others' recollection or understanding of her talk.

The following October *Scientific American* ran a summary of the meeting written by Adriano Antonio Buzzatti-Traverso, of the University of Padua. Buzzatti-Traverso recognized that a complete genetic account of life requires an explanation of "the way genes control the development of the various structural and functional traits of any organism." He saw this explanation coming from "the so-called 'one gene–one function' hypothesis." He noted three distinct approaches to the gene: that of classical genetics, as a material particle that accounts for heredity and transmission of traits; that of mutation—especially radiation—genetics, as the particle of mutation; and that of the emerging concept of the gene as a particle that plays a special role in the biochemistry of the cell. "This same particle, the gene, was thus being attacked from three different angles: no wonder that the conclusions reached by the three types of attacks sometimes do not fit perfectly with one another. The same thing happens when three climbers approach the summit of a mountain from different sides: although it is the same mountain it may look very different to each climber." Goldschmidt's paper, he noted gingerly, "made the audience feel the difficulties encountered when one tries to interpret some genetic phenomena with the assumption that the genes are discrete particles having no interrelationships with neighboring genes." McClintock and Stadler "brought new evidence to the subject," he continued. Their contributions "pointed to the present need of regarding the activity of the gene as a function of the internal organization of the chromosome."[35]

McClintock's paper thus did arouse interest, although it seems it was expressed in small groups, rather than in the formal setting of the symposium session. Clearly McClintock was exaggerating when she said no one understood. How then could she have felt ignored? Interview subjects give a consistent answer, and one supported by documents: McClintock was not primarily talking about transposition. All agree that her talk was long and complex, and that although many had enough familiarity with maize to understand her argument, few had the extensive background required to evaluate it.

The real reason, however, was that so little of her paper was about genetics. This is the way Zinder remembers it: "She really didn't want to speak about transposition. She was talking about regulation and development. And that's what she was pissed off at [us about], and nobody would take that seriously. *Everybody* took seriously the transposition. What she wanted was that transposition was the regulatory control of development. And nobody there, or at least the very bright geneticists in the room, could understand how stochastic processes could really be involved in something as organized as developmental regulatory stages."[36]

Joshua Lederberg agrees. What made her paper difficult were her "epigenetic hypotheses about the significance of gene jumping. And she was never very clear about that. That's still part of her presentation that remains quite vague. How she brings this in as a developmental phenomenon."[37] Similarly, Waclaw Szybalski could accept the transposition, but McClintock's interpretation left him puzzled: "While I was convinced that she had good evidence that things moved, that this has any significance for development I was not convinced, because she did not present anything to me which would convince me at that time or even later."[38]

If McClintock was disappointed about the reaction to her talk, she hid it well. Demerec thought she "thoroughly enjoyed" the meeting.[39] Her letters to Rhoades in the summer and fall of 1951 have none of the despondency that gave a sour flavor to some of her other correspondence, both before and after 1951. In September, she was in high spirits and planning to head out to Pasadena later in the fall, with a stop in Champaign-Urbana. Her summer crop had come in well—"all I could have hoped for."[40] But by October, she was down again, "in a confused and indecisive state." The work was overwhelming her, as was the prospect of preparing her winter greenhouse crop. She was exhausted, and Demerec was concerned about her health. "When I told him I was going to Pasadena," she wrote, "he was distinctly disturbed (as was my mother also) because he felt I should break completely with work."[41] She deferred the trip.

Transposition Redux

Royal Alexander (Alec) Brink was one symposium attendee who certainly did not doubt transposition: he had seen it in his own cultures. Brink was a leading figure in maize genetics, chairman of the genetics department at the University of Wisconsin at Madison. He had a reputation for clean and trustworthy, if not dazzlingly imaginative, experiments. He had been interested in mutable genes for some time; he had received stocks containing the *Dotted* element from Rhoades as early as 1945.[42] In the summer of 1948, Robert A. Nilan, a young Canadian biologist, began doctoral work with Brink in Madison. When Nilan arrived, Brink handed him a fistful of maize papers and told him about McClintock's new work on mutable genes. He told Nilan that his project was to be a reinvestigation of the first known mutable gene, Emerson's *P* locus. Did transposition of tiny inhibitory elements explain the mutations at the *P* locus?[43]

It did. The crucial plants were harvested in the summer of 1950, and by the time Brink and Nilan completed their analysis, in October or November, they were sure that transposition was occurring at the *P* locus.[44] They wrote a quick note for the *Maize Genetics Cooperation News Letter*, an informal annual publication read by all maize workers, and it was published the following March.[45] Nilan and Brink called their new element *Mp*, for "modulator of pericarp." They noted that the patterns were similar to those produced by McClintock's *Ac*. (By the mid-1950s it was strongly suspected that *Mp* is in fact identical to *Ac*; this was later proved.) Nilan finished his thesis by June 1951, when he headed west for a faculty position at Washington State University in Pullman. Brink headed east, for Cold Spring Harbor and the 1951 symposium. In 1952, they published their findings in *Genetics*.[46]

The following year, Peter Peterson, a young corn geneticist from the University of Iowa who had spent time at Cold Spring Harbor in the late 1940s, published on yet another movable element. He discovered it in strains of maize experimentally subjected to the atomic blasts at Bikini atoll. He published a short note in the *Maize Genetics Cooperation News Letter* and also gave a paper at the meeting of the American Society of Naturalists. Peterson named it *Enhancer (En)*, and showed that it operated at the *pale-green (pg)* locus. Both Peterson and Brink acknowledged McClintock's precedence in discovering movable genetic elements.[47]

Transposition, then, was not controversial—but its significance was. Though Brink admired McClintock, he disagreed with her developmental theory. According to Jerry Kermicle, who joined Brink's group in 1957, Brink deliberately avoided use of the terms "controlling units" or "controlling elements." "That's where the nexus of the tension was" between them, he said.[48] James F. Crow, a population geneticist in the department at Madison, lunched with Brink nearly every day. Crow recalled that "her use of the words 'controlling elements' didn't sit well. And most people here, especially [population geneticist] Sewall Wright and R. A. Brink, and by extension me, thought that she was probably wrong on that account."[49] Brink, and soon his lunch companions, used more purely descriptive terms such as "modulating element" and "transposable element" to strip away the functional connotations of control.[50]

Brink understood McClintock's argument. Her interpretation, he wrote in 1960, was that "the various loci are progressively activated during development of the individual by controllers functioning in genetically fixed ways and in a predetermined sequence."[51] But one could "turn McClintock's controller argument around," he said, "without compromising the facts, and assume that what she considers primary causes are, themselves, effects."[52] In other words, transposition did not control development; development controlled transposition.

Though Oliver Nelson told me McClintock disparaged Brink's intellect (in this regard Brink was in the distinguished company of such innovators as Boris Ephrussi and Theodosius Dobzhansky), Madison was, nevertheless, a regular destination for her, perhaps second only to Champaign and Pasadena.[53] The visits, however, came to an abrupt end in 1958, when McClintock gave a seminar to an audience that included Brink, Crow, Kermicle, and Wright. Wright, one of the architects of the modern evolutionary synthesis of the 1930s and a dedicated Mendelian, was in active "retirement." In his late sixties, he was still sharp mentally and just as obstinate as McClintock when convinced he was correct.

According to several accounts, after hearing McClintock explain her developmental hypothesis and watching her diagram it on the blackboard, Wright got to his feet, erased her drawing, and replaced it with an alternative explanation that he favored. Wright's argument was probably similar to Brink's: that she had reversed cause and effect. McClintock responded by erasing Wright's figure and redrawing hers. After two or three cycles of this, McClintock stormed out of the seminar room, vowing never to return to Madison while Wright was still alive. Wright, however, turned out to be the Methuselah of genetics: he died in 1988, at the age of ninety-eight. McClintock never did go back to Madison.[54]

Edgar Altenburg, a drosophilist at Rice University, was one scientist who did accept McClintock's controlling-element theory. In January 1953, he wrote to McClintock, asking her to review a manuscript of his that addressed her mutable-gene work. "Although Goldschmidt's article on your work appeared in the *Am*[erican] *Nat*[uralist]," he wrote, nodding tactfully at Goldschmidt's dicey reputation, "a second article with a somewhat different approach and purpose might be justified." The importance of her work was not being matched by its impact on the field, "perhaps largely because of the difficulty in understanding it, in view of the intricate nature of the problem, and also the necessarily condensed account which you have had to give of your work."[55]

McClintock read the manuscript, praised him for his penetrating insight and objectivity, gave him extensive comments, and encouraged him to publish it (although apparently he never did).[56] She was especially impressed because he had had to work from the primary material she had sent him, aided but little by her published accounts: "The published descriptions of this material, are too inadequate," she wrote. "They have been concerned mainly with generalized descriptions and interpretations." She also sent him a copy of a manuscript she was working on, which she felt nailed down her argument. Her work, Altenburg replied, "heralds a new era in the study of mutation." That other contemporary work in genetics could be explained in light of her results and conclusions, he told her, was the "sort of thing which is arousing such keen interest in your results." The classical gene concept, he believed, "will certainly have to undergo alteration in light of your results." It may not need to be abandoned entirely, "though we may have to equate chromosome and gene."[57]

The Mutable-Gene Business

The article she was writing in January 1953 appeared in *Genetics* the following October.[58] It was her most empirical article on genetic control and her only primary research article on controlling units in a peer-reviewed journal. In it she provided culture numbers, ample tables giving the results of crosses, and explanation of tests for the presence and activity of *Ac*. But her main argument derived from a highly original approach for genetics of the 1950s.

Her primary finding was that she had mimicked Rhoades' mutable *Dotted* by inducing the breakage-fusion-bridge cycle at specific loci. The idea of making new mutables probably had its germ in the first mutable she observed, plant B87-1, which in 1945 she noticed was "like *Dotted*." *Dotted* acted on *a1*, the recessive form of the *A1* gene, which encodes the red pigment anthocyanin. Starting with strains containing a normal, nonmutable *a1* allele, McClintock created new alleles at the *a1* locus that were mutable and controlled by *Ac*. When *Ac* was present, they produced kernels that looked like *Dotted* but were stable in the absence of *Ac*.

She was thus able to select a gene at will and make it mutable. As early as 1950, she was so confident in her ability to make new mutables at any given locus that she wrote to Rhoades, "I think we can go into the business. If any one wants a locus to be mutable, just put in the order and one will be sent the following year!"[59]

Although the 1953 paper was much less theoretical than her 1950 *PNAS* paper, McClintock continued to embed her work in the debate over the nature of the gene. These new results strengthened her conclusion that the chemical-change-in-the-gene model was too limiting. "Mutations need not express changes in genes," she concluded, "but may be the result of changes affecting the control of genic action." The control was manifest in nongenic or extragenic agents carried in the chromosomes, presumably heterochromatin. Here again, however, where she dipped into more speculative arguments, her language and supporting evidence became vague. "A particular type of change in genic expression," she wrote, "might reflect a particular type or types of change in chromatin components at the locus." This statement was bolstered by evidence unstated and unavailable to others for evaluation: "Evidence in support of this assumption has been obtained (McClintock unpublished)."[60] If one went to her laboratory and asked to see the chromosomes and kernels, she would likely have taken an afternoon to explain it; to those outside her inner circle, however, her theoretical conclusions must have seemed highly speculative.

In her paper, McClintock interpreted Brink's and Peterson's results as support for her theory of genetic control. The *Modulator* and *Enhancer* systems, she believed, supported her "generalizations . . . concerning the existence and behavior of controlling units."[61] Brink and Peterson would have agreed that their data supported McClintock's evidence for the existence of movable elements. But they did not see their work as supporting her generalizations about the behavior of movable elements.[62]

In the 1970s and 1980s, McClintock said she had received only two or three requests for a reprint of the 1953 paper.[63] Such a claim is difficult to verify, but indeed the paper was not cited heavily: the *Science Citation Index* lists only three mentions in 1954 (twice by Brink's group) and only twenty-seven more between 1955 and 1964. Since even some of those maize geneticists working on McClintock's transposable elements did not cite the paper, preferring instead to cite her more general 1951 review, it may be that the 1953 paper was seen as too specialized and not directly relevant to others' work.

McClintock, however, considered it the "final crucial experiment" that would convince her colleagues of her developmental theory. She told Altenburg that it would place her work "within a generalized conceptual framework."[64] The evidence suggests it was the reception to the 1953 paper, not the 1951 symposium talk, that really knocked her.

Expressions of Confidence

Immediately after writing the 1953 paper in January and February, McClintock suffered a bout of depression. The response she received from her friends during this period shows the esteem in which some of the nation's most powerful scientists held her. Although McClintock's scientific correspondence with Beadle diminished in the late forties, he continued to be a strong source of emotional and professional support for her. In February, she wrote to him, "after 12 years at Cold Spring Harbor, I have come to the conclusion, finally, that it may not be possible for me to remain here."[65] This was neither the first nor the last time she considered leaving Cold Spring Harbor, but it was the closest she came to actually going.

The problem, she said, stemmed from the logistics of growing her corn. Although she kept her crops small, the field work was still overwhelming and required at least some assistance. The incompetence of the field hands forced her to do much of the maintenance herself. Emergencies such as storm damage overtaxed her. She could never feel comfortable about leaving her plants for very long. Compounding her frustration, she explained, was the lack of scientists at Cold Spring Harbor who understood plants and the "inevitable weariness that comes from remaining too long at one place especially when the environment is so limited both intellectually and spiritually."[66]

Beadle replied quickly and kindly, but he could not mollify her. McClintock thought Demerec's responses to her requests were flippant. So she quit. In March, she sent a letter of resignation to Carnegie president Vannevar Bush. McClintock told Beadle that Bush was "shocked" by her resignation and assured her that all would be made well for her, either at Cold Spring Harbor or elsewhere. Bush flew Demerec down to Washington, where, she said, "the poor man took a beating from Bush." Bush also wanted to meet with her. She had no personal antagonism, she said, "no mads, no serious gripes, just plenty of enough." She did not know where she would go, but freedom held a romantic appeal for her. To go, she wrote, "is an urge which should be obeyed before it becomes a panic!"— as it had done at Missouri. McClintock's toughness and self-deprecating humor kept her comments this side of maudlin: "If sleeping on a park bench would embarrass the department, please don't fail to advise me. I have a nice cozy spot picked out in Florida that could substitute!" Bush's support gave her some relief. "His expression of confidence in me as an investigator and as an intellect removes some of the major causes of my difficulty here."[67]

She was glum about the prospect of finding another academic job, partly because of her age and partly because of the "general anti-female sentiment" in science. Some time in the late 1940s, she told Evelyn Fox Keller in 1979, she had been offered a job at two thirds her current salary. Insulted, she wrote back a terse letter declining the offer. She later used that story as a test, she said: "women were shocked but men just have blank stares." She told Rhoades that Cold Spring Harbor was "the only place I have been where the anti-female bias did not face me much of the time." She would not seek a university position with tenure, teaching obligations, and administrative duties, she said; a research-associate position would suit her better.[68] The position of research associate is a notorious pigeonhole for women scientists. Countless capable women have been offered these nontenured, low-salary positions in lieu of the tenure-track regular faculty jobs had by their male counterparts.[69] McClintock was willing to trade status for freedom.

Instead, she was offered a remarkable post that would have both status and freedom. Beadle suggested she ask for a roving appointment. It would be, not an endowed chair, but an endowed sedan that would bear her wherever her fancy took her. She would come and go from several institutions at will, with no institutional affiliation. It would be a tenured, Carnegie-sponsored version of the National Research Council fellowship on which she had been so productive in the early 1930s. Not surprisingly, McClintock thought this "would be splendid!" She would give Bush an ultimatum: she wanted to be "foot-loose and tied to no place permanently . . . If he does not agree to this, then I am out of Carnegie." Beadle got President Lee DuBridge to endorse his offer of Caltech as one of her bases. Rhoades offered Illinois as another.[70]

Bush did not bat an eye at McClintock's extraordinary proposal. He, Beadle, Demerec, and Rhoades corresponded about the situation. "All of us here have the highest regard for Barbara," Beadle wrote to Bush, "both as a person and as a scientist." McClintock was "a grand person," Bush replied, "and certainly worth all of the aid and support we can give her." He had spoken of the matter with Demerec, who, for all McClintock's complaints, Bush felt was "in accord and just as anxious to help as I am." Bush was willing to grant her request, or any other that would "help her get completely back on the rails." Demerec wrote to Beadle that McClintock had been at a low point recently, although she seemed to be rallying, and that he was "keeping up with efforts to help [her] in her present situation."[71]

In the end, she did not take the appointment. Her emotional state continued to improve, and for the time being she lost the need to leave Cold Spring Harbor. That summer she began growing her corn at Brookhaven National Laboratory, an hour to the east, which seems to have alleviated some of the strain of the lack of field help. The offer was nevertheless a remarkable show of support that illustrates the unique role McClintock played in the community of American geneticists. Five years later, in 1959, she was again apparently unhappy and was given a formal offer from the University of Illinois (again, she would have retained Carnegie funding), which she again declined.[72]

In the 1960s, McClintock could still marshal extraordinary support from leading scientists. By 1964, she had made the short list of candidates for the Kimber Medal. The medal, awarded by the National Academy of Sciences from 1955 to 1970, was the highest honor awarded specifically to a geneticist. It held great prestige, in part because it was awarded by a committee of prominent geneticists. Max Delbrück won it in 1964; in 1965, the award went to Alfred Day Hershey, another founding father of phage research and McClintock's Cold Spring Harbor colleague. McClintock ranked second, above Boris Ephrussi, the Russian geneticist Nikolai Wladimirovich Timoféef-Ressovsky, and three recent Nobel laureates: François Jacob (1965) and James Watson and Francis Crick (1962). In 1966, Timoféef was bumped to the top of the list for political and diplomatic reasons—he was given the prize, though it was never formally announced.[73]

Nineteen sixty-seven, then, was to be McClintock's year. When the nominations came in, she was the unanimous top candidate. To the shock of the Kimber nominating committee, Ernst Caspari, in charge of appointing the committee members, asked McClintock herself to serve. Unaware of her ranking for the award, McClintock accepted. The committee, then, comprised Marcus Rhoades, chairman, Max Delbrück, drosophilist Bentley Glass, James F. Crow, and McClintock. "I do not know how it happened that she got onto the committee," Delbrück wrote to Rhoades, "and I think it is a mean trick if it excludes her from being a recipient. I certainly think she is the most worthy of my three nominees." Glass, too, saw the problem and asked Rhoades what could be done about it.[74]

Rhoades determined that her presence on the committee did not automatically eliminate her from consideration. But it remained a delicate issue. Neither asking McClintock to vote for herself nor asking her to resign from the committee was an attractive option. Rhoades solved the problem by devising two ballots: one for McClintock, listing candidates 2, 3, and 4; and one for the rest of the committee, with McClintock's name heading the list. McClintock made her selection, everyone else voted for her, and she won the award.[75]

Those who maintain that McClintock lacked the respect of her colleagues like to cite the arrogance of one young and prominent geneticist who said McClintock was just "an old bag who's been hanging around Cold Spring Harbor for years."[76] Yet such comments seem to have been rare. Certainly, for some her quirky personality, stellar intelligence, and fierce honesty were off-putting. McClintock could be scathing in her assessments of others' work and general intelligence. But such traits can also inspire loyalty and affection. Her own letters as well as those of friends and colleagues show there were many—among them some of the leading geneticists of the day—who felt such allegiance to her.

A Widening Audience

Though McClintock avoided most scientific meetings, she did travel to give seminars and lectures. Talks she delivered during the mid-1950s brought controlling units to an increasingly wide audience. Scientists debated and often disagreed with her interpretation of mutable genes as evidence for developmental genetic control. During this time, her evidence became more complex. Transposition was easier than ever to understand, but control grew ever more difficult.

In the spring of 1954, she gave a lecture series at Caltech. It was probably arranged the previous year, as Beadle negotiated for her roving appointment. Biology 225 was a graduate "special topics" seminar, whose theme varied as the professor alternated. From the second week in January to the second week in March, McClintock gave two lectures a week to the Caltech biology graduate students, beginning with an overview of maize genetics and progressing through a historical account of her work on mutable genes and developmental regulation. Her lecture notes indicate that the students received the full weight of her evidence and her theoretical speculations. The notes for her last lecture state, "Since these controlling systems have made themselves evident in cells whose nuclear composition has been altered, we infer that they were present in the nucleus before the alteration occurred. Performing in same manner—controlling gene action and doing so at specific times in specific cells during development." Further, if it were accepted that the controlling units are present in cells before the mutations occur, "then they must be functioning in normal nuclei." Thus, she "assumed" (that is, concluded): "(1) They serve to control specific gene actions at certain stages in development," and "(2) That the somatic segregations [of controlling units] likewise occur during normal development."[77]

By 1955, it was clear that transposable or controlling elements were not rare, at least in maize. McClintock could now cite five apparently unique controlling units: *Activator/Dissociator; Suppressor-mutator*, a new element in her cultures (see Chapter 8); Rhoades's *Dotted;* Brink's *Modulator*, and Peterson's *Enhancer.* In addition, M. Gerald Neuffer, at the University of Missouri, had discovered two new *Dotted* loci in South American maize and another new mutable element that seemed similar to *Ac.*[78] These occurred at different sites on the chromosomes. In 1998 Neuffer told me, "Now I wasn't thinking about them as moving around. I was thinking of them as arising on different sites. I thought there was a *Dotted* site and that these other sites arose by translocation to new sites by chromosome breakage. I thought there was an original *Dotted* locus on chromosome 9 and it was duplicated on another chromosome. I was not yet thinking of *Dotted* as transposable."[79] McClintock, though, assumed (correctly) that they were transposable. She believed she had yet another controlling system in her own maize cultures and could point to the work of several other researchers whose experiments indicated (either to them or to her) the presence of controlling elements.[80] The number of scientists engaged in transposition or controlling unit work remained limited, but it was growing.

Thus far, she had referred to these regulatory bits of chromatin as controlling units, to distinguish them from genes. The term "units" recalled William Bateson's "unit character," later dropped in favor of "gene." In June 1955, in a paper presented at the Brookhaven Symposium on Mutation, she wrote that she had called them units because "they may show regular inheritance patterns and thus be followed as 'units' in progeny tests." They were, in short, Mendelian units, but not genes. In that paper, however, she dropped the term "unit," apparently to distance her system further from conventional genes. She began referring to them as controlling *elements*. She asked a question that, characteristically, actually indicated her conclusion rather than her hypothesis: "Are controlling elements present at all gene loci and are many of the so-called gene mutations the consequence of changes occurring to them?" She concluded, "There can be no doubt about the presence in the maize chromosome complement of elements that control the action of genes."[81]

Although her argument was analogical rather than empirical, she suggested that evidence from bacteriophage, bacteria, protozoa, and several plants and animals indicated that controlling elements were widely distributed in nature. She did not discuss heterochromatin; while she still clearly believed controlling elements were different from genes, by 1956 she had stopped speculating about their material nature. Her primary conclusion, however, remained firm: "controlling elements are normal components of the chromosome complement and . . . they are responsible for controlling, differentially, the time and type of activity of individual genes."[82] The real point was control.

Unlike her presentation at the 1951 Cold Spring Harbor meeting, her talk at the 1955 Brookhaven Symposium was followed by extensive formal discussion, among McClintock, Ruth Sager, Rollin Hotchkiss, Hermann J. Muller, and others. Several of the discussants addressed the questions of whether controlling elements were in fact qualitatively different from genes and whether they controlled development in normal cells. This last, especially, seems to have been a matter of some concern, and one difficult for other geneticists to swallow. O. H. Frankel, for example, wondered whether McClintock's phenomena might instead be a "'disease' of the gene." He pointed out that "if it were normal, then it should not be as readily reversible as it appears to be." McClintock reiterated that, on the basis of a large body of evidence, she believed the phenomenon was a general one, "and that it depicts one of the mechanisms responsible for control of gene action during development," though she admitted that extensive studies had been carried out only with maize.[83]

The following April, in a bad mood, she wrote to Caspari about her perceptions of the reaction to her work among her colleagues. Other scientists, she said, did not appreciate the way it undermined a genetics based entirely on the study of mutation, or the way she had been addressing the genetic control of development. Although her material, she told him, "has some basic implications for viewing 'gene mutation' and for effecting developmental processes, the general negative attitude on the part of most geneticists over the past years makes me hesitate to push the story personally. In this respect, I must confess that I am very tired and weary and can just hope that others can take over." Fortunately, the transposable element work of other maize geneticists seemed to be doing that—she interpreted it, as always, as direct support for her developmental theory: "Certainly, the maize geneticists are finding the same story in their own materials and rather generally so. That takes some of the burden from me. However, for me to talk about it is not easy and I don't look forward to the Cold Spring Harbor symposium in this respect." She had been asked to present another paper on controlling elements at that summer's symposium. "I only took the assignment because the topic of the symposium required that the subject be presented by someone and I know more about it than others."[84]

The 1956 symposium was titled "Genetic Mechanisms: Structure and Function."[85] Once again, many of McClintock's friends and colleagues were in attendance. These included many who had heard her 1951 paper, such as Caspari, Harriet Creighton, Rollin Hotchkiss, Edgar Altenburg, Bentley Glass, Boris Ephrussi, Ruth Sager, Edward B. Lewis, Jack Schultz, and Charles W. Metz, as well as some not present in 1951 but who knew her work well, among them Beadle, Sterling Emerson (son of Rollins), and many Cold Spring Harbor staff members. Also present were several young scientists actually working with McClintock's material. Among them were A. H. Ar-Rushdi, of the Scripps Institution of Oceanography; Seaward Sand, of the Connecticut Agricultural Experiment Station; and Drew Schwartz, who had done his degree with Rhoades and gone on to Oak Ridge National Laboratory.

This paper contained more data than did the 1951 paper, including several tables, though none of the photographs that make the 1951 argument for transposition so compelling. But as in 1951, she emphasized controlling elements, not transposition per se, and she focussed on her new element, *Suppressor-mutator.* She began by discussing the distinctions between controlling elements and genes, stressing that controlling elements were a different type of chromosomal element. This distinction, however, was an inference based on their apparent behavior, not a conclusion based on biochemistry. She acknowledged she did not know what the elements were made of, although she believed they were integral to the chromosome. Some argued that hers were free-floating elements that bind to the chromosomes. This "episome" model later became a popular interpretation of controlling elements, and one McClintock never liked.[86]

By 1956, then, several of McClintock's assumptions and hypotheses about controlling elements had changed. Among geneticists generally, heterochromatin faded as a research topic, and McClintock seems to have gradually dropped it as an explanation. Between 1960 and 1967 she adopted the terms "operator" and "regulator" from the bacterial geneticists. She would later also drop the notion that controlling elements were materially distinct from normal genes. As she discarded the heterochromatin hypothesis, so McClintock seems to have given up her speculations about "stickiness" and "repulsive forces."

Two aspects remained constant: that controlling elements determined growth and differentiation in normal organisms; and that the mechanism of transposition was conventional.

Though she published no more data papers in peer-reviewed journals, McClintock exposed many scientists to her ideas, through lectures, courses, and seminars. Her work began to be written up in textbooks. It was cited in major review articles. Geneticists working on maize and other plants began using McClintock's elements in their own research; others considered whether Neurospora, fruitflies, bacteria, or other organisms might carry controlling elements. The main current of genetics during these years was DNA, one gene–one enzyme, and bacteria. Eukaryotic genetics as a whole was on the decline, and complex ideas such as McClintock's controlling elements were not mainstream even in that subfield. That she received as much recognition as she did may be attributed to her reputation for careful experimentation and insight into complex biological phenomena.

According to the *Science Citation Index,* McClintock's 1950, 1951, and 1953 articles received collectively 44 citations by 1954. By 1964, they had been cited 186 times. Her Carnegie reports added slightly to her total. Many of the citations were by maize geneticists, as might be expected, but not all. Several bacterial geneticists cited her, as did workers on a variety of other model organisms, tobacco, alfalfa, various fungi, and mice.

Although some of the citations may be viewed as token obeisances to possible but unlikely mechanisms, a number of workers, particularly younger scientists, took her ideas seriously. Many scientists benefited from direct contact with her. Such contact was easy to obtain, especially if one worked nearby at Harvard, Yale, or Brookhaven, or at her oases, Champaign-Urbana, Madison, and Pasadena. Situated elsewhere, one had only to write her a letter. McClintock was extraordinarily generous in responding to questions and requests. If a correspondent seemed to be well-informed, she took his questions seriously, often writing out many pages of explanation and analysis and tucking in kernel samples.

Lewis Stadler had several people working on mutability at the *R* locus, including Gerald Neuffer and Margaret Emmerling. John Laughnan, a former Stadler student and a colleague of Rhoades at Illinois, was studying mutation and instability in the *A1* locus. Drew Schwartz, by then at Oak Ridge National Laboratory, was by 1960 developing a molecular model for gene instability and later developed molecular mechanisms to explain some of the subtle phenomena McClintock described in *Suppressor-mutator.* His colleague at Oak Ridge, Alex C. Fabergé, published a theoretical article dealing with crossing-over, breakage-fusion-bridge, and control and used some of McClintock's stocks, such as the rearranged chromosome 9 that led her to the breakage-fusion-bridge cycle, to support his arguments.[87] Members of Brink's group, too, benefited from kernel samples and long analytical letters from McClintock. Several people at nearby institutions, such as Elwood Johnson Dollinger, of Brookhaven National Laboratory, received guidance, advice, and samples from her. In her 1956 Brookhaven paper, McClintock wrote that Paul Mangelsdorf of Harvard believed transposition could explain the large number of spontaneous mutants he observed in teosinte.[88] Oliver Nelson, a corn breeder at Purdue and then Wisconsin who moved into biochemical genetics, used *Ac* and *Ds* in his studies of the *waxy* gene.[89] By the late 1960s, Nelson's ten years' work on this gene led him to an interpretation that "the controlling elements may well be normal regulatory elements that in the mutable genes are seen in an abnormal context." He told McClintock, "I don't think this differs substantially from your view, broadly speaking."[90]

Researchers working on other organisms took her work seriously as well. Some, such as Ruth Sager and Patricia St. Lawrence, were personally influenced by McClintock. Others simply absorbed her work and tried to apply it to their systems and may not have had a personal connection with McClintock. They cited her work in botanical journals, genetics journals, general scientific journals, and occasionally even specialized journals, such as *Progress in Biophysics.*[91]

At least four textbooks from the 1950s and 1960s discuss her work; among these is *General Genetics,* by Adrian Srb, Ray Owen, and Robert Edgar, which was widely used in genetics courses across the country.[92] McClintock herself vetted the relevant chapters in another, Francis Ryan and Ruth Sager's 1961 *Cell Heredity.*[93] Her work was also cited in major review articles, such as Stadler's classic 1954 valedictory address and François Jacob's 1959 Harvey lecture. James A. Peters included her 1950 *PNAS* article (as well as the 1931 paper on crossing-over coauthored with Harriet Creighton) in his anthology, *Classic Papers in Genetics.*[94] A collection of articles edited by L. C. Dunn appeared in 1951 under the title *Genetics in the Twentieth Century.* In it Beadle cited McClintock's 1950 paper as an important new direction in gene structure research, and Torbjorn Caspersson and Jack Schultz called that paper "spectacular."[95] Perhaps the most telling citation was one by Guido Pontecorvo, which concluded, "The outstanding work in maize (review: McClintock, 1956) . . . is probably very relevant here. Unfortunately I do not understand the details of this work well enough to put my finger on what may be particularly significant."[96]

McClintock gave numerous talks and lectures during this time. She lectured at Harvard, the University of Pennsylvania, Columbia, Yale, Cornell, Brookhaven, Illinois, Indiana, Wisconsin, and other institutions throughout the 1950s and 1960s. Every one of these talks concerned genetic control, gene action, or developmental regulation; none focused solely on the fact and mechanism of transposition.[97]

"You Couldn't Question Her Results"

During the 1950s and 1960s, McClintock's science was not ignored; nor was she marginalized or ostracized from the scientific community. Yet her work was not wholly accepted. Casting her reception into the rigid molds of rejection and acceptance elides the subtle relationships among competing theories and complex individuals.

By the 1950s, transposition in maize was a well-established fact. Yet how to interpret that fact remained obscure. Not even those maize geneticists working on mobile elements accepted McClintock's interpretation that they executed the plant's developmental program. In their correspondence of 1953, Bush and Beadle concurred that McClintock's current work was of great importance but that she did not seem to have worked out a convincing interpretation. Rhoades, who understood her work better than anyone else, included controlling elements in his genetics lectures. Yet even he questioned her interpretation. In his lecture notes for a survey course on genetics, probably from the late 1950s and 1960s,[98] he left a telling blank:

> *McClintock's mutable genes*
> Formal analysis—beautifully done
> Interpretation—

Drew Schwartz recalled taking the course in which Rhoades presented McClintock's mutable genes. He confirms that Rhoades questioned her interpretation, although, Schwartz said, "you couldn't question her results."[99]

Melvin M. Green is a drosophilist at the University of California at Davis. In the 1960s, he discovered transposition in Drosophila (see Chapter 9). When asked how drosophilists felt about McClintock's mutable genes, the voluble Green blurted out, "Faith! We had tremendous faith!" The subject of her theory of genetic control, however, subdued him. "This was a little touchy. It was very touchy," he admitted.[100] It was touchy because it did not seem supported by evidence. It seemed intuited, analogical, mystical. He and others found it difficult to accept that this scientist they admired so much seemed to be going off on such a speculative tangent. Similar statements were made by corn geneticists, drosophilists, and microbial geneticists. It was not transposition they could not understand; it was control.

Why could they not understand control? The McClintock myth implies that other geneticists were incapable of grasping a revolutionary idea—misrepresented as transposition—that McClintock presented with clarity and a wealth of evidence.[101] McClintock's research notes and published papers suggest otherwise.

She failed to persuade for two types of reasons. First, her control theory contained speculative elements that she could not substantiate. Her argument hinged on pattern and analogy. She never did demonstrate that controlling elements transposed in a coordinated way during the development of the plant. She showed (a) that the elements transposed and (b) that they affected gene action. It was known (d) that genes must somehow be turned on and off during development. McClintock never demonstrated the implied (c), that her system effected that coordinated turning-on-and-off.

Second, where she did have data, she did not present them. In her zeal to shake up the genetics community, to force her colleagues to rethink the concept of the gene, she overestimated their ability to accept on faith a radical argument. Most if not all of those who were convinced of her argument had benefited from extensive personal tutorials from her. She invested extraordinary effort in those she judged sufficiently intelligent and sensitive. As with Altenburg, so with many others: she sent seeds, long letters, pages of diagrams, all in detailed explanation of how controlling elements affected gene action. Yet with the exception of her 1953 *Genetics* paper, every one of her journal articles on mutable genes and controlling elements was a review that summarized data and concentrated on the conclusions. Her Carnegie reports contained somewhat more data, but these too were intended as summaries, partly for nonmaize geneticists, and were seldom cited except by specialists.

Ironically, this may have been a result of *too much* data. McClintock amassed so many "cases" of transpositions and studied them in such detail, and her arguments were such intricate syntheses, that she may have felt it impossible to give all the relevant data to support any of her major conclusions. So she left it all out. She suggested as much in 1950, as she was preparing her first paper on mutable genes for *PNAS:* "There is too much evidence to give all in one dose," she told Rhoades. In 1953, she told Altenburg she had not expected that "geneticists should necessarily consider [her interpretations] objectively or favorably for there is so little published evidence on which one can work." She "refrained from publishing the details," she continued, because she wanted to be sure that she had amassed evidence from all possible quarters. Rather than publishing an incomplete story, she relied on her reputation for license to speculate. When the 1953 paper was not received as the crucial experiment proving her theory, her disappointment kept her from publishing the monograph that might have won her more supporters.[102]

The core of her theory—that the action of the controlling elements was orchestrated by a developmental program—left scientists scratching their heads. No known mechanism could account for such control. McClintock's elements might be a means of introducing new variation into the genome, but few could see how the elements could be part of an organized system that controlled gene activity during the development of the plant.

By the mid-1950s, McClintock began to consider that control could occur without transposition, but transposition, she believed, always indicated control. To discuss transposition was to discuss control. This link explains why she felt ignored. Because she saw her interpretation as inseparable from her genetic data, she felt her colleagues rejected her entire corpus. It also explains why, in the 1980s, scientists were offended by the accusation that they had ignored her. We didn't reject her, they protested. We didn't get that odd bit about gene regulation, which appears to have been wrong anyway, but we didn't doubt her really important contribution: transposition.

Transposition and developmental control thus followed different trajectories. Through the 1960s and 1970s, McClintock pursued developmental, and later evolutionary, control. She responded to the rejection of her theory by isolating herself from all but trusted friends and plumbing "the deep" into which she had dived in 1950. Her arguments grew more complex, more integrative, in some cases more speculative, and ultimately led her to a profound vision of organic change. Although transposition faded from McClintock's studies, it was discovered independently by molecular biologists who saw it as a general genetic mechanism. The full integration of control and transposition in McClintock's work requires that we follow each of these paths in turn.

8

RESPONSE

The integrative process undoubtedly has remodeled organizations and interrelations in the higher organisms.

—BARBARA MCCLINTOCK

Primitive corncobs were no bigger than your thumb. They bore a dozen or so stony kernels. Prehistoric Americans from Appalachia to the Andes discovered and cultivated the giant grass that produced them. They may have crossed the plant with teosinte, a related species, to get a better yield or hardier plants (biologists still argue over which plant is the ancestral form). Maize proved adaptable to domestication. Different peoples bred maize for different qualities, leading to a tremendous variety of types of corn: pod corns, with each kernel wrapped in a husky sheath; popcorns, in which the seeds are encased in a tough shell that can be burst with heat; dents, which look like water-worn molars; tough-kerneled flints. Kernels may be tiny, round, and tightly packed, like a cluster of freshwater pearls, or long and triangular, inverted pyramids splayed radially from the cob. In color they range from milk through sunflower, orange, and blood, and on to blue jay, grape, and walnut. They may be tiger-striped, leopard-spotted, birds-egg speckled. The cobs may be shaped like baseballs, baseball bats, scimitars, pine cones, or dowels. From the first tiny, stony-kerneled cobs we have bred the juicy, foot-long, twenty-rowed marvels that grace late-summer suppers. This is more than a feat of agriculture; it is a tribute to corn's genetic diversity. Yet nearly all the varieties are immediately recognizable as corn. Under the riot of colors and shapes, the mind picks out a basic corn pattern.

As young seedlings, these different varieties look much alike. Differences become apparent as they grow. One grows a little faster, another produces more ears. Green seedlings begin to express pigments of red, purple, brown, or yellow. Each variety has a characteristic developmental pattern. In a growing plant, subtle alterations in pattern can have dramatic effects. A shot of growth hormone at a key moment, a squirt of pigment in an early stem or leaf cell can propagate through development and change the fate of whole tissues. Timing is everything.

187

In the 1950s, McClintock remained concerned with the production of variety, with the ways developmental programs could be executed and altered. She discovered *Suppressor-mutator,* a new controlling system that made the *Ac/Ds* system seem simple. Then, at the end of the decade, she embarked on a research project unlike any she had undertaken before. She analyzed and compared the chromosomes of many varieties of maize from across the Americas. So much variety led her to consider the underlying theme—a sort of Ur-pattern—that made corn, or any organism, what it was. Through the 1960s, this project competed for her time with her controlling-element research.

In the 1970s, she integrated these two concurrent studies, one concerned with reversible changes during development, the other with permanent, structural change over evolutionary time, into a profound and speculative vision of organic change. She extended her observations to other plants and even to animals. She contemplated the species-pattern, how it changes when a tadpole becomes a frog, or how a stick insect can mimic the pattern of a twig. Her overarching goal was to understand the complexity encoded in the maize plant and in nature generally.

Biographers have paid scant attention to McClintock's findings and ideas after 1951, and not without reason. *Suppressor-mutator* is baroque in its complexity; variables pile up on one another to the verge of chaos. Her comparative study appears to be a detour, irrelevant to the main trajectory of her thought.[1] Her late explorations of wildflowers and natural mimicry seem a meandering coda to a highly productive scientific life. In short, McClintock's late work has seemed to comprise the extensions of earlier experiments and the lightweight diversions of a scientist in active retirement.

Ignoring this work, however, comes at heavy cost. These research projects illustrate important developments in the themes of freedom, integration, pattern, complexity, and control. The complexity of the *Suppressor-mutator* system gave McClintock the confidence that controlling elements could supply the subtlety of control needed to regulate development. The comparative maize study catalyzed her interest in evolution and the development of her theory of organic change. Her studies of wildflowers and mimicry, far from being some retirement hobby, show her seeking pattern and control outside the laboratory and beyond the cornfield, through development into evolutionary time. When McClintock revisited earlier work, it was to apply established techniques in new arenas or to integrate old findings within a wider setting. Her later work shows that as she grew older, the natural themes in her work intertwined and took on new and deeper meanings.

Most important, this late work gives coherence to McClintock's entire career. It shows how cytology, genetics, mutation, transposition, development, and evolution were all elements in her lifelong study of the landscape of the chromosomes. As she integrated these elements, she came to a broad understanding of the whole of nature, which she expressed for the first time in mature form near the end of her life. Consideration of this last third of McClintock's career, then, ensures that we are not misled into thinking that transposition, or even developmental control, were endpoints on her intellectual trajectory.

Though McClintock prided herself on her autonomy, her late work was partly a response to her perceived reception among the genetics community. I have recounted how, by the mid-1950s, McClintock's colleagues had accepted transposition but largely rejected developmental control. Because the two were inseparable in her mind, she began to feel rejected. Her work on the races of maize took her out of the scientific community that revolved around Cold Spring Harbor and placed her in a new one, in which her authority was unchallenged.

Yet like the genes she studied, McClintock was not an independent pearl on a string. Concurrent with the comparative maize study, she linked her controlling-element work to one of the most exciting developments in genetics. François Jacob and Jacques Monod's operon model of genetic regulation drew her back into a debate central to the concerns of bacterial geneticists. At the end of her career, she addressed at last the question that has always bothered geneticists about her work: where does the control lie? Her answer was that it lies in the organism's response to the environment.

Suppressor-Mutator

In 1950, McClintock had begun a project to create new instances of Rhoades's *Dotted* mutable. These experiments led to her quip about going into the mutable-gene business. She then borrowed some of Rhoades's *Dotted* stocks and bred them to try to confirm her theory. Among these, one and then two more new mutables appeared at the *a1* locus.[2] They turned out to harbor a new controller.

The *A1* locus is one of four then known to influence the color of the kernel's aleurone, the tissue beneath the pericarp, or outer covering. It lies on chromosome 3 and produces the red pigment anthocyanin. A second anthocyanin locus, *A2*, lies on chromosome 5. The *C* locus on chromosome 9 is a third, and the *R* locus, on chromosome 10, was the fourth. The anthocyanin loci, *A1* and *A2*, offered the advantage that they act in both kernels and plant. McClintock thus had access to both halves of the maize life cycle, allowing her to ask whether the same controlling element had different effects in the gametophyte—the embryo and kernel endosperm—and the sporophyte.

Although McClintock's new mutable *a1* resembled *Dotted*, when she crossed the controller out of the stock and crossed *Dotted* in, the *a1* locus was no longer controlled. This result indicated that *a1* had come under the influence of a new controller. By 1950, she had deliberately produced three mutable *a1* loci by crossing an existing allele to controlling-element stocks. She named the alleles *a1-m1*, *a1-m2*, and *a1-m3*, but she laid aside the latter two for the time being.[3]

At the end of March 1952, she described the new controlling unit in a staff meeting. In contrast to *Ac*, it exerted two kinds of action on a locus. Early in the development of the ear, anthocyanin production was completely suppressed; when she bred out other kernel color alleles (*C*, *R*, and the other *A*), the kernels were colorless. Later in development of the ear, some dots or spots of pigment appeared. The new element underwent or caused mutations that reactivated the anthocyanin genes. These could be of several types, leading to a graded color series, from pale to normal red. In 1953, she named the new controller *Suppressor-mutator*. She first abbreviated it *Sm* but, by 1954, was using *Spm*.[4]

Like the *Ac/Ds* system, the *Spm* system had two elements; however, she never named the *Ds* homologue, referring to it simply as "the element residing at *a1*" or some variant. By 1954, she had found that *Spm*, like *Ac*, transposed. This hardly surprised her, naturally. But in the ensuing years, McClintock scarcely worked on the transposition of *Spm*. She used the fact of *Spm*'s transposition in conducting tests of dosage effects and other experiments, but its transposition per se was never a significant part of her investigations.

In one sense, the action of *Spm* was simpler than that of *Ac*: it did not (at first) show dosage effects. One *Spm* element in the nucleus seemed to produce the same effect as two or three. But *Spm* produced a much wider range of responses than *Ac*; the size, number, and intensity of spots on kernels varied greatly. Also, its two-part action was intriguing. Finally, the affected locus showed distinctive behavior even in the absence of *Spm*. Description of *a1-m1* behavior in the presence of *Spm* was no longer sufficient to characterize a state of the system. She now had to account for *a1-m1* activity in the absence of *Spm* as well.

Suppressor-mutator was an even stronger pattern generator than *Ac/Ds*. As she pursued the various alleles of the two *A* loci affected by *Spm*, she found rules, then exceptions to the rules, then exceptions to the exceptions, and contingencies, and interactions. The complexity of the system not only increased, it accelerated. Each state or condition, initially clearcut and distinct, under further investigation shattered into a spectrum of shaded, dotted, and streaked kernels and plants. No rule seemed absolute. In molecular terms, she was working with a mutational system that worked not only on other genes, but on itself as well. But in McClintock's steadfastly nonmolecular frame, it was the complexity of the patterns it produced that drew her on.

In February 1955, McClintock summarized for herself the similarities and differences between *Ac* and *Spm*. Where *Ac* showed many states, *Spm* showed none. Similarly, dosage effects were "clearly expressed" with *Ac*, whereas for *Spm* there were "none noticed." With *Ac*, transposition "depends on dose. 1 dose = frequent[;] higher doses, delayed until after meiosis." *Spm* worked the same way. Both controllers could be found in many different positions in the chromosomes, and both were inherited as single genetic units and showed normal linkage patterns (except, of course, when they transposed).[5]

Almost immediately, this neat characterization began to break down. The first to go was the observation that *Spm* did not show multiple states. In her Carnegie report for 1955, she described two distinct types of modifications resulting from the mutator action of *Spm*. "One effects a stable mutation, and the mutants so formed give rise to a series of alleles that differ from one another both quantitatively and qualitatively," she wrote. "The second type of modification, of rarer occurrence, leads to a change in the controlling element at *A1*—a change in state." Like alleles, states were stable and heritable. The classical genetic concept of a mutation is a change, presumably a chemical change, from one allele to another. A change in state, however, was not a mutation; rather, it was another sort of change that affected the pattern of future mutations. "In the presence of *Spm*, these modifications are expressed by changes in the kinds of mutations that occur, their frequencies of occurrence, and their times of occurrence during the development of the tissues. These modifications of state also affect the degree of action of the genic materials at the *A1* locus in the absence of *Spm*."[6]

She distinguished the various states on the basis of pattern, both in the presence and in the absence of *Spm*. In each case, the behavior of *a1-m1* in the presence of *Spm* led to spots, dots, or blotches, and in its absence to uniform kernels, but the pattern and intensity of pigmentation varied. In the presence of *Spm*, the original state of *a1-m1* produced early mutations, leading to a Holstein-like pattern of large pigmented blotches on a colorless background. In the absence of *Spm*, the kernel was uniformly pigmented. Other states produced fewer, later mutations, rather like a high state of *Ac*, leading to many small dots of color; in the absence of *Spm*, the kernel might be colorless to moderately and uniformly pigmented. Still other states produced a fainter version of the original Holstein pattern, with less intense pigmentation in the blotches on a background of light to moderate color. Each state also produced characteristic patterns in the leaves and stem. These were not necessarily parallel to a given state's effect on the kernel. In one state, for example, in the absence of *Spm* the kernels were only faintly colored, but the plants were darkly pigmented.[7]

Spm itself occurred in different forms. In one plant, McClintock identified an element in which the suppression action of *Spm* was reduced; she called it *Spm-w* (for "weak") and now distinguished it from *Spm-s* (for "strong" or "standard"). Although in her Carnegie reports she mentioned only these two types of *Spm*, in her research notes she identified several others, including *Spm-D* and two forms of *Spm-delayed action*. The delayed-action forms showed cycles of *Spm* action, resulting in nested patterns of color and no color. Depending on when the *Spm* changed, suppression of anthocyanin might be complete, partial, or sporadic.[8]

She added another level of complexity in 1956, when she described modifiers at *a1-m1*. The states of *a1-m1* were quantitative: they formed a graded series of timing and intensity of mutations. In some stocks, however, plants and ears showed patterns characteristic of the next higher state. She saw state 2 patterns in plants known to be genetically state 1. There seemed to be one or more modifying alleles that could bump up the state of *a1-m1* one level. The first modifier seemed to act only in the presence of *Spm*, although later modifiers acted in its absence.[9]

McClintock also had a mutable allele of the *A2* locus, *a2-m1*. It had arisen in her broken-X-broken experiment in the summer of 1944. At that time, she had not followed it up, engaged as she was by the broken chromosome mutable. When she returned to it in the early 1950s, she noticed it was associated with a controlling element that was not *Ac*.[10] Most notably, *a2-m1* showed dosage effects. In some respects, the *a2-m1* mutable seemed to be controlled by *Spm*, but in other ways not. A puzzling aspect was its frequent mutation from colored to pale (but not colorless). She toyed with the idea that it was *Spm* plus a new modifier, but she wrote little on *a2-m1* in 1955 and 1956, though she continued to experiment with it.[11]

In 1957, she recognized that *a2-m1* was indeed controlled by *Spm*, but in a more complex manner than *a1-m1*. She described two states of *a2-m1*. The first was similar to *a1-m1*. In the absence of *Spm*, it gave kernels with weak red color, indicating some gene action but less than normal *A2*. In the presence of *Spm*, color was completely suppressed, except after occasional mutations, which resulted in full *A2* red.

The second state, however, showed dosage effects. One *Spm* gave many pigmented spots, both large and small (a single dose of *Ac*, recall, gave only large spots). Two *Spms* gave essentially only small spots. Three or more *Spms* gave kernels that were either completely colorless or mostly colorless with a small region containing a few small spots. The control systems were expanding rapidly.[12]

An important difference between *Spm* and *Ac* was that in the *Spm* system the affected locus had a strong influence on the pattern. The *Ds* element affected all loci in essentially the same ways: it either inhibited gene action or caused chromosome breaks. Working on *Spm*, McClintock began to view the controller at the locus and the locus itself as a unit. Both states of *a2-m1* had many variants, which she now grouped into "class 1" states and "class 2" states. Class 1 states gave no gene expression in the absence of *Spm*. Different class 1 states gave different amounts and kinds of suppression and mutation in the presence of *Spm*. They could also cause the gene locus to alter permanently, giving rise to other states, either class 1 or class 2, or to mutants, which were insensitive to *Spm*. Class 2 states gave gene expression in the absence of *Spm*—the amount varied, depending on state—but did not give permanent alterations or changes of state. They also showed dosage relations with increasing numbers of *Spm* elements.[13]

During the quiet months of March and April 1958, McClintock experienced an intense period of thinking that led to many "integrations." The *a2-m1* locus continued to vex her. In the presence of *Spm*, pigmented areas appeared on a colorless background. But within those dark spots, colorless areas appeared that were indistinguishable from *a2-m1* in the absence of *Spm*. In her research notes, she wrote, "turning on gives no color. Turning off gives *A2* color. Within *A2* areas, colorless areas appear due to turning [on] of Spm. Within these colorless areas, *A2* specks may appear from turning off of Spm activity." It seemed as though *Spm* was no longer acting in the cells. But puzzlingly, later in the development of the kernel, *Spm* action was again visible.[14]

She explained the patterns by assuming that *Spm* cycled between active and inactive "phases." Alternation of the phases led to nested patterns: tiny dots of color within colorless areas that punctuated colored spots on a colorless background. She referred to the off phase as "inhibited" and then "inactive" *Spm*. She came to see inactive *Spm* as the "rate controller"—the element that set the rate at which mutations occurred. Active *Spm* behaved differently in the presence of an inactive *Spm* than it did alone. Inhibition of *Spm* was released gradually, stepwise, leading to graded pigment intensities in the kernels. *Spm*, then, not only controlled the modified gene loci, it controlled *Spm* itself.[15]

Phase changes, McClintock emphasized, were under precise control. In her Carnegie report for 1959, she described how they could and did occur in any part of the plant, at any stage of development. She continued to assume that the regulation differed at different developmental stages because the elements *determined* the stages. This assumption forced her to investigate the actions of *Spm* in all tissues, at all stages of development.[16] At last she seemed to be realizing her goal of understanding the genetic control of the development of the entire plant. Many details remained fuzzy, but her system was growing sufficiently complex that she could begin to see it as a model for the execution of a developmental genetic program.

In 1961, turning her attention back to the second mutable allele at the *a1* locus, she realized that the reason it had confused her earlier was that it responded to *Spm* in the opposite way as *a1-m1!* Gene suppression occurred in the absence of *Spm*, while mutations and transpositions occurred in the presence of active *Spm*. The reason for this, she believed, was that in *a1-m2*, *Spm* lay very near the *a1* locus. The effect of *Spm* seemed to be invertible, simply by its transposing from far away to snug up against the locus on which it acted. A few years earlier, she might have called this a new sort of position effect, but by this time she no longer used this language.[17]

To try to make sense of the diversity of these patterns, on an August day I went up to Cold Spring Harbor Laboratory's Uplands Farm, eleven acres on a hilltop about a mile from the laboratory. Cold Spring Harbor's maize geneticists have grown their corn at Uplands Farm since 1985.[18] Two cornfields front the road; the stalks were tall and capped with floppy

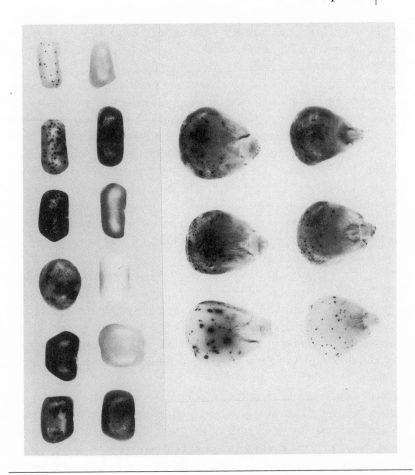

Figure 8.1. One of McClintock's prints demonstrating the complex patterns produced by different states of *Spm*. Reprinted with permission from the American Philosophical Society Library.

brown tassels. Behind the fields are greenhouses and a barn with field equipment, microscopes, and a cold room. My guide was Robert Martienssen, wiry, with thinning blonde hair and hawklike features, a senior staff scientist at Cold Spring Harbor. Martienssen is a molecular biologist who studies both maize and a little weed called Arabidopsis, a workhorse of plant molecular genetics. He contributed significantly to a collaborative project to develop maize controlling elements as "gene traps," powerful tools for identifying simultaneously the structure, function, and location of new genes. Following McClintock's death in 1992,

Martienssen helped catalog her enormous corn crib, packing up most of it and sending it to the Illinois seed bank, an enormous, priceless agricultural archive. He knew McClintock as both a colleague and a friend, has studied her controlling-element papers and discussed them with her. He is one of the three or four people who best understand McClintock's work.

Martienssen led me into the Uplands Farm cold room, a giant refrigerator lined with shelves containing steel bins full of old corncobs. On a top shelf were a couple of bins containing ears of McClintock's corn, each ear rolled up in Saran wrap. When he packed her corn off to Iowa, he saved just a few ears for Cold Spring Harbor, he said. Shivering in the refrigeration, we pulled out some ears, then went back out into the heat of the barn to a lab table. Small price tags with string piercing the end of the cob bore each ear's culture and plant number in McClintock's handwriting.

We arrayed the corn on the table. I recognized some of the lower culture numbers as being from the late 1940s. Expecting big, voluptuous ears of supermarket decorative corn, I was a little dismayed at the runty, gap-toothed artifacts before me. At first, the patterns on the kernels looked chaotic, just random dots. They crystallized as Martienssen explained them, pointing out individual kernels with subtle variations. He showed me examples of *Ds* at different loci, different doses of *Ac* acting on *Ds*.

Then he pulled an ear from the late 1950s. "Here's a case of *Spm* at the *A* gene," he said. "This must be *a1-m1* in a class 1 state. No wait a minute, that can't be right. Maybe it's *a2-m1* with a modulating element. See this speckled area with the clear spot in it? That's caused by . . . ah . . ." It was no use. Martienssen smiled and shrugged.

The more McClintock crossed her plants to one another, the more diversity they threw at her. She kept the distinctions straight and grouped them as best she could. The complexity continued to increase.

Spm could suppress by being present and release by being absent, or vice versa. It could be active or inactive, but when it was inactive it could make an active element cycle through phases of activity. The nature of that activity depended on several variables, including the state of the *Spm* element and that of the element at which it resided. The distance between *Spm* and the element on which it acted could affect and even reverse its action. The only constant seemed to be that *Spm* had to be present in order to affect gene action.

The pattern generator—the controlling elements, continuously altering both the genes and themselves—exploded even this constraint. In 1964, she showed that *Spm* could also control genes when it was not there at all—it merely had to have been there once. At *a1-m2, Spm* seemed to continue to have an effect even after it had been removed from the nucleus. Controlling elements, she said, act like a "genetic clock." (That year, Linus Pauling and Emile Zuckerkandl published their hypothesis of a "molecular clock"—the suggestion that if mutations occur randomly and at a constant rate, divergence in DNA or protein sequence might be used as a measure of evolutionary distance.) The *Spm* element, McClintock postulated, sets the rate of the clock, with genetic events—turning the gene on and off, or modulating its activity—being ticked off at a regular rate.

The element residing at the locus could advance or retard the clock to produce more subtle effects. It could also "preset" the clock to behave in a certain way; the clock would continue at the rate given by the element residing at the locus, even in the element's absence, until another element came in and altered it. Such "preset patterns," she said, occurred in plants grown from kernels from a plant that carried *Spm*—but only in those kernels that had lost *Spm*.[19] The presence of *Spm* in the parental, or even grandparental, generation could modify the action of the genes, such that a pattern characteristic of *Spm* appeared even after *Spm* had been lost. Presetting and subsequent setting of a locus to give a specific pattern could also be "erased," such that the effects no longer occurred. In 1965, she used another mutable *A1, a1-m3,* and a mutable *Waxy* allele, *wx-m7,* both of which are controlled by *Ac,* to demonstrate that *Ac* also showed phase changes. In one kernel, she showed that the pattern of *Waxy* expression under the influence of active *Ac* was superimposed on the pattern produced by inactive *Ac*.[20]

Little wonder other geneticists had a difficult time following her work. The complexities she was finding were not artifacts—molecular analyses begun in the 1980s have explained many (though not yet all) of these effects in terms of the often-bizarre mutations that *Spm* wreaks upon itself (see the Appendix). But her explanations, framed entirely within classical cytogenetics, strained the scientific language she used. She studied *Spm* with essentially the same techniques and analyses she had applied to the breakage-fusion-bridge cycle: clever genetic crosses, acute observation, and hard thinking. Forced to invent new terms and concepts to accommodate the snowballing complexity of her material, she resorted to descriptions that sound ad hoc.

Despite these limitations of language and technique, McClintock kept pace with her pattern generators. She could look at a chaotically pigmented kernel and tell an elaborate story of gene action and control. Her theory provided an impressive toolkit for explaining nearly any pattern or variant that she saw. The different varieties and phases of *Spm*, the classes and states, and the modifiers could produce any imaginable pattern. Every action of each element was contingent on modifiers and interactions with other elements, and most were capable of graded action.

Such complexity was precisely what she required. Controlling elements could not play the role she ascribed to them unless they could produce the kind of graded variation one actually saw in nature. She intended her theory to explain the development of the organism. She knew that a developing plant must execute a genetic program far more complex than that which she was observing in kernel and leaf pigmentation. The apparent complexity of her system was a limitation of the human mind, not an argument against her theory.

Until almost the end of her career, she resisted efforts to understand her system on the molecular level. In 1962, Oliver Nelson was working with *Spm* at the *waxy* locus. He offered to collaborate with her on an exploration of the biochemistry of genetic control at *waxy*. McClintock answered his letter two weeks later but did not address his offer.[21] In 1965, she attended a symposium on the genetic control of differentiation at Brookhaven National Laboratory. Her paper was titled "The control of gene action in maize."[22] In discussion following her presentation, Robert S. Bressler, of City College of New York, asked McClintock whether she would be willing to speculate on the molecular basis of controlling systems. "I believe that we are not yet ready to construct any model at the molecular level," she replied. "We are still too ignorant of the composition, at the molecular level, of chromosomes in higher organisms and also of the various types of gene-control mechanisms that may operate in these organisms."[23]

Molecular Control

The rest of genetics, meanwhile, was going molecular. While McClintock was probing *Suppressor-mutator*, the most pressing questions in genetics concerned the physical nature of genes and the mathematical problem of how information was encoded in nucleic acids. These were questions that could not be asked with corn, or even fruitflies. Bacteria make flies look like corn—they reproduce faster and are easier to work with. In addition, they have but a single chromosome, just a couple of handfuls of genes. In bacteria, the tasks of unraveling the structure of the gene and discovering how genetic information guides the making of proteins was like locating a set of books in a well-ordered though uncataloged community library. With corn, it would have been like searching for those books in Manhattan, mapless, in a hurricane.

In 1951, McClintock's work had been at the center of a major debate among geneticists over the nature of the gene. As the decade wore on, however, bacterial geneticists arrived at a radically different and extremely powerful vision of the gene—indeed, of the entire genome and its mechanism of transmitting genetic information. A decade later, the two visions of the genome seemed to coincide. The microbial geneticists seemed to come around to many of the general views McClintock had been voicing for years. The similarities, however, were mainly by way of analogy. They did not, as McClintock believed for a time, signal a reintegration of the various approaches to heredity—at least not along the lines she championed.

The 1951 Cold Spring Harbor Symposium marked not only the ascendancy of microbes over large complex higher organisms, but also the beginning of the eclipse of the bacteriophage system of the 1940s. Max Delbrück, spiritual leader of the American phage group, had decreed that all work with phage be done with the T-series phages. The T-series phages are known as virulent phage: on infection they lyse, or burst, their host cell and release more phage into the surrounding broth. Delbrück knew, as others knew, of experiments that had shown the existence of another, "temperate," type of phage, which need not lyse its host. Bacteria carrying temperate phage were called lysogenic, because although as a whole the culture appeared normal, small numbers of phage particles were generated, which were capable of infecting and lysing related bacterial strains. In the 1940s, this strange genetic property was poorly understood. For years, Delbrück thought lysogeny nonsense; as late as 1953, he seemed to think the whole phenomenon tainted with mysticism.[24]

Not everyone belonged to Delbrück's phage church. Notable apostates included the American Joshua Lederberg and the members of a burgeoning school of virology and bacteriology under André Lwoff at the

Institut Pasteur in Paris. In 1950, Lwoff showed that lysogenic strains of bacteria contain the viral genome, integrated into the bacterial chromosome, its genes replicated right along with those of its host. He called this form of virus the prophage. With Louis Siminovitch and Niels Kjeldgaard, Lwoff irradiated lysogenic bacteria with ultraviolet light, which rapidly "induced" the production of phage particles from lysogenic bacteria. The virus, in other words, existed in two states: either the "vegetative" state, in which it produced phage particles and burst its host, or the lysogenic state, in which it lay quiescent, integrated into the host cell, and acted as part of the bacterial genome.

The next year, Esther Lederberg, then wife of Joshua, found that *Escherichia coli* strain K-12, with which her husband had demonstrated the existence of sex in bacteria, was lysogenic. She named the virus it contained lambda. Joshua Lederberg discussed lambda, as well as the recent finding by his student Norton Zinder, of transduction, or virus-mediated gene transfer, in his six-hour talk at the 1951 Cold Spring Harbor meeting. Since *E. coli* was the *Drosophila melanogaster* of bacteria, with far more known about its properties and genetics than any other species or strain, lambda and K-12 quickly became the preeminent model system for the study of lysogeny.[25] By the middle to late 1950s, lysogenic phages superseded lytic phages as the dominant phage model system.

Lytic phages, however, still had a few secrets to reveal. In 1952, the year McClintock first hypothesized a new controller system at *A1*, Alfred Day Hershey and his technician Martha Chase, working downstairs from McClintock in the Carnegie department's Animal House, used phage T2 to demonstrate the nature of the genetic material. Hershey and Chase labeled phage protein and DNA with radioactive tracers, let the phage infect bacteria, and then knocked the phage particles off the bacteria by whirring the mixture in a Waring blender. When they examined the bacteria for the tracers, they found to their surprise that only DNA had been taken up. They showed that DNA was sufficient to allow bacteria to make new phage. DNA was the genetic material.[26] The Hershey-Chase experiment— not as clean as that of Avery, MacLeod, and McCarty in 1944, but it used the tools and language of phage genetics and came eight years later—convinced virologists that DNA was the stuff of genes.

The following year, 1953, Delbrück organized the Cold Spring Harbor Symposium on viruses. The meeting was held next to McClintock's laboratory, in a new Carnegie-built auditorium named for Vannevar Bush. Another new building with ample laboratory space had been constructed next door, and McClintock had just moved into a spacious lab on the lower floor. At the last moment Delbrück inserted a paper by James Watson and Francis Crick, who had been working at the Cavendish Laboratory at Cambridge University. Watson, not quite twenty-five, in T-shirt and shorts, looked like a teenager, but the opening words of his paper possess maturity, pride of bearing, bravura: "It would be superfluous at a Symposium on Viruses to introduce a paper on the structure of DNA with a discussion on its importance to the problem of virus reproduction. Instead we shall not only assume that DNA is important, but in addition that it is the carrier of the genetic specificity of the virus (for argument, see Hershey this volume)."[27] The structure—with its twin chains of nucleotides, one the mirror image of the other, fused at the bases like a gently twisting ladder—was, as Watson said, undeniably "pretty."[28] But its great significance was that it instantly clarified the nature of the gene and the principles, if not the details, of how genetic information was passed on, from cell to cell and from organism to organism.

The nature of the gene was transformed from a particle to a codescript. The rungs of the double helix ladder are paired molecular groups known as bases. DNA contains only four—adenine, thymine, cytosine, and guanine—and each pairs only with one other: adenine with thymine, cytosine with guanine. Each strand of the double helix therefore specifies the other. Watson and Crick recognized that if the strands could be unzipped, each could be used as a template to make a new double helix, thus duplicating the genetic material. Genetic information was coded in the sequence of bases and was transmitted, through generations of cells or generations of elephants, by this simple copying mechanism. Black boxes, metaphorical speculations, and mysterious forces were no longer necessary to explain genetic properties. One could now ask specific, molecular questions about the passing of genetic material from cell to cell and the precise nature of individual genes.[29]

The double helix gave the theory of the gene as a unit of mutation new power. In 1955, Seymour Benzer extended mutational analysis of the gene to the interior of the gene itself with what has been called "the archetype of modern genetic analysis."[30] He used clever genetic technique and the super sensitivity of bacterial genetics to detect recombinants within the phage T4 *r* gene. In so doing, over five years he produced a mutation map of the interior of the gene. Genes were indeed divisible, but not into anything so simple as identical repeating subunits. In 1958, Guido Pontecorvo compared the pre-Benzer gene to messages in Chinese, where each character is simultaneously a unit of structure, function, and copying. Now, he said, the gene was only a unit of function emerging from linear arrangements of letters. "Miscopying has now become misspelling: a mistake in letters or in their order, not usually a mistake in words."[31]

Bacteria, however, continued to yield strange new phenomena that did not fit the developing paradigm. In 1957, Joshua Lederberg and Tetsuo Iino described "phase variation" in Salmonella. Different serological types of Salmonella can be distinguished on the basis of proteins called antigens. Since 1922, it had been known that a certain antigen could be expressed in one of two forms, or phases, and that these could alternate, with bacterial colonies expressing first one antigen and then the other. The switching was called phase variation. Lederberg and Iino linked the phases to two genes, *H1* and *H2*. They showed that phase variation could be explained with a model in which the *H2* gene existed in two "states," active and inactive. In the active state, *H2* dominated over *H1* and the bacteria expressed the *H2* protein. With *H2* in the inactive state, *H1* predominated and its protein was expressed.

This model is remarkably close to the phase changes McClintock described for *Spm*. Lederberg and Iino recognized at least part of the parallel in their discussion, in which they compared phase changes in Salmonella with the effects of transposition of McClintock's *Ds* element. They noted that "the suppression of phenotypic effects of transposition of heterochromatin (E. B. Lewis 1950) or of the *Ds* and *Mp* elements (McClintock 1956; Brink and Nilan 1952) may also be viewed as local changes of state, which although they depend on structural modifications nevertheless suggest how gene action may be regulated. These may therefore be considered as conceptual, if not mechanical, models of phase variation."[32] In a recent interview, Lederberg explained what they were getting at. "This turns out to be programmed in one sense. That is, there's a site-specific recombinase [enzyme], that when it acts on a DNA loop, switches the sense and gives you an inversion of the intervening sequence, and it turns one gene on and turns the other one off.

"Now I can't call that a developmental phenomenon, in that I cannot verify that it is a developmentally programmed operation," he continued. "It's the way that [Salmonella] switch surface antigens. But what has never been elucidated is whether that's a purely random event, occurring spontaneously, or whether it could be environmentally induced," in other words, whether there was a source of control. "If we could add the element of environmental responsivity, it would be a very good substantiation of what McClintock was talking about. And we sort of said so in this remark."[33]

Meanwhile, the bacterial geneticists at the Institut Pasteur were pursuing two sets of studies. The first line of work concerned phage lambda. Like other lysogenic bacterial strains, *E. coli* K-12 could be induced to release lambda phage by treating it with ultraviolet radiation. In 1954, François Jacob and Elie Wollman discovered a new type of induction. When bacteria lysogenic for lambda conjugated with nonlysogenic, or sensitive, bacteria, the prophage was injected into the sensitive bacteria. There it was induced, produced phage particles, and lysed its new host. This was called zygotic (or, more titilatingly, "erotic") induction.[34]

The second line of work concerned the *lac* gene, which governed the digestion of the sugar lactose. It was well-known that bacteria could adapt their enzyme production so as to be able to digest the type of sugar that happened to be available. Grow them on galactose, for example, and they produce galactosidase; grow them on glucose and they do not. Bacteria raised on glucose, then, cannot grow on other sugars, such as lactose, because they lack the enzyme to digest that sugar. In 1951, Joshua Lederberg isolated the first "constitutive" mutants in *E. coli* K-12. Such mutants *can* grow on lactose, even when raised on, say, glucose. Lederberg's work focussed attention on the enzyme beta-galactosidase, necessary for the digestion of lactose. André Lwoff, and his younger colleague Jacques Monod, began to study mutations of the genes involved in beta-galactosidase production. They found they could "induce" beta-galactosidase production by adding galactosides, the enzyme's substrate, which was referred to as "inducer."[35] In 1956, Monod discovered the permease, an enzyme that pumps beta-galactosides into the cell for digestion. The *lac* system now contained three adjacent loci: *i*, for inducibility; *z*, for production of beta-galactosidase; and *y*, the permease.

In 1957, Arthur Pardee, an American, joined the Pasteur group and undertook an investigation with Jacob and Monod that has been canonized as the PaJaMo experiment. In brief, they measured galactosidase production while mating bacteria with the galactosidase genes to bacteria that lacked them. Contrary to expectation, they found that production of beta-galactosidase was not turned on by an inducer, but was rather turned *off,* by a molecule they named the repressor. The system operated by negative, rather than positive, feedback.[36]

Around the lab, the scientists had joked about having given the same name—induction—to such two different processes as virus production and gene action. In 1959, Jacob recognized that it wasn't a joke. The realization came, he wrote in his memoir, suddenly, in an intuitive flash. He had been trying to write a talk for the Harvey Society, in whose prestigious lecture series he had been asked to participate. The writing was not going well. He went to a movie. It was a bad movie. His mind wandered:

> Slumped in my seat, I dimly perceive in myself associations that continue to form, ideas for proceeding. An indistinct hullabaloo, whose unfolding I make no attempt to grasp. Shadows move on the screen. I close my eyes to heed the extraordinary things going on within. I am invaded by a sudden excitement mingled with a vague pleasure. It isolates me from the theater, from my neighbors whose eyes are riveted to the screen. And suddenly a flash.[37]

The flash was that zygotic and enzymatic induction were fundamentally the same. In both cases, a repressor is inactivated, by ultraviolet rays in zygotic induction, by lactose in enzymatic induction. But in the case of lambda, the entire phage genome—perhaps fifty genes—is induced or derepressed. The only way, he thought, to achieve such regulation was for the repressor to bind to the DNA itself. It was a molecular approach to the problem of gene regulation.

In his Harvey lecture, Jacob described the repressor as a molecule that floated free in the bacterium and was capable of binding to the chromosome to alter gene action. The phage genome, he wrote, "brings into the cell a plan for the synthesis of new molecular patterns." This plan is executed by the cell. Phage genes could be seen to "control the activity of the adjacent DNA molecule," making the DNA "not only a unit of replication, but also a *unit of activity*" (emphasis his). He even invoked position effects, noting that if the regulation operates on the entire DNA molecule, "certain chromosomal rearrangements would produce a disturbance in the functioning of the displaced determinants (position effect by chromosomal rearrangement)."[38]

Jacob did cite McClintock, but not in this passage. Later in the paper, he described the prophage as a molecule that could exist in two states, either floating free or integrated into the bacterial chromosome.

He called molecules with this two-state property episomes and noted that the term encompassed a number of phenomena that had been described in bacteria and other organisms. He wrote that "in maize the factors described as 'controlling elements' (McClintock, 1951) also appear to behave as episomes, since these agents are not always present, but when present they are added to certain chromosomal sites and can move from one site to another, and even from one chromosome to another."[39]

Thus did McClintock's comet shoot into Jacob's mental orbit, only to be hurled off again by the force of his own gravity. Jacob knew of McClintock's work, but he understood it as an example of episomes, not of genetic regulation. Nor did he and Monod—who had known McClintock since the 1930s—acknowledge or cite her work the next year, when they sketched the first draft of their simple, elegant model of gene regulation.

The paper appeared in 1960. At the end of the year, they submitted a much longer, more thorough version, in English, to the *Journal of Molecular Biology*.[40] The papers laid out a model of the *lac* locus in which both beta-galactosidase and permease were produced by "structural" genes, whose action was controlled by two adjacent genes. (See Figure 8.2.) One, the operator, was the binding site for the repressor molecule. The other, the regulator, synthesized repressor. If the regulator produced repressor, the structural gene would be off. If the regulator shut off, repression would be lifted, and the gene would turn on.

The authors seemed deliberately to invoke the two poles of classical genetics, Thomas Hunt Morgan and Richard Goldschmidt—and in so doing, to position themselves in the vanguard of molecular biology. They concluded,

The hypothesis of the operator implies that between the classical gene, independent unit of biochemical function, and the entire chromosome, there exists an intermediate genetic organization. The latter would include the *units of coordinated expression (operons),* comprising an operator and the group of genes for structure which it coordinates. Each operon would be, by way of the operator, under the control of a repressor whose synthesis would be determined by a regulator gene (not necessarily linked to the group). The repression would be exercised either directly at the level of the genetic material, or at the level of "cytoplasmic replicas" of the operon [emphasis theirs].

Controlling Elements

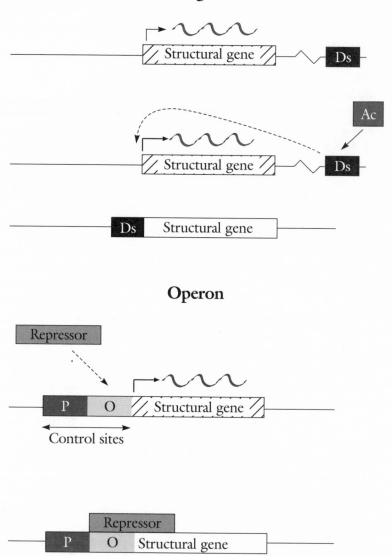

Operon

Figure 8.2. Controlling elements and the operon. In McClintock's model, a gene would be turned off when a *Ds* element transposed into or next to it. In Jacob and Monod's operon model, a structural gene was regulated by regulatory genes immediately upstream. A gene would be turned off when a repressor molecule bound to the operator; no transposition was involved.

Monod told historian Horace Freeland Judson that their failure to cite McClintock in this paper was an "unhappy oversight." Oversight or not, it indicates that Jacob and Monod, though aware of her work, thought of it in terms of transposition, not gene regulation.[41]

McClintock, however, saw the parallels in terms of regulation. When she read their 1960 paper, she "went through the ceiling" with delight. She recognized the fundamental similarity between their work and hers: both were talking about gene regulation. Decades later, in an interview with Will Provine and Paul Sisco, she explained that transposition had nothing to do with the similarity between the operon and controlling elements. "Oh, there wasn't any transposition" in the operon, she noted. "The main thing was control. They had mechanisms that were—they showed that there was a control."[42]

She immediately gave a Cold Spring Harbor staff meeting on the parallels between the bacterial operon and maize controlling elements, and then wrote an article for *The American Naturalist* that appeared in early 1961.[43] It was to the operon what Goldschmidt's 1950 *American Naturalist* piece was to her own work: an effort to bring attention to the author's research by linking it to work of widely recognized importance.

The bacterial operon, she wrote, controlled enzyme production in response to changes in the cellular environment. Maize controlling elements could operate either this way or by bringing about permanent developmental or evolutionary changes in gene action. She adopted the bacterial nomenclature for her elements: the controlling element residing at the gene locus she called the operator, and the independently located element (*Ac* or *Spm*) she called the regulator. She emphasized that it was immaterial that the controllers transposed in maize but not in bacteria. In the *Spm* system, the operator was also fixed in location near the gene, and the movement of the regulator, while possibly connected to some of its regulatory functions, was not essential to functioning of the system.[44]

The operon was accepted immediately; it won a Nobel Prize for Jacob, Monod, and Lwoff in 1965, just five years after initial publication. Referring to controlling elements as operators and regulators was in large measure McClintock's effort to establish her precedence in gene regulation—partly for credit, perhaps, but certainly also to show other geneticists that gene regulation was widespread, that it occurred in higher organisms, and that maize still had some things to teach.

If the 1953 Cold Spring Harbor Symposium is the double helix symposium, then the 1961 meeting is the operon symposium. Jacob and Monod gave a long research paper on the operon and presented the concluding statement of the meeting. McClintock, as usual, was in the audience. In their summing up for the symposium, Jacob and Monod seemed to give McClintock unqualified credit:

> Long before regulator genes and operators were recognized in bacteria, the extensive and penetrating work of McClintock (1956) had revealed the existence, in maize, of two classes of genetic "controlling elements" whose specific mutual relationships are closely comparable with those of the regulator and operator . . . Although, because of the absence of enzymological data in the maize systems, the comparison cannot be brought down to the biochemical level, the parallel is so striking that it may justify the conclusion that the rate of structural gene expression is controlled in higher organisms as well as in bacteria and bacterial viruses, by closely similar mechanisms, involving regulator genes, aporepressors, operators, and operons.[45]

In their research paper, they began by acknowledging McClintock's control argument. "Evidence for the existence of genetic systems of control has already been reported in maize (see McClintock 1956; Brink 1960)." They then cited McClintock's article in *The American Naturalist*. "As pointed out by McClintock (1961), such dual systems of control in maize may, in some respect, be compared with the regulator gene-operator system of bacteria: it is conceivable that the maize system also operates through the intermediacy of specific, cytoplasmic molecules, acting as signals like bacterial repressors."[46]

But Jacob and Monod did not accept McClintock's interpretation. In contrast to bacterial operators and regulators, controlling elements "seem to be dispensable and able to move from one chromosomal location to another. In this respect, the maize controlling elements behave like certain particular types of genetic elements in bacteria, called episomes." Transposition makes controlling elements "dispensable" and therefore more like foreign bits of DNA or like viruses than like integral, essential regulatory mechanisms. "That such episomic elements may interfere with some regulation system of the bacterial cell is clear in the case of prophage in lysogenic bacteria."[47]

To McClintock, the idea that controlling elements were foreign particles implied they had no useful purpose in the cell. It undermined the most important part of her theoretical argument: these elements were essential to the normal functioning of the plant. In her 1963 Carnegie report, she argued against the foreign particle idea, presenting evidence from her own and others' work and insisting that controlling elements were "true chromosomal components of present-day maize." In a 1965 paper, she argued that controlling elements were part of the genetic complement of normal plants, though in her plants they were disrupted in a way that enabled her to visualize them. Complexity provided the evidence. Considering only the simpler *Ac* system, she acknowledged, might lead one to conclude that controlling elements were foreign particles that influenced genes in uninteresting ways, merely as a side-product of their invasion of the host. But the various and subtle forms of control exerted by *Spm* surely were too elaborate for some viral invader.[48]

By 1968, she had backed off the parallels between controlling elements and the operon. In a letter that year to maize geneticist Oliver Nelson, she admitted that comparing her work to the operon had linked it more with episomes than with genetic control. "I have never thought that the episome model could be applied directly," she wrote. She did not doubt that such particles had been incorporated into the genomes of higher organisms, or that the regulatory processes of microorganisms were present in higher organisms. But complex organisms such as maize must have other regulatory tools not available to bacteria. Evolution itself was a kind of integration, or occurred by means of integrations. "The integrative process undoubtedly has remodeled organizations and interrelations in the higher organisms. Thus, one cannot always draw a direct homology between modes of operation in microorganisms and those in higher organisms, even though many of the basic mechanisms must be the same. Thus, I do not believe that controlling elements in maize may be homologized with models based on the current dogma of the operon."[49]

Races of Maize

In the late 1950s, McClintock became involved in a parallel research project that, on the surface, appears to have had little to do with her controlling-element work. Employing the cytological techniques that she had created decades earlier, she used her skill with the light microscope to trace the origins of cultivated maize and the divergence and distribution of the many races grown throughout the Americas. She made one conceptual discovery, in the second year of the project, and developed a novel representation of the data. The project competed for her time with her research on controlling elements; she spent the next twenty-three years immersed in exhaustive data collection and analysis, academic politics, administration, and publishing details.

The races-of-maize project was intimately connected to her previous and ongoing research. It took her back to the cytological techniques and analysis with which she had made her early career. She did little cytology with *Spm,* so methodologically the two projects complemented one another. Even more important, the races-of-maize project stretched her intellectually. In the races-of-maize work, McClintock compared stable differences in chromosome structure across the entire distribution of the species. These two qualities—stability and breadth—took her beyond developmental change and into the realm of evolutionary change.

She remembered its beginning romantically, like a scene out of an old movie. In early 1957, the maize geneticist Paul Christof Mangelsdorf had been visiting her laboratory at Cold Spring Harbor. At the end of the visit, McClintock recalled in 1979, she drove him to the train station. "Just as we were getting near the station he said that he would like to have somebody in Peru trained in cytology, and asked me would I be interested. And I said, just as we got to the station, 'Yes,' and the train was nearly pulling out, and on the steps of the train he said, 'Do you really mean it?' and I said, 'Yes.'"[50]

Mangelsdorf was one of the leading figures in what had become a classic and thorny problem in maize genetics: the origins of cultivated maize. He had trained under Edward Murray East at Harvard and then worked with Donald F. Jones, a former student of East, at the University of Connecticut Agricultural Station. In 1940, he returned to Harvard as a professor, where he concentrated on the origins and evolution of maize.

Unlike Drosophila or other genetic model organisms, maize has no wild type; it is a cultivated species. Since at least the 1930s, maize geneticists have argued over its evolutionary origins, in particular its relationship to its nearest relatives, teosinte *(Euchlaena)* and Tripsacum. Most maize geneticists today agree with George Beadle's 1939 argument that maize was developed by intense artificial selection of teosinte. Mangelsdorf, however, articulated an alternative, the "tripartite" hypothesis, in which teosinte is the descendant of a hybrid of ancestral maize and Tripsacum.[51] This "corn war" still rages, with proponents of both sides marshaling evidence from fossils, hybridization experiments, and DNA analysis.[52] The stakes are high: the two theories imply different strategies for agricultural improvement.

Mangelsdorf was a member of the Committee on Preservation of Indigenous Strains of Maize, cosponsored by the National Academy of Sciences and the National Research Council (NAS-NRC). The NAS-NRC committee was an early genetic diversity project that comprised academically trained geneticists who collected and analyzed the varieties of maize. The committee worked closely with the Rockefeller Foundation, which since the 1930s had been promoting Latin American agriculture through research and education. Mangelsdorf had also been involved with the Rockefeller's Mexican agricultural program since 1943.[53] Roughly speaking, the Rockefeller Foundation put up the money and the NAS-NRC committee provided the scientific expertise.

The Rockefeller Foundation established field stations in several Latin American countries and maintained seed centers at laboratories throughout Latin America: La Molina, Peru; Medellín, Colombia; Chapingo, Mexico; and Piricicaba, Brazil. It installed American-trained scientists at these field stations to conduct research and train local workers.[54] These scientists had been attempting to characterize the races of maize on the basis of morphology, the form and structure of the plant. The work had been going badly. Mangelsdorf wished to stimulate the project by adding cytological analysis. McClintock, as the founder of maize cytogenetics, was the natural choice.

In late June 1957, McClintock received an official invitation from Ralph Cleland, who had moved from Johns Hopkins to Indiana University and was chair of the NAS-NRC committee. In 1933, Cleland had done McClintock a favor, by encouraging her to apply for a Guggenheim fellowship. He now asked her for one: to "help train someone who can carry on maize cytological studies" at the Rockefeller Foundation's field station in La Molina.[55] McClintock left for Peru on 5 December. She spent about six weeks at La Molina and at the main laboratory in Lima, examining the chromosomes from strains of maize that had been collected by the Peruvian team.[56] She also trained a young Latin American scientist, Ulises Moreno, in her cytological techniques.[57]

The gist of the work was to prepare chromosomes, taken from pollen grains from the tassels of corn plants, stain them with a cytological dye, and examine their structure under the microscope. The primary measures she used were the size, shape, and position of chromosome knobs and centromeres, and the ratio of arm lengths of the chromosomes. The goal was to characterize each strain cytologically and, if possible, to infer evolutionary relationships among them. After she returned to the United States in January, Alexander Grobman, the director at La Molina, wrote to thank her for her time. He also implied that she had treated the Latin American workers harshly: "I understand your exacting requirements, and know that you would hardly feel satisfied of the progress of people who even trying would not be up to your standards."[58]

In her controlling-element work, McClintock considered knobs as regulators of gene action. At La Molina, she concentrated on knobs as taxonomic characters. Knobs were found in many places besides the tip of the short arm of chromosome 9. As she told Franklin Portugal in 1980, "There are 22 places in the chromosomes where these big, deep-staining structures occur. They could be absent or they could be present and when they are present they can be various sizes: little ones, bigger ones, bigger ones, and very large ones."[59] Her task at La Molina was to train Moreno and others to prepare the chromosomes and learn to recognize the different characteristics of knobs, and then to characterize the various races of maize according to each one's constellation of knobs. The work is difficult. Simply learning to identify the different chromosomes can take weeks. Recognizing the knobs and reliably distinguishing "bigger ones" from yet bigger ones requires both talent and experience.

Back in the United States, in February 1958, Mangelsdorf met with her to discuss problems with the races-of-maize project at other Rockefeller stations. Pleased with her work, the following June he asked her to go to the Rockefeller station at Medellín, Colombia. While in Colombia, she could jaunt over to La Molina to check on progress there. In July, Cleland issued a formal offer.[60]

McClintock agreed and began making arrangements for another winter trip. One of the reasons was that she thought a comparative study of maize knobs might shed light on controlling elements. Gerald Neuffer, one of Lewis Stadler's students, had recently discovered *Dotted* elements in a Brazilian yellow flint and a purple-kerneled race from Peru.[61] "The experiences in Peru," she wrote later in 1958, "made me realize that through such studies, we might get some clue to the action of knobs in developmental processes."[62]

On 9 December, her Avianca flight left New York for Bogotá.[63] When she arrived she set up her equipment, and the local researchers began bringing in samples of maize they had collected from Ecuador, Bolivia, Chile, and Venezuela. She and her assistants examined specimens of a given race, tabulating for each chromosome the number, position, size, and shape of each knob. At first, she went about the task mechanically.

Bottles of preserved sporocytes—spore cells—would be brought to her, she said, "and I'd just take one bottle after another, I was not really interested. But I was doing the work, and I could do it fast and carefully." But it "wasn't more than two weeks or so that I realized that I'd discovered something very very important."[64]

"All I had given to me were these bottles that had the collections in them so that I could look at the chromosomes. I had the name of the race, the country from which it came, and the elevation [at which they had been collected], that's all." This was enough information for her to see the pattern. "I was recording everything as it went along, and then I began to realize that everything that was up in the very high elevations, with two exceptions only, was all the same."[65] In nine of ten highland races from Ecuador, in eleven of twelve highland races from Bolivia, and in all ten highland races from Chile, she found a consistent pattern of knobs.[66] "But the things that were down low were very different." Lowland races differed widely in number, location, and size of knobs. She realized that "for 2,000 miles or more there was one genotype only—cytologically the same. But as soon as you got down to the lower elevations, something very different."[67] This made her realize that "there was something going on there."[68]

She had discovered that she could trace the origin of the races of maize through their chromosomes.[69] This was the major conceptual leap of the project. The knobs were not only distinguishing markers for a given race. They told the history of maize. Since each race had a distinctive pattern of knobs, she could use cytology to trace the spread and evolution of the races by the knobs, as a comparative anatomist draws up phylogenies on the basis of variations in the shape and orientation of bones.

McClintock characterized each race as she had done maize as a whole in 1929, by drawing idiograms (see Figure 8.3). Each idiogram stylized the ten maize chromosomes and represented the presence and form of knobs characteristic to a given race. Incremental changes in idiograms illustrated the divergence of races from one another. She then gave each race a unique symbol and superimposed their distributions upon geographic maps (Figure 8.4). Often, a given race had a geographic area of concentration but sent out fingers into other regions, where it abutted or graded into another race. In this way, she collapsed the four-dimensional distribution of maize through time onto a two-dimensional map. Distribution maps of maize races had been constructed before, but not distribution maps of maize chromosome patterns. Thorough collection of corn

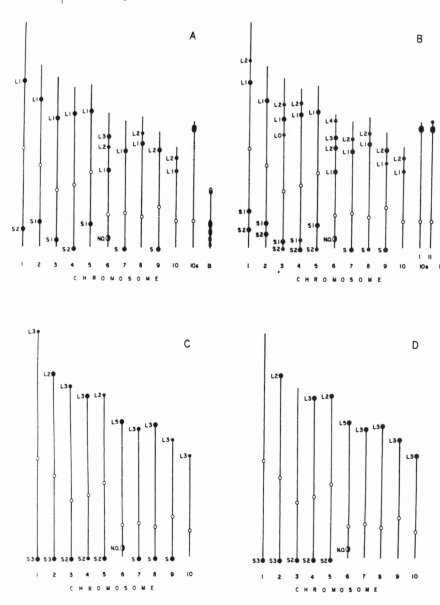

Figure 8.3. Idiograms of four different races of maize. Open circles represent centromeres; filled circles represent knobs. From Barbara McClintock, T. Angel Kato Y., and Almiro Blumenschein, *Chromosome Constitution of Races of Maize: Its Significance in the Interpretation of Relationships between Races and Varieties in the Americas* (Chapingo, Mexico: Colegio de Postgraduados, Escuela National de Agricultura, 1981), p. 24.

throughout the Americas and careful construction of the maps would allow her to reconstruct the spread of maize throughout the Americas. These maps became the primary representation of the data.

"I came to the conclusion then, from studying this, that these knobs were highly conserved," she said. "They'd gone through generations without changing. Then I decided that if we had all the maize throughout all the Americas we'd be able to find out where they came to a focus."[70] Summarizing her data for the Carnegie Institution *Year Book*, she concluded that it should be possible to use knob constitutions to trace the origins and migrations of maize races and draw inferences about hybridization in the development of various races. She also asserted that present-day maize may have derived from several different centers, with migration followed by hybridization.[71]

McClintock's involvement in the NAS-NRC project continued for more than twenty years. Edward J. Wellhausen, the head of the Rockefeller Foundation's bureau in Mexico City, persuaded her to extend her study into Mexico and Guatemala. From 1962 to 1964, she became involved in a Rockefeller-funded program to train Latin American scientists in cytogenetics at North Carolina State College (now University) in Raleigh. She and her colleague William Brown, a research scientist (later president) at Pioneer Hi-Bred, the company that developed commercial hybrid corn, designed the fellows' research programs and oversaw their progress.

The collaboration was often tense. The genetics faculty at North Carolina State, which was administering the program, asked her to submit her plans, more for information than critical review. McClintock responded haughtily: "Since I have never before been required to defend the appropriateness of a research project that I had outlined for a graduate student, this method of review was received by me with surprise and some discomfort." H. F. ("Cotton") Robinson, chairman of the department, backed off immediately, assuring her, "If you feel that you would not like to have such discussions in the future, then there will be none."[72]

Two of the fellows, Almiro Blumenschein and Takeo Angel Kato Yamakake, stayed on the project, and McClintock continued as their informal but absolute supervisor. This was the longest lasting of McClintock's numerous mentoring relationships. She then became involved in the founding of an international center for the improvement of maize and wheat (known as CIMMYT), established in 1965 in Mexico and extant today as a renowned agricultural research center.[73] One of her stated goals

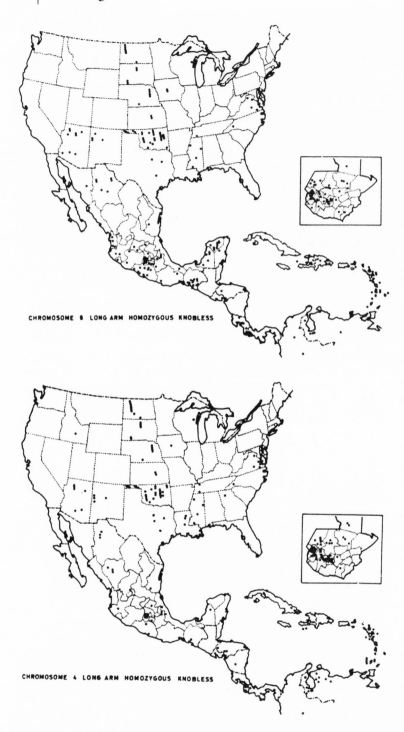

CHROMOSOME 8 LONG ARM HOMOZYGOUS KNOBLESS

CHROMOSOME 4 LONG ARM HOMOZYGOUS KNOBLESS

was sociopolitical: "To produce an effective integration of efforts and objectives among investigators in different countries of the Americas."[74]

Through the 1960s and 1970s, she guided Blumenschein and Kato through the completion of the project, alternately praising and excoriating them, and with them she ultimately produced a major monograph summarizing the chromosome studies. When in 1970 Kato presented to her an analysis of some strains that diverged from her system, she scolded him as though he were a schoolboy, outlining her methods and reasons—which Kato knew perfectly well—and concluding, "I should be pleased to know your reaction to them. If you cannot agree that these points-of-view have value for you, I should be grateful to be so informed."[75] He begged her forgiveness and restored himself to her good graces. Soon thereafter, she went to great lengths to get him into a doctoral program. She wrote letters for him, sought funding, identified an advisor. She even asked Beadle, by this time a Nobel laureate, to write a letter of recommendation. (Recently "retired" to the presidency of the University of Chicago, Beadle had rekindled his interest in maize and had worked with Kato one summer in Mexico.)[76] Kato got into graduate school, with a fellowship, and received his degree under Walton Galinat at the University of Massachusetts at Amherst in 1975.[77] He wrote large sections of the book manuscript, which McClintock rewrote, as she did all of Blumenschein's contributions. Publication was further delayed by administrative tangles; the book finally appeared in 1981.

Patterns and Programs

One of McClintock's favorite pastimes was to walk the Cold Spring Harbor Laboratory grounds. Rain or shine, from at least the early 1950s through the 1980s, alone or with a friend, she hiked through the woods or strolled down Bungtown Road to the Sand Spit and back. Evelyn Witkin recalled that "every couple of feet she would see something growing that she would point out to me and give me a little lecture about it."[78] The walks, though relaxing, had a serious component. Out in nature but near

---

Figure 8.4. Geographic distribution of maize showing two different knob constitutions. From Barbara McClintock, T. Angel Kato Y., and Almiro Blumenschein, *Chromosome Constitution of Races of Maize: Its Significance in the Interpretation of Relationships between Races and Varieties in the Americas* (Chapingo, Mexico: Colegio de Postgraduados, Escuela National de Agricultura, 1981), p. 70.

the lab, these walks led to a string of minor discoveries that sparked her synthesis of the races of maize and controlling elements. On these walks, she integrated heredity, development, and evolution.

In the late 1960s, McClintock noticed in the meadows around the laboratory patches of touch-me-nots *(Impatiens biflora)*. Touch-me-not flowers come in clusters, and each flower bears a constellation of pigmented dots on a colored background. The patterns vary widely, but the flowers on any one plant have the same pattern. A given pattern may recur on different plants. The dot patterns reminded McClintock of the patterns on her *Spm* kernels. Their regularity led her to assume that a gene-control system must be responsible for producing them.

In Queen Anne's lace *(Daucus carota)*, a wild carrot, she found a different sort of pattern. Look at a hundred plants of this common wildflower and you will see on each a cluster of tiny white flowers, called an umbel, with a single dark purple flower in the center. Look at thousands, as McClintock did, and you may see variants on this pattern. The color of the central flower may be red, or even yellow. Rarely, it may be variegated with stripes of purple on a light background. (See Figure 8.5.)

Most significant, she found examples of Queen Anne's lace in which several central flowers were pigmented, and even some where a small cluster of pigmented flowers formed further out in the umbel. She interpreted these variants as errors in a presetting mechanism. Usually, the presetting occurred so as to make only the central flower purple. If it occurred early, a cluster of purple flowers resulted; if late, perhaps only part of a petal of the central flower would be pigmented. If the presetting mechanism was triggered at an inappropriate place and time, the off-center cluster of purple flowers would result. Such patterns did not prove the existence of a genetic controlling system, and she did not pursue genetic studies with Queen Anne's lace or touch-me-nots. Rather, she shot hundreds of black-and-white pictures with her camera and macro lens, developing them in a darkroom in her laboratory. The patterns were sufficient proof. Of Queen Anne's lace, she wrote, "There is no reason to doubt that the variegated expression results from some modification of the system normally operating to control pigment production in the floral parts of the central region of the umbel."[79]

A program is a pattern that develops through time. It is also a set of instructions, like a computer program. McClintock began to speak more and more of developmental programs and genomic programs. In 1967, she attended the annual meeting of the Society for Developmental Biology. In a paper published in the proceedings of the meeting, she wrote that "during development, all components of the genome must undergo sequential stages of programming, which must continue until the final stage of differentiation of the cell."[80] By "programming," she meant the execution of patterns of gene activity through time. The programs, she argued, were carried out by controlling elements.

Figure 8.5. A demonstration slide of Queen Anne's lace from the 1970s. The top left image shows many pigmented and partially pigmented flowers in a cluster that normally would have just one. Reprinted with permission from the American Philosophical Society Library.

As McClintock came to see nature in terms of programs, she began to extend her vision of pattern from phenotype—the study of form in the context of genetics—to morphology, the study of form in the context of natural history. Morphology became simply a complex pattern, which had to be produced by a developmental program. In this light, well-known phenomena of natural history, such as metamorphosis and mimicry, took on new significance for her. A caterpillar and the moth it becomes have the same genome, yet are dramatically different looking creatures. During metamorphosis, the animal shifts from a caterpillar pattern to a moth pattern. Controlling elements, which altered gene expression without altering the genes themselves, could, if properly coordinated, produce such shifts in pattern. She grew fascinated by examples of mimicry, for example a clam with a lure that resembles a fish. The tissue composing the lure expressed a fish pattern. Plant galls, vegetative tumorlike structures formed in trees when infected by an invading insect parasite, indicated to her that external species could alter the developmental program of its host.[81]

In her emerging vision of organic change, developmental programs in all tissues were connected and interacting. Returning to the meeting of the Society for Developmental Biology in 1977, she discussed the development of the maize kernel as revealed by controlling elements. Extending her 1940s observations of twin sectors, she discussed the clonal sectors of tissue within the kernel. A clonal sector was a region of tissue derived from a cell with a given developmental pattern. The various patterns communicated with one another to ensure the coordinated development of the ear. Development, she wrote, was a form of integration. "During development of the kernel, events are so regulated that each tissue adjusts its developmental progression to coordinate with all others. The endosperm is only one such tissue. These integrations are exceptionally well described and illustrated by Randolph (1936)." Controlling elements were a good candidate for the execution of such patterns because they were flexible and reversible. Finally, she integrated the races-of-maize study—on which she contributed a chapter to the volume *Maize Breeding and Genetics*, published the same year. She noted that while she had tried to illustrate the basic "theme" of maize endosperm development, "The selected examples . . . do not explore the many variations on this theme that should be encountered, especially among kernels produced by some of the races of maize." She did not attempt a molecular explanation, though by now she accepted that one was needed and hoped her work would "project the responsible mechanisms at the molecular level."[82]

Development, then, used a more or less fixed set of genes in an infinite variety of ways. The genome was like a set of organ pipes, with each gene modeled to produce a unique note. The qualities of any note could be varied by the player, but most important, the organ could produce anything from a Bach chorale to rhythm-and-blues, depending on the combinations, order, intensity, and timing with which the keys were pressed. Development, she thought, was often, usually, or even exclusively governed by reversible modulations of gene action. Evolution, however, must result from permanent, structural genetic changes.

In 1971, McClintock asked an audience at Cold Spring Harbor, "If all organisms have the same basic parts and can produce basic tools, why do we have so many different types of organisms?"[83] Her answer was that different organisms simply expressed different patterns. In an interview with Will Provine and Paul Sisco in 1980, she explained. "Years ago I began to realize that any genome can make any other, practically." The genome, she imagined, was a set of interchangeable genetic parts, "like an erector set." By using different combinations of genes in different arrangements, one could build almost any organism from a single set. These potentials reflected the "extraordinary integration of the genome." All organisms, she said, had an "overall pattern that's built into the genome somewhere so that it builds itself."[84] Pattern is built into the genome; if pattern shifts, the organism changes.

It is a small leap from an insect or frog genome that can make two different organisms to the production of new species. McClintock began to see speciation as a shift in pattern. This idea dates back to the turn of the century, to Hugo de Vries and his *Mutationstheorie*. From observations of local populations of evening primrose, de Vries proposed that new species formed suddenly, in saltations, or jumps, dramatic reorganizations of the hereditary material. Many biologists at the time saw de Vries's results as contradicting Darwinian gradualism. McClintock's longtime friend and ally Richard Goldschmidt was a great advocate of this saltational theory of speciation. He argued that the creation of new species and higher taxonomic orders, which he termed macroevolution, proceeded by large, systemic mutations in which the total composition of the chromosomes was altered. Most such mutations produced "hopeful monsters," but those well-adapted to their environment became new taxa.[85] Today a variant on the theory continues in the "punctuated equilibrium" of Stephen Jay Gould and Niles Eldredge, tagged by critics as "punk eke" or "evolution by jerks."[86] Though not by any means canonical, saltational evolution continues to be taken seriously by scientists on both sides of the debate.

Asked in 1980 whether her races-of-maize studies shed light on evolution, McClintock replied, "Yes. Macroevolution undoubtedly." Since at least 1951, she had agreed with Goldschmidt's thinking that chromosomes, rather than genes, were the proper unit of heredity. As the interview continued, she described knob patterns in maize and its relatives and how they reflected evolutionary history. She concluded, "The main changes in evolution are regulatory. They have to be."[87]

She had long favored the idea that a gene's action depends on its chromosomal context. She now realized that if developmental programs were determined in part by chromosome structure, then they could be remodeled by rearranging the chromosomes. A given gene's neighborhood—other genes, certain promoters, controlling elements, heterochromatin—could determine its action. Putting a gene in a new neighborhood could change its timing or strength of expression, could expose it to new modulators, operators, or repressors. At the level of the genome, she reasoned, speciation could be structurally related to metamorphosis. If so, then in theory all organisms could be derived from a single set of genes. Macroevolutionary changes could come about without chemical changes in genes.

In her own earlier work, McClintock readily found a candidate mechanism for rearranging the genome. The breakage-fusion-bridge cycle was a self-perpetuating cycle of chromosome rearrangements, and she knew it could produce dramatic effects: the first controlling elements had appeared in stocks undergoing BFB. She now returned to the BFB cycle, uniting it with her knob studies and controlling-element work to develop a genetic theory of speciation.

In 1978, she used chromosome knobs as indicators of large-scale "unorthodox types of chromosomal rearrangements."[88] Her original controlling-element theory involved microscopic bits of heterochromatin. But by the late 1970s, she had made her detailed study of chromosome knobs. Knobs and centromeres, she argued, were involved in large-scale chromosomal changes such as inversions.[89] In at least one instance, chromosome breaks occurred at a knob and a centromere, so that the knob ended up at the centromere position and vice versa. Such arguments echoed her findings from Colombia, described in her 1960 Carnegie report. There she described several inversions as well as an "abnormal chromosome 10" with an extended segment added to the end of its long arm.[90] Her conclusions in 1978 concentrated on "resetting" and "restructuring" the genome, as

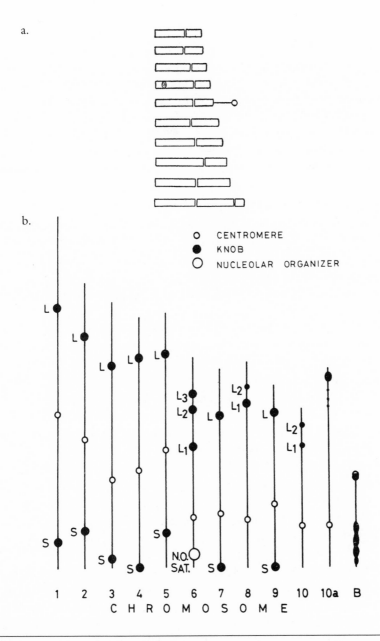

a.

b.

○	CENTROMERE
●	KNOB
○	NUCLEOLAR ORGANIZER

C H R O M O S O M E

1   2   3   4   5   6   7   8   9   10   10a   B

Figure 8.6. Idiograms of maize, from 1929 *(a)* and 1981 *(b)*. From Barbara McClintock, "Chromosome morphology in Zea mays," *Science* 69 (1929): 629, copyright 1929, American Association for the Advancement of Science; Barbara McClintock, T. Angel Kato Y., and Almiro Blumenschein, *Chromosome Constitution of Races of Maize: Its Significance in the Interpretation of Relationships between Races and Varieties in the Americas* (Chapingo, Mexico: Colegio de Postgraduados, Escuela National de Agricultura, 1981), p. 15.

initiated by "stress," either cytological or environmental. She expressed her macroevolutionary perspective in the last sentences:

> I believe there is little reason to question the presence of innate systems that are able to restructure a genome. It is now necessary to learn of these systems and to determine why many of them are quiescent and remain so over very long periods of time only to be triggered into action by forms of stress, the consequences of which vary according to the nature of the challenge to be met.[91]

In this paper, McClintock provided a schematic diagram of the maize chromosomes showing the positions of knobs. The knobs were the breaking points, the sites of potential reorganization, the substrate for Goldschmidtian macroevolution. The diagram was an idiogram: rotated ninety degrees, it was identical to those used in her races-of-maize work, which she was at that time writing up for publication, and to her original maize idiogram of 1929.[92] (See Figure 8.6.)

The knob patterns in the races of maize, then, were not simply markers; they were agents of control, arbiters of pattern. When McClintock integrated controlling elements and the races of maize, she derived macroevolution. The twenty-two knob positions determined the possible large-scale recombinations—precisely the kinds of rearrangements Goldschmidt's theory required. The patterns of the knobs reflected the evolutionary history and potential of maize.

These ideas of restructuring and reorganizing the genome brought her to the final question of her career, concerning the seat of the control. What "told" the controlling elements when and where to act? What guided genome restructuring and reorganization? Throughout the 1950s and 1960s she had been vague on this point, apparently taking the existence of control on faith. It was on this subject, the source of the control, that McClintock sounded most mystical. The organism seemed to have "knowledge of every cell, where it belongs in the system," she told Provine and Sisco in 1980. "It's the overall thing that dominates. What is that overall thing?" One interviewer noted that this was exactly Hans Spemann's great concern: the overall controller—that for which he was accused of vitalism. "Sure!" she answered. "But that is basic." Where it gets complex, "where you get all mixed up," she said, is where one considers, not simply one tissue inducing another, but the development of the entire organism. She and a few others had a special quality of mind, she believed, that enabled them to see the patterns and their relation to control. That quality was freedom. "This is all part of the freedom from constraint," she said, freedom from preconceived notions and accepted dogmas.[93]

McClintock's last writings suggest that the environment is the ultimate source of control. The genome, in other words, responds to the environment. Her Nobel lecture, in 1983, concerned the response of the genome to conditions of stress. In a simple way, this idea recalled her work on the breakage-fusion-bridge cycle. As Stadler's X rays had disrupted the chromosomes in some way so as to stimulate the BFB cycle, so she believed the BFB cycle had shaken loose the controlling elements such that they acted out of their normal context, which enabled her to see them. This theory helped explain not only developmental responses to environmental change but (macro-)evolutionary responses as well: "stress, and the genome's reactions to it, may underlie many formations of new species." She called for greater attention to the genome, "with greater appreciation of its significance as a highly sensitive organ of the cell that monitors genomic activities and corrects common errors, senses unusual and unexpected events, and responds to them, often by restructuring the genome." She linked controlling elements with phenomena even more disparate than crown galls and mimicry: heat shock, in which heat activates a set of genes that enable an organism to tolerate heat; and the "SOS" response, in which the bacterial genome recognizes DNA damage by activating a cascade of genes that allow the cell to survive that otherwise lethal insult.[94]

By the end of her career, then, McClintock had developed a sweeping, integrated vision of organic change that united development and evolution. She saw both as changes in pattern: one within the individual, the other among individuals. In McClintock's vision, the expression of pattern is a response to the environment, as in Darwinian evolution. Yet she was not a throwback to nineteenth-century biology. Her vision was manifestly that of the late twentieth century: like her, other biologists in the 1960s began reintegrating the problems of growth and form. They did so, however, under the paradigm of molecular biology, which McClintock never adopted. Beginning in the late 1960s and continuing through the end of the twentieth century, molecular biologists discovered in McClintock's vague, almost poetic language echoes of a new understanding of the genome. This is what I call McClintock's renaissance.

9

RENAISSANCE

She shook the tree of chromosomal stability.
—JOSHUA LEDERBERG

Nature, Mr. Allnut, is what we are put in this world to rise above.
—ROSE (KATHARINE HEPBURN), *THE AFRICAN QUEEN*

By the 1960s, Barbara McClintock had slipped into the category of venerated but largely irrelevant older scientists. Most geneticists—save a fraction of those who had come into microbial genetics from physics—knew of her early contributions from textbooks. She ranked with George Beadle, Marcus Rhoades, Milislav Demerec, Calvin Bridges, Alfred Sturtevant, perhaps even Rollins Emerson and Thomas Hunt Morgan as a past master. By the beginning of the 1960s, all but the first three were dead; Demerec would pass away in 1966. McClintock had not published a primary research article in years. She of course remained active, but her ideas grew more complex and obscure. They seemed increasingly off-topic to molecular and bacterial geneticists.

In an informal interview in 1970 with Charles Burnham, whom she had known since the early 1930s, and several students, McClintock described how assumptions—preconceptions—colored one's ability to accept something new. "A person doesn't know they have a basic assumption [but] they certainly use it to make something clear to them," she said. To accept a new vision of nature, one must be ready to recognize and to reject one's assumptions. Of her colleagues, she said, "They're not ready." She then related a story of telling her mother that men might go to the moon—this must have been in the early 1960s. "I told her about it and she had heard it was being gently talked about . . . and she said, 'Oh, I'm so glad I'm not going to live, I don't think I could stand it.' Not because she was against it, it meant the whole reorganization of her whole idea of what this earth was [about]."[1]

McClintock, born the year before the Wright brothers' first flight, not only saw men walk on the moon, but also endured the reorganization of her idea of what genetics was about. In the 1970s, recombinant DNA technology and protein and gene sequencing opened the black boxes of genes and gene action. The molecular level of resolution demonstrated that no mystical or vitalistic forces, no omniscient controller need be invoked to explain the wonderful complexity of living systems.

Under this new vision of nature, McClintock was reborn scientifically. She was invited to meetings on strange new topics, lavished with awards, prizes, and publicity. She became moderately wealthy and famous. It wasn't that the scales finally fell from scientists' eyes and they recognized she had been right all along about genetic control. Rather, McClintock's central idea was cut out and transposition, the mechanical fact of movable genetic material, was recast as her major contribution.

The process occurred in stages. From Jacob and Monod's initial association of controlling elements with episomes, bacterial and viral geneticists began to tie controlling elements to transposition and to divorce them from control. In bacterial and viral genetics, Drosophila genetics, and yeast genetics, genetic mobility was discovered and parallels were drawn to McClintock's elements. In studies of gene structure, viral infection, antibiotic resistance, even immunology, genetic rearrangement emerged as a central principle. Transposition was linked to many kinds of mobile nucleic acid, including some kinds of elements that McClintock would have preferred not be associated with her controlling elements. Redefined, transposition grew into a major new subfield of molecular biology. McClintock became its figurehead.

Phage Mu

The prodigious body of work that emerged from the Institut Pasteur in the late 1950s and early 1960s focussed bacterial geneticists' attention on the linked problems of lysogeny, episomes, and the genetic control of biochemical pathways. The discovery in the 1950s that lysogeny was a state in which a virus became genetically part of the host chromosome began to dissolve the notion of the genome as a self-contained unit. Transduction showed that phage could transfer bacterial genes from one bacterium to another. Episomes reminded geneticists that genes need not lie on chromosomes. These results did not negate powerful generalizations like the string-of-pearls gene model or Francis Crick's "central dogma," which said that once information got into protein it could not get out again.[2] But they did show that these rules had exceptions. By the 1960s, then, genetic information was coming to be seen as fluid, if viscous.

In the early 1960s, in the immediate wake of the operon theory, a few microbial geneticists linked their work to McClintock's. One important question was precisely how the viral prophage integrated into the host chromosome. In 1959, François Jacob suggested that it "hooked to" the chromosome.[3] In 1962, Allan Campbell argued that the prophage actually inserted into the genome, and compared them to McClintock's controlling elements.[4] The Campbell model of episome integration was an extension of Jacob and Monod's assumption that controlling elements were in fact a kind of episome. In substantiating Jacob and Monod's view, Campbell cited McClintock's *American Naturalist* article on the parallels between controlling elements and the operon: "It is their transposability which justifies our inclusion of these [controlling] elements as episomes. Their functional aspects make them more similar to bacterial regulators and operators, for which transposability has not been found (cf. McClintock 1961)."[5] Campbell thus separated controlling elements' structure from their function, and emphasized their structure. Transposition began to denote the insertion of DNA into DNA.

A second connection came with the discovery of a new bacteriophage. In 1961, Austin Lawrence Taylor, a graduate student of Edward A. Adelberg at the Berkeley campus of the University of California, discovered it as a chance contaminant in a widely used strain of *E. coli*.[6] He had direct evidence that the mutations occurred in a host bacterium when the phage inserted into it, and indirect evidence that one of the mutations was genetically linked to the insertion site, which in this case was the galactose operon. In 1999, Taylor told me, "I had not yet proved that there was absolute linkage, that in fact there was cause and effect, that in fact the prophage in this particular strain was inserted into one of the galactose genes."[7]

Taylor submitted his thesis in the spring and then went to Brookhaven National Laboratory on Long Island, for a year's postdoctoral fellowship with Milislav Demerec. (Demerec had retired from the Carnegie Institution in 1960 and taken a post at Brookhaven.) "He of course knew Barbara McClintock very well, he knew Cold Spring Harbor very well, and he introduced me to both."[8]

During his year at Brookhaven, Taylor found that his phage inserted into multiple, seemingly random sites in the bacterial chromosome. When it did so it caused mutations. It was probably during this time that he renamed his virus—which had had the unpoetic name X-373—mu, "for mutagenic, not because mu comes after [phage] lambda in the Greek alphabet." He and McClintock "used to talk about phage mu to some extent, and about its behavior," he said. "Her reaction was one of great interest and great enthusiasm," he recalled, but he gave no credence to the anecdotal report by muologist Ariane Toussaint that McClintock in fact suggested the name "mu."[9]

From Brookhaven, Taylor went to the National Institutes of Health in Bethesda, Maryland, where he wrote up his discovery of phage mu. In October 1963, Demerec, a member of the National Academy of Sciences, communicated Taylor's paper to the National Academy's *Proceedings*. In the paper, Taylor was cautious but did not waffle in comparing mu, which he optimistically called Mu1, to controlling elements: "Phage Mu1's dual ability to occupy many chromosomal sites and to suppress the phenotypic expression of genes with which it becomes associated resembles the 'controlling elements' of maize more closely than any previously described bacterial episome." The comparison came about, he said, as a result of his conversations with McClintock. "It was evident to me that there could be a similarity between the apparent mutational changes occurring in maize under the influence of controlling elements" and those caused by mu insertion. "I did not have a notion about transposability," he told me. "The similarity was in the ways mu was behaving." Taylor had no evidence for either mechanistic similarities or any suggestion of a common evolutionary role in mu and controlling elements.[10]

In 1966, biologists began to investigate phage mu with the electron microscope, revealing a classic phage structure resembling the Apollo Lunar Module or a water tower. Between 1968 and 1970, a small cadre of scientists coalesced and began to exploit mu in addressing some of the problems of DNA excision and insertion.[11]

In 1970, one of Taylor's graduate students, a Pakistani émigré named Ahmad Bukhari, came to Cold Spring Harbor for a postdoctoral fellowship and brought mu with him. Bukhari had written his thesis on the biochemical genetics of the synthesis of diaminopimelic acid, an amino acid found only in bacteria—not on mu, though he used mu in his doctoral work. At Cold Spring Harbor, however, Bukhari began working on the biology and biochemistry of mu insertion. He engaged McClintock's attention, and she certainly shaped his thinking on the parallels between mu and controlling elements. By the early 1970s, Cold Spring Harbor had become a center for mu research and discussion. Phage mu became molecular biology's strongest claim to intellectual descent from controlling elements.

In 1972, Bukhari was promoted to the scientific staff and got his own lab. In July, he organized a workshop on phage mu at Cold Spring Harbor.[12] It brought together a small yet diverse and far-flung crowd and helped define the field.[13] Conveniently, it overlapped with the bacterial genetics course (successor to the phage course started by Max Delbrück in 1945). Taylor was a visiting instructor in Delbrück's course and also participated in the mu workshop. The members of the mu community were young and idealistic, and they shared a passion for open collaboration and collegiality. Around Cold Spring Harbor they called themselves the mukaryotes.

Bukhari began to think of mu as a transposable element in a viral package. In so doing, however, he redefined transposition. Mu rubber-stamps itself onto another site in the host chromosome. The original copy stays put, and the number of copies of mu thus multiplies. This process became known as replicative transposition; the original form of transposition was given the retronym "conservative" transposition. By the 1980s, the mukaryotes had identified many mu genes and characterized both the physical virus and its genome in some detail.

Among its most unusual features was the G segment, a chunk of DNA that flip-flopped within the viral chromosome, existing equally happily in the forward, or flip, direction, or the reverse, flop. The G segment was first detected by electron microscopy. Flop cannot base-pair with flip. When a DNA strand carrying flip G encounters a strand carrying flop G, they will zip up into a double helix—except at the G segment, which loops out into a single-stranded bubble. In the mid-1970s, mukaryotes discovered that the G segment was flanked by special sequences similar in structure to another class of movable element, discovered in the late 1960s, called insertion elements (see following). The mu phage itself could also insert into the host genome either forward or backward. Mu, then, was a transposable element within a transposable element.[14] Phage mu continues to be a major model system for molecular investigation of transposition. But the romantic era of mu ended in 1983 with Bukhari's tragic death by heart attack at age forty-two.

Drosophila Controlling Elements

In 1965, Melvin M. Green, working at the Davis campus of the University of California, discovered transposition in fruitflies. Although in an interview in 1996 he downplayed his discovery's significance, it marked the first instance of transposition outside corn. As did Bukhari's work on phage mu a few years later, it stretched the definition of transposition. "The only thing which I did—well I didn't do it, the flies did it—the only thing which I uncovered which was supplementary from what McClintock found, was that I actually described transpositions of a gene under the direction or influence of a controlling element. Because maize wasn't made to order for this. Flies were."[15]

In McClintock's corn, the genes themselves stayed put; only the controlling elements moved. In Green's flies, the *white* gene moved from the X chromosome to the third. He wrote to McClintock and discussed his work, outlining to her a theory of mutable genes in Drosophila that involved controlling elements. "And in this connection I wrote a paper, which was published in *Genetics,*" in 1967, and cited McClintock.[16] This too was a new form of transposition—the movement of actual genes rather than their controllers. Accordingly, Green linked his research not only back to McClintock's work but also to more recent studies on viruses such as phage lambda, which integrate into the host chromosome. Perhaps controlling elements were, as Jacob had suggested, a form of episome. In 1968, Green wrote McClintock that he had made this link "in part to see whether one can stir up interest apart from the few people who have a parochial interest in controlling elements and in part because this equating of the two allows one to do a few more experiments."[17]

By 1967, however, McClintock was beginning to find the bacterial analogy too restrictive. Reminding Green of the diversity in her own material, she said she "found it difficult to relate all of these modifications to simple episomic infection. The patterns expressed are too distinctive and too diverse in type."[18]

Green's 1967 paper drew little interest among geneticists. "I didn't get many reprint requests, and this bothered the hell out of me. Because I thought this was a rather interesting and even important contribution to genetics." Frustrated, he called McClintock, saying he would be traveling to Washington, D.C., and asking if he could come visit. "She says, 'Sure, come on in.' So I went down to Cold Spring Harbor and met with Barbara and said, 'Barbara, what's wrong with this paper? I don't get any reprint requests for this paper. This is transposition!' It just wasn't one, it was four independent transpositions I had discovered. And she said, 'There's nothing wrong with it. People aren't ready for this yet.'"[19]

Most microbial geneticists still were not listening. Part of the problem was that the younger bacterial geneticists read much less outside their field than the classical Drosophila and maize geneticists had done. Arrogance, perhaps in some cases, but in defense of the youngsters the volume of literature was exploding—simply staying current in one's field was becoming a Sisyphean chore. Besides, bacterial geneticists were too caught up in problems of genetic control to be concerned with transposition in flies.

Insertion Elements

During the 1960s, the pieces of one of biology's greatest puzzles were laid out and assembled. The puzzle was the controlled flow of genetic information, the molecular mechanism by which instructions coded in the genes are read out and direct the building of proteins. Genes, encoded in the language of DNA, are transcribed into messages in the related dialect of RNA by enzymes called polymerases. These attach to a site on the DNA called the promoter, discovered in 1964, and then slide down the DNA, reading off the genes and spinning threads of messenger RNA. Which genes are read when, and in what amounts, is determined at the operons by whether repressor protein is bound to the operator, adjacent to the promoter. The RNA messages then encounter ribosomes, through which they are pulled like magnetic tape across a reading head. The ribosomes read the message three nucleotides at a time. Each triplet, or codon, signifies one amino acid, the basic unit of protein. When the ribosome reads out a codon, yet another species of RNA comes into play: transfer RNA. A molecule of transfer RNA has a complex clover-leaf shape. One leaf bears an anticodon, complementary to one RNA codon; another acts as a grappling hook for an amino acid. As a given codon is read out, the corresponding transfer RNA docks in a pocket in the ribosome, binds to the RNA message, and its amino acid is welded on to the growing protein chain. In 1961, the first codon was associated with an amino acid, and by 1966 the entire genetic code, the cipher by which nucleotide triplets are translated into amino acids, was solved. The four nucleotide bases—adenine, cytosine, thymine, and guanine, referred to by their initials, A, C, T, and G—can make sixty-four different triplets. Since there are only twenty amino acids, there is room for both redundancy and, critically, punctuation. Three of the codons encode a stop signal rather than an amino acid.[20]

This, tersely, was the emerging story of genetic control. It was a molecular story, one based on the time-tested principle of a basically stable genome with regulatory and structural genes grouped into functional bunches. It had little to do intellectually and nothing to do technically with McClintock's vision of movable controlling elements.

In 1964, James Shapiro was an idealistic midwesterner with a bachelor's degree in English, pre-med aspirations, and a fellowship that took him abroad to study at Cambridge University. He wound up working under the "sporadic, mercurial, and unofficial tutelage of Sydney Brenner," he wrote in a brief memoir in 1992.[21] Brenner's group at that time was focussed on mutations that interfered with the reading of the gene into protein. In 1964 and 1965, they discovered a new kind of mutation, which they called "ochre," which stopped transcription. Like the previously discovered "amber" mutations, these were mutations that resulted in an abbreviated protein chain. A stop signal is read in the middle of a gene.[22] Mutations that result in the substitution of one amino acid for another are "missense" mutations; those that result in a stop are "nonsense" mutations. A stop signal will crop up eventually as a result of any mutation that changes a gene's reading frame. The DNA sequence *GTA ACG*, for example, translates as the amino acid pair valine-threonine. But if a mutation bumps the reading frame down one base, so that G*TA AC*G is read as a codon, a stop signal will result. Brenner directed Shapiro toward a study of the effects of transcription on mutation, which he commenced, using the galactose *(gal)* operon. In 1966, Shapiro went to the Postgraduate Medical School of London to finish his dissertation under the great bacterial geneticist William H. Hayes.

Among numerous spontaneous *gal* mutations, Shapiro found an unusual class of so-called polar mutations. Polar mutations were first identified by Jacob and Monod. Rather than disrupting a single gene, polar mutations suppress the activity of all of the structural genes in an operon. Initially, Jacob and Monod had thought the mutations lay in the operator, but later work showed them to be nonsense mutations in the first structural gene downstream of the operator. Polarity could be strong, with gene expression mostly or completely suppressed, or weak, with gene expression merely reduced. Curiously, the further upstream the mutation, the stronger its polarity. Shapiro learned that Sankhar Adhya, a graduate student at the University of Wisconsin, was working on a similar problem, and they decided to publish their results together.

Shapiro's mutations were more strongly polar than amber and ochre mutations—even though they occurred downstream of known amber and ochre mutations. Shapiro quickly ruled out DNA deletions as the cause of the mutations because his mutants regularly reverted to wild-type. A second mutation restores a deleted segment of DNA as often as a second power surge restores lost files to a damaged computer hard drive. Other tests and observations ruled out other possibilities. "Sometime during the writing of my thesis, it occurred to me that the properties of these puzzling mutations could be explained as the results of extra DNA inserted into the *gal* operon," Shapiro wrote in 1992.[23] "Moreover, this idea could be tested." The following year, as a postdoctoral fellow in the laboratory of François Jacob, he did the experiment.

The test was to weigh bacterial chromosomes with and without the mutation and see which was heavier. Shapiro transduced *gal* genes into virus particles and then centrifuged the viruses in tubes containing a gradient of cesium chloride, a heavy salt. Pulled outward by the force of spinning and pushed back by the ever-increasing density of the cesium chloride, the viruses migrated through the solution and stopped where their buoyant density matched that of the surrounding solution. The viruses containing mutated DNA migrated further and were therefore heavier than those with wild-type DNA. He concluded that his spontaneous mutations were "the consequences of the random insertion of large pieces of foreign DNA into a structural gene."[24]

Meanwhile, at Cologne, Germany, Peter Starlinger, together with graduate student Heinz Saedler and postdoctoral fellow Elke Jordan, had also found a set of strongly polar mutations at the *gal* operon, similar to those found by Shapiro and Adhya. Shapiro, then at the Institut Pasteur, visited the Cologne group, and he and they shared data and hypotheses. Later, Shapiro was surprised to find that the German group had performed the same experiment to test the insertion hypothesis. They too found the mutated chromosomes to be heavier than wild-type chromosomes, and they concluded that a thousand to fifteen hundred DNA nucleotides had been added to the chromosome. The Starlinger group published their results in 1967.[25]

Shapiro took a leave of absence from science, spending eighteen months in Cuba and getting involved in antiwar politics. "I was in a 'Bolshie' phase," he said, and generously distributed his mutants to interested researchers.[26] He gave his mutants to Michael H. Malamy, at Tufts, and to Waclaw Szybalski, a former Cold Spring Harbor staff scientist, now at the University of Wisconsin. Malamy and Szybalski explored the insertion mutations, generated many more, and characterized them fully. They found that the segments of DNA that inserted could do so not only in the *galK* gene, but also at other sites in the *gal* operon, the *lac* and other operons, and in phage lambda. They showed that insertions at different locations contained DNA of the same size. Using the electron microscope, they saw that complementary DNA strands from different insertions could base-pair with one another, indicating that the sequences were identical. Cesium chloride density gradients showed that different insertions had the same buoyant density. In short, segments of DNA of the same size and same nucleotide sequence could insert into many different operons, as well as viruses and episomal DNA called plasmids. The authors identified a total of three different sequences, which could insert into a chromosome either forward or backward.[27]

In 1972, Malamy, Szybalski, and Michael Fiandt coauthored a pair of papers describing their results and proposing the name "insertion sequence" or IS elements, for the inserted DNA. The authors noted that "genetic elements were found in higher organisms which appear to be readily transposed from one to another site in the genome. Such elements, identifiable by their controlling functions, were described by McClintock (1956) in maize." The authors then triple-hedged a speculation on McClintock's elements: "It is possible that they might be somehow analogous to the presently studied IS insertions."[28]

The same year, the Cologne group guardedly acknowledged McClintock as a predecessor. In a review article, Starlinger and Saedler briefly discussed her controlling elements, attributing to her, ironically, the idea that controlling elements were episomes.

> Nothing is known at present at the molecular level about the occurrence of insertion mutations in higher organisms. It should be recalled, however, that McClintock established in Zea mays the presence of elements, which can be associated with and removed from genes, which thereby loose [*sic*] or change their function (McClintock, 1956, 1961). McClintock discussed the possibility that episomes might be responsible for these observations. It would be very interesting to see whether these hypothetical episomes bear some resemblance to the insertion mutations described in microorganisms.[29]

As I have noted, McClintock did *not* suggest that her elements were episomes. But interpretation in science is a collective process. Insertion elements were soon recognized as a distinctive and fundamental type of genetic element. In the ensuing years, insertion elements or IS-element-like sequences were found at the ends of many puzzling mobile genetic elements. Among the first problems they helped solve was that of the transfer of antibiotic resistance.

Antibiotic Resistance

One of the disturbing realizations to come from bacterial genetics following the discovery of penicillin was that bacteria could become resistant to antibiotics. Worse, bacteria could transmit antibiotic resistance to other bacteria. In 1962, Tsutomu Watanabe discovered that this resistance was transmissible on plasmids, independently of the chromosome.[30] By the early 1970s, the molecular tools to discover the basis of the resistance—and therefore potentially to foil it—seemed to be in hand.

The discovery came in pieces. In 1973, Phillip Sharp and Norman Davidson, at Caltech, together with Stanley N. Cohen at Stanford, looked at the DNA of several drug-resistance plasmids as well as the bacterial F, or sex, factor under the electron microscope. In several plasmids, they saw one or more stem-and-loop, or lollipop, structures. Each lollipop indicated an inverted repeat, a segment of DNA that occurs in one position and then again, some distance away, in reverse sequence; the similar sequences base-pair, bowing the intervening sequence into a loop. Most important, they found that the factor that conferred resistance to tetracycline was bracketed by an inverted repeat. The next year, Robert W. Hedges and Alan E. Jacob, at the Royal Postgraduate Medical School in London, showed that the gene for ampicillin resistance was carried on a plasmid called RP4. The resistance gene could be transferred from plasmid to plasmid, or from plasmid to chromosome.[31]

In 1975, these and other pieces were united in a remarkable series of coincident findings. That year, four groups published papers describing transposition of drug-resistance genes. At the University of Washington, in Seattle, Fred Heffron, Craig Rubens, and Stanley Falkow found that ampicillin resistance could translocate, or transpose, from one plasmid to another. The ampicillin-resistance gene was flanked by an inverted repeat; the whole unit resembled an insertion sequence element with the ampicillin-resistance gene inserted into the middle. At MIT, David Botstein, with postdoctoral fellow Nancy Kleckner and graduate students Russell Chan and Bik-Kwoon Tye, showed the same for the gene for resistance to tetracycline. In Switzerland, Douglas E. Berg, Julian Davies,

Bernard Allet, and Jean-David Rochaix described similar transposition of kanamycin resistance. And at the National Institutes of Health, Michael Gottesman and Judah L. Rosner described transposition of chloramphenicol resistance.[32]

Each of the groups designed and executed their experiments independently and more or less simultaneously. But—bacterial genetics was still a small enough field—all of the groups had personal connections with one another. Falkow and Botstein became friends during this period. Heffron befriended Chan when Chan moved to Seattle for a postdoc, and the two shared data and conversation. Heffron also sent electron micrographs of DNA to Jacob, Hedges, and Naomi Datta in London. Gottesman's wife, Susan, was a postdoc in Botstein's lab. "There is no question that each of the groups knew what the others were doing from both meetings and personal interactions," Botstein told me. The Berg paper was published back-to-back in the *Proceedings of the National Academy of Sciences* with that of Falkow's group. The description of transposition of drug-resistance genes, then, was one of multiple independent discovery and collective interpretation.[33]

The interpretation that emerged was that many drug-resistance genes move by virtue of having been picked up by a pair of IS elements and incorporated into a plasmid. If an insertion element cut the DNA to either side of itself, it excised from the DNA. It became clear that the IS element could also cut to one side only. Two IS elements bracketing another DNA segment, if they cut only to the "outside," would thus lift out the intervening segment, perhaps to reinsert it on another plasmid or chromosome. Insertion elements, then, are employed in many cases where a chunk of DNA—a gene segment, a gene, a plasmid, a whole virus—is taken out of one site and inserted into another. Since they were found in viruses, the best evolutionary argument was that IS elements had evolved as a way to insert a virus into a host chromosome—in other words, as part of the mechanism of lysogeny. But it was easy to imagine that IS elements left behind in the host chromosome had given bacteria an evolutionary advantage, and that they had since been put to a variety of uses.

None of the five groups investigating antibiotic resistance cited McClintock in their first transposition papers—not that she wasn't asked. David Botstein spent the 1974–75 academic year at Cold Spring Harbor, where he met McClintock and discussed his results with her as he was writing up the paper. She read and commented on the paper, and Botstein asked her if he might cite her. But Botstein said McClintock thought that "knowledge of her work didn't contribute in any way" to the interpretation of the paper, so she insisted that she not be cited. It would have been a political citation, she told him, not an intellectual guidepost to the history of the research. And Botstein consented, though he told me in 1999 that, given a second chance, "I'd've just done it."[34]

Yeast Mating-Type Switching

Another transient connection to McClintock's work was made in the early 1970s, this time in yeast, in 1971. Two Japanese researchers, Yasui Oshima and Isamu Takano, developed a controlling-element model to explain the peculiar phenomenon of mating-type switching.[35] It had been known since the late 1950s that in certain genetic constitutions, yeast can alternate spontaneously between the two mating types, *a* and *alpha*. Oshima and Takano proposed that a controlling element might explain the patterns: the *a* and *alpha* genes might be the same physical unit, with alternative states determined by the action of a controlling element. This model competed for some years with a "flip-flop" model that proposed a gene that could be read in either direction along the chromosome, with one direction indicating the *a* mating type, the other indicating *alpha*. The flip-flop model, developed by Richard Egel at Freiburg, also involved transposition, of the replicative type, and also invoked McClintock.[36]

In 1976, Ira Herskowitz, together with Jeffrey Strathern and James Hicks at the University of Oregon, proposed a "cassette" model of mating-type switching.[37] Their model contained transposition but not controlling elements. In it the *a* and *alpha* genes were distinct and transposable "cassettes" that inserted in front of a regulatory gene. Depending on which cassette was inserted into the regulatory site, the yeast cell would assume an *a* or an *alpha* mating type. Hicks, Strathern, and Herskowitz recognized Oshima and Takano's citation of McClintock but mentioned it themselves only to argue against the controlling-element model.[38]

Strathern and Hicks then took their yeast to Cold Spring Harbor, where they were joined by Amar Klar, another new staff scientist who had also been working on mating-type switching, in Seymour Fogel's lab on the Berkeley campus of the University of California. Nine months of the year, they did their experiments in a teaching lab. Summers, they were relegated to a steamy trailer, which shook visibly from the centrifuges whirring inside. From 1978 to 1980, the yeast group cloned the mating-type genes and clarified many aspects of the complex recombinations and rearrangements involved in mating-type switching. Thinking of both mating-type switching and controlling elements in terms of DNA rearrangements, the yeast group scientists engaged McClintock's interest and had many discussions with her.[39]

Insertion Elements, Plasmids, and Episomes

By 1976, then, scientists knew that insertion elements were found in episomes, plasmids, viruses, and chromosomes and that functional genes could be transferred by means of their cutting and pasting activity. They had seen phage mu jump in and out of the host genome like a transposable element. They recognized yeast mating-type cassettes as being similar to transposable elements. These all began to be seen as varieties of transposition. Further, transposition could be either conservative or replicative. It included movement of genes of known function, like antibiotic-resistance genes, and even insertion of foreign DNA not originally part of the chromosome, such as a virus. In premolecular days, these diverse phenomena could never have been related to one another. The commonality among them was at the molecular level. A shared DNA motif—the insertion element and particularly the inverted repeat—linked these disparate fields and suggested a common mechanism.

In 1976, these lines of research were brought together at Cold Spring Harbor, at a meeting entitled "DNA Insertions." Inspired by conversations at a meeting on plasmids at the Squaw Valley ski resort near Lake Tahoe, California, the previous year, the meeting was organized by Jim Shapiro, Sankhar Adhya, and Ahmad Bukhari.[40] Shapiro had been teaching genetics in Cuba and said that one impetus for his role in organizing the meeting was a realization that his early work was taking on broader significance. "I felt a need to reestablish my place in the insertion element field [because of] an absence of a few years," he said in 1999. McClintock was called in to advise, quite likely by Bukhari. "McClintock was there as a kind of icon," Shapiro said.[41] Few if any of the other participants grasped her vision of the genome. A few other classical geneticists who worked with movable elements were invited, such as Mel Green and Peter Peterson, probably at McClintock's suggestion. McClintock is listed on the program and is thanked in the preface of the published volume, but Shapiro did not recall whether she actually gave a talk, and she did not contribute a paper to the volume.[42]

At the meeting, a team of bacterial geneticists, led by Allan Campbell, met to codify the definition and nomenclature of mobile elements. David Botstein recalled there was "basically a weeny war of nomenclature."[43] While other geneticists had informally adopted the term "transposon" for mobile elements, since 1975, Botstein and Stanley Falkow had been using the term "translocon" (or "translocation element"), thinking it less interpretive. Botstein chaired the session "Drug resistance translocation elements," in which Jacob, Hedges, and others presented a paper entitled "Characterization of transposons determining resistance" to various antibiotics. Botstein allowed himself to be persuaded that "transposable" was really no more loaded than "translocatable." Transposable elements stuck, and in a sentimental gesture, the bacterial geneticists credited McClintock with coining the phrase.[44] It had actually been used in 1952 by Royal A. Brink, as a term less interpretive than McClintock's "controlling elements."

The group defined transposable elements as "DNA segments which can insert into several sites in a genome."[45] The definition is striking both for its generality and for its mechanistic connotations; a transposable element was now simply a movable piece of DNA. As shown by the hair-splitting debate over whether to call them transposable or translocatable, great effort was spent in ensuring that no interpretation seeped into the names and definitions. Of course, that itself was an interpretation. In making transposition a hot topic in molecular biology, the bacterial geneticists stripped McClintock's controlling elements of the functional connotations that were for her their most important quality.

Mobilization

The renaissance of McClintock's elements was gradual. Acknowledgments accumulated, connections were formed between disciplines, and along the way the McClintock revival reached several thresholds. Winning the Kimber Award in 1967 was certainly one such threshold. When the award was to be announced, Marcus Rhoades, the award committee chairman, asked geneticist Tracy Sonneborn to review the proposed press release. Sonneborn, a long-time McClintock admirer if not a close colleague, suggested inserting "that her studies of controlling systems of gene action were the precursors of the famous regulator-operon theory that won for its promulgators the Nobel Prize and that their thinking was probably much influenced by Barbara's notion of 2-member interacting controls."[46] Though Rhoades had reservations about her developmental regulation theory, he inserted the text. As I have shown, Jacob and Monod's thinking was not at all influenced by McClintock's work. The citation was an example of benevolent myth-making. Many more such examples were to follow.

In 1973, the work on transposition of antibiotic-resistance genes was yet to be published, but McClintock's star was clearly on the rise. Her old friend Ernst Caspari wrote to her in June, "I completely see that you are going to have many calls for many activities. You are, after all, very famous now, and it is rather astonishing how your fame has increased at a time of life where in most people it is on the way down." In Caspari's opinion, this interest showed that her ideas were far from obsolete and, in fact, that the younger generation of biochemists and molecular biologists were most receptive to them.[47]

Another threshold was a major review article on maize controlling elements, published in the widely read *Annual Review of Genetics* in 1974. The first author was John Robert Stanley Fincham, from the University of Leeds. Fincham had worked on unstable genes in snapdragons *(Antirrhinum)* and was familiar with McClintock's work. He apparently wrote McClintock in 1973 to ask her permission to write a review of controlling elements in maize. She recognized that her summary reports to the Carnegie Institution had in some measure failed. "The time has come when a more detailed exposition could be effective," she replied. "At least it would indicate the quantity and range of the studies that have contributed to summary statements appearing in my reports." She liked Fincham: he integrated. "Your capacity to comprehend and to integrate is conspicuously demonstrated in your Antirrhinum studies. They are sharply focused, clearly executed, and they come right to the point—no fuzziness!"[48]

Fincham's coauthor was G. R. K. Sastry, who had worked on maize controlling elements since the early 1960s and had been at Leeds since 1971. McClintock was less kind to Sastry. In 1971, she delayed ten months in replying to his request for seed. In the end she turned him down, writing that she was happy to send out seed "if I am assured that their use will lead to supportable conclusions. In this regard, I have had some trying, frustrating, experiences in the past. In these instances, the persons using my stocks have not paid attention to the published accounts of their modes of operation. Had they done so, they could not have made such unpardonable errors of interpretation."[49] If easy to incur, such ire rarely endured. Sastry soon returned to McClintock's good graces; after the 1974 Fincham and Sastry review she was friendly to him. Her irritation likely stemmed from the same quality in Sastry that makes the Fincham and Sastry review perhaps the clearest summary of McClintock's work: a tendency to simplify.

"The subject of this review is often regarded as complicated, difficult, and bizarre," the article began. It proceeded to lay out the wealth of findings, by McClintock and many others, on maize controlling elements. It is a long, masterly review that tackles not only *Ac* and *Ds* but also *Spm* with all its phase changes and modifiers and presetting and erasure. Fincham and Sastry demonstrated a deep understanding of the previous twenty-five years of McClintock's work, distilling it into clear principles. Yet they did not shy from critical appraisal. This was particularly evident in the section on whether controlling elements were native components of the genome or episomic invaders. They assumed controlling elements were selfish DNA, probably the result of viruses becoming integrated into the host genome. They did not deny that such invaders could lose their capacity for making infective particles and subsequently become incorporated in the apparatus of normal genetic control. But they conceded that such events should be rare and that "controlling elements ought only to play a rather peripheral role in regulation. Whether they are, in fact, involved in essential developmental processes is a question that cannot be answered at the present time, but the balance of argument seems rather against the idea."[50]

In 1976, McClintock was nominated for the Nobel Prize in Physiology or Medicine—doubly an honor for a plant geneticist. Her nominator was Judson John van Wyk, an endocrinologist at the University of North Carolina at Chapel Hill. Van Wyk had nominating privileges as a result of having spent the previous four years at the Karolinska Institute in Stockholm, which administers the prizes. He told me that in 1975 he had been having dinner with Gordon Sato, a geneticist who had spent several summers at Cold Spring Harbor in the 1950s, when Sato suggested to him that McClintock ought to have a Nobel Prize. "I didn't know that much about McClintock," he said, but he was familiar enough with her work that when Sato made the case, he agreed to submit the nomination. In the same nomination, he also put forward the name of Seymour Benzer, whose fine-structure mapping studies cracked open the bacterial gene in the late 1950s. Sato, van Wyk recalled, wrote an evaluation of McClintock's work that accompanied the brief formal nomination.[51]

The evaluation exaggerated her early influence, crediting her with "demonstrating, for the first time, that chromosomes possess distinctive morphological features that make them individually recognizable."[52] The individuality of the chromosomes, of course, had actually been known since the first decade of the century and was a cornerstone of Drosophila genetics. Beyond such overstatements, however, the evaluation demonstrated a subtle understanding of McClintock's work, especially her findings on controlling elements. It detailed her published results, noting the ability of controlling elements to influence gene action. Her work, the nomination read, "led to the first clear enunciation of the distinction between structural and regulatory genetic elements."[53] In short, she was nominated for her discovery of control—but not developmental control. In the wake of the operon, control and development had become separate. Her work, the evaluation continued, "laid the conceptual groundwork for the historic achievements recognized in the honor conferred on Professors Jacob and Monod by the Nobel Committee in Physiology or Medicine."[54] After initial review, however, the nomination was turned down. McClintock had not elucidated genetic regulation with the same precision or certainty as Jacob and Monod had done for bacteria.[55]

Over the next three years, transposable-element research broadened further. Cornell University yeast geneticist Gerald Fink discovered a transposon that affected the activity of a histone gene and published the results in 1979 and 1980. He called the unit Ty1, for "yeast transposon one."[56] Fink wrote in 1992 that his search for transposable elements in yeast was suggested by McClintock.[57] Other work from farther afield also tied in. Retroviruses, for example, enter the host as membrane-bound RNA. They immediately reverse-transcribe themselves into DNA, which is capped with a long repeated segment at either end, concluding in an inverted repeat. This DNA, which by 1980 was being compared to antibiotic-resistance transposons, is then inserted into the host genome.[58] Also, immunologists had discovered a complex system of gene rearrangements that led to the antigen-specificity of immunoglobulins. These too were considered transposable elements.[59]

Slowly, the big complex organisms of classical genetics became amenable to molecular analysis. Drosophila began to make a comeback, as did maize and other plants, this time as molecular model systems. As molecular techniques were developed for animal and plant models, the vast banks of genetic data on flies and corn made these organisms immensely valuable. Details of the molecular mechanism of transposition and the structure of mobile elements began to be gathered. Efforts were under way to clone and characterize McClintock's controlling elements, research carried out notably by Nina Fedoroff, a former student of Norton Zinder, who by the 1980s was running her own lab at the Carnegie Institution of Washington's Department of Embryology in Baltimore, and by Starlinger and Saedler in Cologne.

By 1980, the umbrella of transposition stretched so wide that transposable elements could not be ignored as a major new field of biology. A Cold Spring Harbor Symposium on transposable elements was "virtually unavoidable," wrote James Watson, now director of Cold Spring Harbor Laboratory.[60] Near-bankruptcy in the years following Demerec's retirement had forced the laboratory to remodel the symposium, once a forum for the vanguard in genetics, into a showcase for sexy fields guaranteed to draw large crowds and generate income. The 1980 meeting was titled "Movable Genetic Elements."

Mobile elements were by this time mainly a subfield of microbial genetics. The Cold Spring Harbor program included seven sessions on bacteria and phage, one and a half sessions on yeast, one on retroviruses, one on plants including maize, and half a session on Drosophila. The talks included all types of genetic rearrangement. They touched on important medical considerations, including antibiotic resistance, cancer (via retroviruses), and immunology. They even connected to the commercially and politically explosive area of recombinant DNA technology: man-made transposons were now ferrying genes from one species into another, making genetic chimeras and creating the possibility of biological factories for insulin and other medically important chemicals. Transposition had become a hot topic. It was also completely divorced from developmental control.

"The Katharine Hepburn of Science"

In 1980, McClintock, now pushing eighty years of age but, as ever, abreast of the current literature, linked her work to bacterial transposons. In the proceedings of the Miami Winter Symposium, she compared maize controlling elements to bacterial insertion elements, antibiotic-resistance transposons, phage mu, Ty1 in yeast, and the ends of DNA fragments in the ciliate protozoan Oxytricha.[61] There was no developmental control in the comparison. The parallels in this case were strictly mechanistic. McClintock, then, was responsive to the work being done in other fields. She quite likely wanted to associate her elements with trendy terms and concepts, but she also kept up with current research and allowed it to modify her thinking. Though she liked to think of herself as having worked all alone and having no influences, in fact she regularly drew upon her extensive knowledge of genetics and molecular biology, making connections between her work and that of other scientists.

Nineteen eighty-one was McClintock's big year. In June, she was made an honorary member of the Society for Developmental Biology and received the Thomas Hunt Morgan Medal of the Genetics Society of America. In October, the Wolf Foundation, based in Israel, awarded her their $50,000 prize in medicine. On 17 November, the John D. and Catherine T. MacArthur Foundation named her their first MacArthur laureate, a plum for senior scientists, since discontinued, which provided McClintock with $60,000 a year for life, tax-free. Earlier in the month, Marcus Rhoades, asked to evaluate her nomination, wrote the foundation that "Barbara McClintock is one of the intellectual giants of her time. I know of no individual more deserving of a MacArthur Prize Fellow Award. Her selection would be applauded by the scientific fraternity."[62]

The next day, the Albert Lasker Basic Medical Research Awards were announced, and McClintock was one of two winners (the other was Louis Sokoloff of the National Institutes of Health, for developing positive emission transverse tomography). Ability to predict the Nobel Prize is a staple of public affairs for biomedical research prizes, but the Lasker has perhaps the best average: in 1981, they boasted thirty-three subsequent Nobelists among seventy-one Lasker winners since the award's establishment in 1946. Indeed, in 1981 McClintock was nominated for the Nobel once again, this time by Stanley N. Cohen and Howard Temin. But the 1981 prize went to neurobiologists Roger Sperry, for his work on right- and left-brain specialization, and David Hubel and Torsten Wiesel, for their discoveries relating to the processing of visual information in the brain.[63]

Nobel or no, this spate of prizes made McClintock a celebrity. She, other scientists, and the press collaborated to create an appealing public myth. Newspapers wrote that she had "revolutionized genetics." She became a "geneticist whose work went largely unappreciated by the scientific community for many years." Interviewed by Lawrence K. Altman for the *New York Times*, McClintock said of her colleagues in the 1950s, "They thought I had flipped my top," but she had not been deterred. Altman reported that "she had persisted in her research because she knew she was 'going in the right direction' and that 'sooner or later I would be all right.'" Although surely nothing could have been further from her mind, medical and business applications were trotted out, to demonstrate the significance of her work: Transposition "contributed to the burgeoning recombinant-DNA technology," and "transposable genes" had "other important medical implications. They help explain how white blood cells can make antibodies so quickly in response to infection, and they may offer new insights about cancer." Her work was even presented as "the basis of today's research in gene exploration, such as gene splicing and human engineering."[64] Not one of the scientists involved in recombinant DNA, immunology, or cancer research claimed her work as a direct influence.

She proved an irresistible media figure: the boyish little old lady, dragged out of the solace of her laboratory, blinking up at the klieg lights, saying everyone had thought she was crazy, she knew all along she was right, she hated publicity, had no need for money. As the reporters fished for quotations, Kathleen Teltsch of the *New York Times* landed a good one and interpreted it correctly. Quoting McClintock as saying, "I'm 79 and at my age I should be allowed to do as I please and have my fun," the author explained, "By 'fun' she means the freedom to work as she always has, alone and independently, to eat and sleep as she chooses, or not at all, but most important, to experience the joy that comes with solving a scientific riddle."[65]

By the time the 1981 Nobels were announced, McClintock's friends and advocates were already mounting a new and wider campaign for the following year. One of the initiators was Herschel Roman, a geneticist at the University of Washington who had been a graduate student at Missouri in the late 1930s when McClintock was there. In September, apparently unaware of the previous nominations, he wrote to Marcus Rhoades, "It seems to several of us that Barbara McClintock should get or share in the Nobel Prize, now that transposons have been shown to be so widespread and so important. I just had a letter from Beets [George Beadle] and he agrees and would support a nomination. He suggests that you lead the charge."[66] Rhoades felt the letter would be stronger if it came from a Nobel laureate (in fact, unless Rhoades's home institution, Indiana University, had been designated as a nominating institution that year, Rhoades would not have been qualified to submit a nomination) and suggested James D. Watson.

Though a logical selection administratively, the choice of Watson had its problems. He and McClintock had had their differences. Watson became director of Cold Spring Harbor Laboratory in 1968, six years after sharing the Nobel with Francis Crick and Maurice Wilkins. He and McClintock lay at opposite ends of the spectrum of genetic style and temperament. Young enough to be her son, Watson epitomized the style of early molecular biology: bold, simplifying, and unsparing intellectually, gregarious yet intensely competitive socially.[67] Like McClintock, Rhoades, and Beadle at Cornell in the 1920s, in the 1940s and 1950s Watson's cohort cockily rejected the scientific tradition in the textbooks—including much of the cytogenetic tradition McClintock had helped establish. Everyone in the phage group respected McClintock, Watson recalled, but in a distant, detached way, as though she were no longer relevant.[68]

By the 1970s, Watson was nominally McClintock's boss. "Barbara was never at ease with directors," he said. "You never felt you were doing the right thing. I would ask her if we could come paint her apartment: 'No!' She would complain about having no secretary, but she certainly wouldn't let anyone read a letter she wrote." The secretary issue went on for years. McClintock would complain about having to do all her own typing and filing, but when she was offered a secretary, she refused. "I used to call her the Katharine Hepburn of science," Watson said. Indeed, the parallels are several: the temperamental disposition, the passionate commitment to her work—even the slacks. "You couldn't win with Barbara."[69]

Watson declined the invitation to nominate her, saying he was already committed to other candidates.[70] His demurral cannot be taken as lack of support—as Joshua Lederberg told me in 1999, "the weight of one's persuasion is going to be diluted if it's divided among several candidates."[71] Yet he might well have felt relief to be absolved of that awkward duty.

Lederberg, however, agreed to support McClintock's nomination. A Nobel laureate since 1958, he had admired McClintock since 1941, when he was a sixteen-year old college freshman and had sought her out on a cytological staining question. He had been astonished to receive some twenty pages of reply.[72] Also, he had cited her work in 1956 in his paper with Tetsuo Iino on phase variation in Salmonella. Roman wrote him on 1 October 1981, noting that both Beadle and François Jacob had agreed to "support the nomination in the event that transposons are recognized for the prize." Lederberg replied that "Barbara for Nobel is of course an excellent idea although my own view is that Ac/Ds was not even her most important work. I would be glad to endorse and forward a nomination that you and Marcus prepared." Other affirmations of support came in from Salvador Luria, Alfred Hershey, and Francis Crick. Rhoades wrote a summary appraisal of McClintock's research, which Lederberg attached to his nomination letter.[73]

Lederberg's nomination, informed by Rhoades's evaluation, noted the importance of "her work on the mechanism of control of gene action in maize, involving the action and interaction of two independent loci. This system was the forerunner of the regulator-operon theory of Jacob and Monod. She found that the transfer of controlling elements from one location to another in the chromosomes brought about a modification of gene action. This has opened up new vistas regarding the way in which chromosome structure may be altered and gene action controlled."[74] In the late 1990s, Lederberg did not believe McClintock's work was the forerunner of the operon theory and did not recall having said it was.[75] It is likely he did not believe it in 1981. The forerunner-of-the-operon argument was initiated by Tracy Sonneborn and propagated by Marcus Rhoades.

Other nominations came in, from Crick, Stanley N. Cohen, Heinz Saedler, and others. Several nominators argued that McClintock deserved a Nobel for her pretransposition work alone. Thus, they supported a sort of lifetime-achievement award. Although a science Nobel is supposed to be given for a single contribution, others—notably Max Delbrück and Salvador Luria, who shared a prize with Alfred Hershey in 1967—had won such lifetime-achievement Nobels.[76]

One scientist close to the nomination said he was not persuaded by either the lifetime-achievement or the forerunner-of-the-operon argument. He noted that McClintock's linkage of controlling elements to the operon was one of several attempts she had made to tie her work to current theories, from heterochromatin position effects in 1950 to bacterial and yeast transposons in 1980. Such arguments, he stressed, "should not receive more attention than the facts presented in McClintock's analyses." She had postulated, he said, "that the turning off, the turning on of genes and their modulation in expression by transposition of controlling elements in maize *might* signify normal events in differentiation," but that "neither she nor other maize geneticists have been able so far to come up with hard evidence for this" (emphasis his).[77]

Nevertheless, the reviewers were convinced that transposition was real and important. Somewhere amid the confidential lobbying, negotiations, and evaluations that precede Nobel Prizes, the argument switched from that which McClintock had been making for over thirty years to that which microbial geneticists had been making for eight. We will not know precisely how this happened until after 2033, the year McClintock's Nobel files will be opened. A good guess, however, is that it was a generational matter. The scientists making the forerunner-of-the-operon argument seem to have been those roughly of McClintock's scientific generation (though Lederberg is of the younger generation, he was the mouthpiece for a nomination initiated by Roman, Rhoades, and others). It may be that the bacterial molecular geneticists marshaled the more convincing argument.

In his nomination letter, Heinz Saedler wrote perceptively that McClintock's contribution "lay in the introduction of a completely new concept. In it, chromosomes are no longer rigid structures, but show flexibility and thereby allow reorganization of the genetic material." Shifting then to a discussion of transposable elements in bacteria, Saedler continued, "Such a reorganization of the ordering of genes is catalyzed by transposable elements. These genetic units not only jump from one place in the genome to another, they can also influence the expression of other genes; because of this McClintock at first named them 'controlling elements' . . . The 'controlling elements' were the first model system for the control of gene activity."[78]

Saedler's letter accomplishes several things. It renders McClintock's work reasonably accurately, describing transposition, influence on gene action, and McClintock's recent emphasis on genomic rearrangements. However, it omits all mention of developmental regulation by controlling elements, which by this time had been discredited. In so doing, it provides a logical basis for viewing McClintock as the forerunner of the bacterial conception of transposition.

The nomination letter, dated 26 January 1983, submitted by Ira Herskowitz and Bruce Alberts, reflects the completion of the shift in strategy for securing McClintock a Nobel. Both Alberts and Herskowitz were on the biochemistry faculty at the University of California at San Francisco. Herskowitz led the lab group that developed the cassette model of yeast mating-type switching; Alberts, today head of the National Academy of Sciences, used bacteriophage to study the mechanism of copying the DNA. The nomination form contains a box labeled "The nomination is based on the discovery of." In the box is typed "Transposable genetic elements." The nomination begins, "Barbara McClintock discovered that discrete pieces of genetic information are capable of movement from one chromosomal location to another and that this movement of 'transposable elements' can lead to major alterations in gene expression." There is no mention of gene regulation or of the control of development of differentiation. Herskowitz and Alberts mention recent work that had "shown that transposable genetic elements are present in essentially all organisms." They go on to note transposition's implications for cancer research, yeast genetics, and genome evolution. In short, the nomination presents genetic transportation as universal, of basic importance to organisms, and of great significance to medicine. It has little in common with what McClintock herself had been arguing now for decades.

One may hypothesize, then, a scenario in which McClintock's first Nobel nominations were dominated by her friends and peers, who rather anemically adopted McClintock's own strategy of 1961, linking controlling elements to the operon and attempting to establish her as the founder of genetic regulation. By the early 1980s, younger molecular geneticists came to influence the nominations more. They adopted a new strategy, leaving out the weak link to gene regulation and emphasizing transposition and genomic rearrangement.

The Prize

McClintock nearly shared the prize with an immunologist. In the mid-1970s, Susumu Tonegawa, then at the Basel Institute for Immunology in Switzerland, began working on the problem of antibody diversity. Antibodies are complex protein molecules that are the source of the body's immunity to foreign toxins and infections. Located on the surface of the B cells of the immune system, each antibody responds to a single antigen, or foreign substance. Vaccines work by stimulating production of antibodies against a given germ. The riddle of the immune system was that it seems able to produce antibodies to any antigen—yet the potential antigens in the world far outnumber the genes possessed by any animal, even if the entire genome coded for nothing but antibodies. How does the genome encode antibody diversity?

The answer, strangely enough, is to have many genes encode each antibody. Antibodies are modular: each comprises two heavy chains and two light ones, and each chain contains a variable region and a constant region. In the mid-1960s, William Dreyer of Caltech and J. C. Bennett of the University of Alabama School of Medicine recognized that a huge number of different antibodies could be made by a relatively small number of genes if these different regions could hook up in any combination.[79] A decade later, Tonegawa found the first direct evidence supporting this combinatorial hypothesis. Through the late 1970s, several groups pursued this problem, but Tonegawa's group repeatedly positioned crucial pieces of the puzzle. Tonegawa presented their model at the 1980 Cold Spring Harbor Symposium on mobile genetic elements. His published paper did not cite McClintock.

Genetic rearrangement was the key to Tonegawa's model. The antibody genes are located near one another on the chromosomes. During the maturation of a nascent B cell, the genes rearrange into new combinations. The genes are not picked up and moved, in the manner of controlling elements; rather, the intervening sequences are cut out and the DNA is stitched back together. The number of variants for each segment ranges from several to hundreds, making for tens of thousands of possible light and heavy chains out of a small number of genes. Since any light chain may associate with any heavy chain, the possibilities escalate exponentially. The number of possible antibodies is in the hundreds of millions.[80]

Tonegawa did link this research to work on insertion elements, and also to RNA splicing. Splicing was discovered simultaneously in 1976–77 by Richard Roberts and colleagues at Cold Spring Harbor and by Phillip Sharp and colleagues at the Massachusetts Institute of Technology. Roberts and Sharp recognized that DNA genes are often much longer than the messenger RNAs produced from them. Intervening segments, called introns, are snipped out of the RNA, bringing together the other regions, known as exons. Tonegawa connected his work to the splicing and insertion-element literature but not to controlling elements. The linking by Nobel nominators of McClintock's controlling elements with Tonegawa's immune system rearrangements demonstrates how mechanistic scientists' interpretation of her findings had become, and how genetic control was swept aside in the campaign to award her a Nobel Prize.

Tonegawa was nominated to share the Nobel with McClintock in 1982 but by the final decision his name had been removed (he had to wait five more years). The 1982 prize went for prostaglandins, but McClintock's nominators tried again the following year. This time, success. On 10 October 1983, McClintock received a telephone call notifying her that she had won that year's Nobel Prize in Physiology or Medicine. The citation read "för hennes upptäckt av rörliga strukturer i arvsmassan"—for her discovery of mobile genetic elements.[81]

McClintock thus joined Marie Curie (1911, chemistry) and Dorothy Crowfoot Hodgkin (1964, chemistry) in the superelite of women who have received unshared Nobels in science. McClintock is further distinguished by the small number of nominations she received before winning—probably only four. In 1987, the week before Tonegawa was announced as that year's Nobel laureate in Physiology or Medicine, *Science* ran a story on the Nobel archives and reported that many eventual winners were nominated dozens of times, often spanning a period of decades, and that many of the most-nominated never won. Once scientists had recast

McClintock's major contribution as transposition rather than control, she won science's most prestigious honor quickly and easily.[82]

The morning of 10 October, she gave a press conference in Cold Spring Harbor's Bush Auditorium, the room where Watson had presented the double helix in 1953. The cameras gave most Americans their first glimpse of McClintock, eighty-one years old, with salt-and-pepper hair, in sturdy shoes, khaki slacks, plain white shirt, and sweater. She endeared herself to many when asked what she would do with the prize money. "I don't even know what the award brings in," she said. It was approximately $190,000. "Oh, it is," she said, surprised. The room rippled with laughter. "No, I didn't know and I'll just have to get to one side and think about this."[83]

Why did she think it took so long for her to win the award, they asked her. "Oh, it was quite all right. It took long, I think, because nobody had the experience I had." She continued with a quick sketch—not very clear—of her discovery and then said, with characteristically quirky syntax and grammar, "Now, the whole point of my going on with it was the regulation part, not so much the transposition, but the regulation, because at that time we had no DNA to work on and we had no regulation about genes. We weren't concerned with gene regulation at the time. We weren't ready. But, this was telling about gene regulation, so I put the emphasis on these and called them controlling elements."[84]

Re-evaluation

By 1983, transposition, the existence of which was never in serious doubt, was of uncontested importance. Geneticists had shown, first, that transposition occurred in organisms widely separated by evolution—corn, flies, yeast, bacteria. This diversity suggests either that transposition evolved many times and therefore is essential to life, or that it arose very early in evolution and thus was a basic property of most or all organisms. Second, biologists could now explain transposition in molecular terms. Repulsive forces and sticky chromosomes had long been replaced with inverted repeats and sequence-specificity. Third, transposition, broadly defined, was clearly a property of genomes that could be put to many uses.

The old vision of the genome was as a dusty genetic archive, from which volumes were periodically drawn as reference materials for the active patrons of the cytoplasm. Geneticists were realizing that the genome is more like a bustling train station. It is full of switches and turnstiles, with proteins coming and going at all hours, picking up and dropping off information to and from innumerable destinations. Molecular biologists found transposition to be a powerful, flexible means for orchestrating these comings and goings.

Yet as Brink argued in 1960, as the Nobel committee insisted in the 1980s, as interview subjects recalled in the 1990s, McClintock never showed that the control processes she observed occurred normally during plant development. The biological community never underwent a change of opinion on McClintock's elements; they had always accepted transposition and rejected McClintock's developmental interpretation.

The generality of transposition, Joshua Lederberg said, enhanced the significance of McClintock's work and "enabled it to be disentangled from the regulatory-theory baggage that went along with it."[85] In this light, the question of why it took so long for McClintock to win the Nobel Prize for transposition has an almost trivial answer. Her Nobel followed the publication of insertion elements by fourteen years—about the average interval between a discovery and its Nobel Prize.

James Shapiro, who has a deep understanding of McClintock's late career and personality, believes the Nobel was bittersweet for her. "She told me she was fortunate this happened to her late in life," he said. For her personally, the Nobel "meant a tremendous disruption of the very private life she had constructed for herself." Though of course she was glad for the scientific recognition, Shapiro believes she was pained by the fact that the recognition missed her central point. He recalled, "She told me she wasn't interested in transposition. She was interested in regulation."[86]

McClintock recognized that her history was being rewritten. She knew her colleagues still did not accept her vision of genome regulation. She disliked being made into a feminist icon of intuitive or holistic (read: nonrational) science. She hated publicity and crowds. Through it all, however, she maintained her wry, self-effacing humor. A photograph taken at a Cold Spring Harbor Laboratory function not long after the Nobel shows her milling about the crowd "incognito"—with Groucho Marx nose, glasses, and mustache.

Publicity following the Nobel Prize referred to McClintock as a "modern Mendel."[87] But analogy and metaphor—reasoning by pattern—operate at several levels. In which sense, precisely, does McClintock warrant comparison with the father of genetics? In chronology, the parallels are strong. Gregor Mendel died in 1884, two years after Charles Darwin. His rediscovery came in 1900, thirty-four years after publication of his paper on hybridization in peas. Had he survived even a few years into the twentieth century, he might well have won a Nobel Prize. McClintock won the Nobel thirty-three years after publication of her 1950 *PNAS* paper, or thirty-five after the first mention of transposition in print, in the 1948 Carnegie Institution *Year Book*.

In social terms, the parallels break down. Mendel seems genuinely to have been ignored and misunderstood. His work was rarely cited and was, until after 1900, unknown to the Darwinians, the dominant group of scientists studying heredity in the mid- to late nineteenth century. It was not grasped even by Carl Nägeli, to whom Mendel described his findings in detail. Moreover, when Mendel's work was rediscovered, it recovered intact. Had he lived and won a Nobel Prize, say "för hans upptäckt av de lagar som styr ärftliga egenskapers utklyvning och oberoende nedärvning" (for his discovery of the hereditary laws of segregation and independent assortment), he would have recognized the citation as an acknowledgment of his most important work.[88] The same cannot be said of McClintock.

Many of McClintock's colleagues, especially Rhoades and his students, but also R. A. Brink, Peter Peterson, Oliver Nelson, and others, clearly understood her argument. In the early and middle 1960s, among the mainstream in heredity studies (that is, bacterial genetics), all of the major players knew of her, owing in part to her presence at Cold Spring Harbor and participation in the annual symposium. By the end of that decade, however, and into the next, many young geneticists working on movable elements read her papers only after doing the experiments that showed parallels with them. McClintock was no more, and possibly less, forgotten than the other great cytogeneticists of the 1930s and 1940s. Her renaissance, moreover, resulted directly from a new generation looking at her work with fresh eyes. They recognized transposition as amenable to dissection with molecular techniques.

Unlike Mendel, McClintock participated directly in her own renaissance. With Larry Taylor and later Ahmad Bukhari working on phage mu, at the Cold Spring Harbor meeting on DNA insertions, with the Cold Spring Harbor yeast group, and with Gerry Fink and the yeast Ty elements, McClintock discussed and suggested parallels between her work and theirs. In most cases, this was simply because the young investigators involved happened to be where she was. The centrality of Cold Spring Harbor in the molecular biology community, as both a research and a conference center, can hardly be overemphasized as a factor in the McClintock renaissance.

McClintock may not have been a modern Mendel, but McClintockism is a modern Mendelism. Mendelism as it is taught today consists of two clear "laws," segregation and independent assortment, and the principle that each parent contributes one copy of each trait to the offspring, who express the trait in accordance with the principles of dominance and recessiveness. Mendel clearly understood segregation and assortment. But the package of Mendelism, as an experimental and teaching tool, was created by de Vries, Correns, von Tschermak, Bateson, Johannsen, and others.[89]

Similarly, the discovery of the mechanism of genetic recombination for which McClintock is primarily remembered is largely the work of Shapiro, Adhya, Malamy, Saedler, Starlinger, and Bukhari. Like Correns and de Vries, they discovered the phenomenon before they discovered their predecessor's papers (or at least before realizing the relevance of those papers to their own research). And like them, they rapidly drew in colleagues and built an intellectual edifice based on their own findings, then graciously elevated their predecessor to the status of forerunner.

McClintock was not the first scientist to receive a Nobel Prize for work other than what she considered her primary contribution. Contrast with other examples is revealing. The 1903 prize in physics went to Pierre Curie and Marie Curie, "in recognition of the extraordinary services they have rendered by their joint researches on the radiation phenomenon discovered by Professor Henri Becquerel." The Curies had discovered radium, a new chemical element with properties so startling that it transformed branches of both physics and chemistry as well as medicine and industry. Pierre Curie was a physicist, well-known to the physics community, and was proposed alone. When notified he was being considered for the prize, he gently insisted it would be "more satisfying, from an artistic point of view," if he shared the prize with his wife. But radium was so important that chemists and physicists both wanted a piece of it. Chemists in the Swedish academy argued that radium was a chemical discovery. Giving the physics prize to the Curies for their researches on "the radiation phenomena" left the door open for a second prize for radium (and polonium), in chemistry, which Marie won eight years later, after Pierre's death. In Marie Curie's case, then, the citation was hedged, not to deny the significance of what Curie and everyone else knew was her major achievement, but to spread among two communities the privilege of honoring her discoveries.[90]

Albert Einstein's case is more complex. Einstein won the 1921 prize (awarded in 1922) "for his services to Theoretical Physics, especially for his discovery of the law of the photoelectric effect," not for relativity. Einstein had been nominated every year but two since 1911. He articulated his theory of special relativity in 1905, the photoelectric effect shortly thereafter. General relativity was not enunciated until 1915, and the bending of light during a solar eclipse, the first prediction of relativity to be borne out, was not observed until 1919. Thus, unlike McClintock's theory of developmental regulation, relativity was still new at the time of Einstein's prize. In 1921, the Nobel committee solicited two evaluations of his work, one for relativity, one for the photoelectric effect. The former was poorly written and inaccurate; the latter, written by Svante Arrhenius, the scientist with the greatest influence over the outcome of the physics prize, was concise, politically expedient, and sensitive to the benefits of the physics community. On 10 November 1922, Einstein received a telegram notifying him that he had been awarded the 1921 prize in physics, "but without taking into account the value which will be accorded your relativity and gravitation theories after these are confirmed in the future." The telegram clearly leaves room for a second prize, for relativity (though it never came). The physics community thus acknowledged the import of relativity, but the Nobel committee—being a famously conservative body and feeling pressure to award Einstein a Nobel—honored him for the better-established discovery.[91]

Like McClintock after him, Einstein received the prize for universally admired work that preceded what he himself considered his major contribution. Yet in a sense, Einstein's case is the mirror image of McClintock's. Few in 1922 would have denied the significance of relativity. Einstein was and remains known principally for relativity, whose implications touched philosophy, psychology, and popular culture. In other words, Einstein did not receive the prize for his best-known work, while McClintock did.

Perhaps the closest analogy is to Peyton Rous, the Rockefeller Institute pathologist who in 1911 discovered a "filterable virus" that caused tumors in chickens. At the time, the idea that cancer could be induced by an infectious agent seemed a throwback to a long-held theory of tumors as inflammations, caused by infection. In the 1910s, the flush of the germ theory of disease, viruses were thought of as tiny bacteria; bacteria caused infections. Cancer pathologists had recently come to the view that tumors were internal in origin, overgrowths of cells dividing out of control. Rous's colleagues rejected his findings, not as wrong but as anomalous and therefore trivial, because they undermined a powerful and broadly true generalization.

In the 1930s, viruses began to be understood as genetic particles. When the link was made between viruses and cancer as a genetic disease—by W. E. Gye, of the National Institute for Medical Research, in London, Richard Shope, of the Rockefeller Institute for Medical Research, and others—Rous's tumor virus began to be resurrected. By the 1950s, virus research had become one of the hottest areas in microbiology and genetics. Renato Dulbecco's development in the late 1950s of techniques for culturing animal cells, as well as other new techniques, opened the door from bacterial viruses to animal viruses as a genetic system. Viruses in general entered the limelight through events such as the Salk and Sabin polio vaccines. Cancer viruses became acceptable too when other viruses were discovered that caused tumors. In 1964, the National Cancer Institute initiated a virus tumor research program and spent nearly $4 million on virus research, while the American Cancer Society spent another $1.8 million. Meetings, courses, and symposia were held on tumor viruses, and new workers were brought into the field.[92] Virology and the genetic theory of cancer, once at odds, merged. It is now standard for textbooks and introductions to collections of research papers to include a reference to Peyton Rous.[93]

Both McClintock and Rous, then, presented results that few doubted but that seemed anomalous in the context of a sound genetic model. In both cases the model has proved robust, but exceptions and complexities have been allowed in. The tidal wash of molecular genetic theory has dissolved the apparent anomalies, embracing both viral cancer and transposition as illuminating exceptions to genetic stability. Such shifts, from what a researcher believes her work means to the interpretation that becomes canonized as established knowledge, may be more common in science than is generally acknowledged. Science is as ruthless as nature in selecting what is useful and rejecting what is not. McClintock's case is simply a particularly clear example of a process that is likely quite common.

10

SYNTHESIS

It's all one.

—BARBARA MCCLINTOCK

Although molecular studies have not confirmed McClintock's theory of developmental regulation by controlling elements, some studies point to an emerging molecular view that is consonant with her vision of the unity of nature. That vision, arresting both in its sweep and in its flouting of conventional biological thought, speculative, unfalsifiable, fundamentally unscientific, establishes McClintock as one of the few biological thinkers who have grasped the connections among growth, heredity, and evolution.

Developmental Control

McClintock's body of research is remarkable in the extent to which molecular biologists continue to mine it for new ideas and experiments. Recently, researchers have found surprising connections between disparate parts of McClintock's corpus. One such connection is between transposition and telomeres, the tips of chromosomes. In her investigations of ring chromosomes and the breakage-fusion-bridge cycle, McClintock discovered the special nature of the telomere. A broken end, she found, tends to fuse with another broken end; a natural chromosome end does not. Hermann J. Muller gave the telomere its name in 1938, deriving it from the Greek for "end-segment."[1]

Since the 1970s, molecular biologists have learned an enormous amount about telomeres and telomere function. Because of a technicality in the mechanism of copying DNA, the very end of the telomere fails to be copied. This results in a progressive shortening of the telomere as the cells divide. Evidence suggests this shortening may act as a cellular clock, ticking off the cell divisions in a given cell's life span, until it reaches a point known as the Hayflick limit and stops dividing. Some organisms and cell types, though, maintain telomere length, thanks to an enzyme called telomerase, which adds telomere sequences back on to the end.[2]

Drosophila, however, uses transposable elements to maintain telomere length. Mary-Lou Pardue, a drosophilist at the Massachusetts Institute of Technology, and others find that mobile heterochromatic elements called HeT-A and TART accumulate at the chromosome tips to form telomeres. One element, HeT-A, accumulates at broken chromosome ends and thus appears to be related to chromosome healing in Drosophila.[3] Thus, in flies (though not in maize) the chromosome healing McClintock described in the late 1930s and early 1940s is explained by tiny movable bits of heterochromatin similar to what she described in 1950.

This is just one example of an emerging set of connections between mobile elements and heterochromatin. In 1950, McClintock believed controlling elements to be made of heterochromatin. Today, evidence suggests that much heterochromatin seems to be made of mobile elements. Together, mobile elements and heterochromatin compose from 15 to 30 percent of the genome in flies and humans. It is now well-established that mobile elements tend to collect in heterochromatin. At Cold Spring Harbor, Robert Martienssen, Richard McCombie, and colleagues recently sequenced a complete chromosome knob from the plant Arabidopsis. In the knob heterochromatin, they found *Ac, Spm,* and mu transposons.[4]

In 1950, McClintock would have explained this result by supposing that the mobile elements were naturally part of the heterochromatin; when she found mobile elements elsewhere, she assumed they came *from* the heterochromatin. Most researchers today, however, explain the accumulation of mobile elements in heterochromatin as a side effect. Because heterochromatin has few genes, random insertion of a mobile element into it is less likely to cause a lethal mutation than if the element hops into gene-laden euchromatin. The elements need not have a preference for heterochromatin. In short, in the minds of most molecular biologists today, Fincham and Sastry were right when they wrote in their 1974 review that controlling elements are selfish DNA.[5]

Some scientists today, however, point to evidence that at least some mobile elements may indeed have a preference for heterochromatin. Further, heterochromatin is not the junk DNA it was once thought to be. It does contain some genes, and it is important in vital cell processes such as chromosome pairing and segregation. What is more, mobile elements have been shown to play a role in genome rearrangements. It is at least plausible, then, that mobile elements are essential in rearranging heterochromatin. This could be a means of differentiating new species from one another, by making hybrids infertile or inviable. Other workers have suggested recently that transposable elements may contribute to the formation of knobs in maize chromosomes, and that the knobs may shape "meiotic drive," the subversion of meiosis such that certain genes are transmitted preferentially to progeny.[6]

Corroboration even for some of McClintock's more farfetched assertions is coming in from comparative genome studies. In the 1970s, she began saying that any organism can make any other. In October 1998, Nicholas Wade, writing in the *New York Times,* reported on results discussed at a recent meeting at the Field Museum of Natural History in Chicago. Humans and chimpanzees differ in just 2 percent of their DNA; a Colorado company argues that as few as a couple hundred genes may distinguish us from them. Further, many chimp genes differ from their human counterparts by only a few nucleotides out of thousands; others, though they may have less overall homology, clearly are related evolutionarily to human genes. In other words, it is only a modest exaggeration to say that a chimp genome contains the information to make a human being.[7]

The differences between us and them, some researchers say, are mainly in the regulation of the genes, not the gene structure. For example, David L. Nelson, a biologist at the Baylor College of Medicine, has asserted that the discrepancy in brain size between us and chimps could be accounted for simply by a mutation that allowed one further doubling of the developing brain cells. Most startling, at least some of the differences may have come from position effects. In a 1998 paper, Nelson and Elizabeth Nickerson showed that five chimpanzee chromosomes contain regions identical to human chromosome segments—but inverted. "Genes on the flipped portion would find themselves in a different environment and might be more or less active than before," Nelson explained. If these were regulatory genes, whose protein products control other genes, they could have a large effect. "Even a small difference in the activity of a high-level regulatory gene could produce significant effects in the developing animal."[8]

Regulation, position effects, the idea that one organism can make another—the story is strikingly close to McClintock's vision of developmental and evolutionary change. Allowing for the hyperbole McClintock often used just to be provocative, and for the refinements of interpretation that come with deeper understanding, McClintock had a remarkable batting average. Certainly, no serious biologist today would disagree that the genome is indeed dynamic, as Goldschmidt and McClintock began arguing in the late 1930s, that genes interact with other chromosomal elements, that the chromosomes interact with the cytoplasm, that both cell and cytoplasm are influenced by the world outside the cell. Having dissected the apparatus and flow patterns of biological information, biologists are now reassembling them, finding connections among them; some of the most exciting work in molecular biology today consists in synthesis, integration.

On Growth and Form

Compared with the reductionist, simple-systems approach to genetics that characterized the mid-twentieth century, this view of life is so different that it may seem revolutionary. In fact, it has resulted from a gradual evolution toward increasing complexity. Modern molecular genetics is constantly describing new gene-gene, gene-cell, and gene-cell-environment interactions. When these converge, resulting in stories like those above, contemporary biology seems to echo the unity of nature expressed by nineteenth-century naturalists. Charles Darwin closed the *Origin of Species* with this poetic passage, summarizing the view of life that emerges from his theory of evolution:

It is interesting to contemplate an entangled bank, clothed with many plants of many kinds, with birds singing on the bushes, with various insects flitting about, and with worms crawling through the damp earth, and to reflect that these elaborately constructed forms, so different from each other, and dependent on each other in so complex a manner, have all been produced by laws acting around us. These laws, taken in the largest sense, being Growth with Reproduction; Inheritance which is almost implied by reproduction; Variability from the indirect and direct action of the external conditions of life, and from use and disuse; a Ratio of Increase so high as to lead to a Struggle for Life, and as a consequence to Natural Selection, entailing Divergence of Character and the Extinction of less-improved forms. Thus, from the war of nature, from famine and death, the most exalted object which we are capable of conceiving, namely, the production of the higher animals, directly follows. There is grandeur in this view of life, with its several powers, having been originally breathed into a few forms or into one; and that, whilst this planet has gone cycling on according to the fixed law of gravity, from so simple a beginning endless forms most beautiful and most wonderful have been, and are being, evolved.[9]

Darwin's choice of metaphor, the entangled bank, with all the elements of nature interwoven, expresses elegantly his effort to understand the whole of living nature. Further, his choice of language—the laws, especially growth with reproduction, inheritance (almost implied by reproduction), variability arising in part from use and disuse—reflects Peter Bowler's description of nineteenth-century evolutionary thinking as characterized by an analogy with growth. In this tradition, development (change within the organism) is used as a model for evolution (change within and among species and higher taxa).

Bowler points out that this analogy leads to the false conclusion that, just as exercise leads to bigger muscles and better individual fitness, so learning, strengthening, and stretching must lead to better evolutionary fitness. In other words, characters acquired during an individual's lifetime may become heritable, part of the evolutionary legacy. Also, the analogy with growth suggests that, as an embryo follows a largely fixed program through development, so a species must evolve according to a fixed program. These two concepts, the inheritance of acquired characteristics and progressive evolution, are typically associated with the pre-Darwinian evolutionist Jean-Baptiste Lamarck. The analogy with growth persisted in the minds of some of the nineteenth century's most celebrated evolutionists, not only Darwin himself but August Weismann, who has been credited with "completing" the Darwinian revolution.[10]

Ernst Haeckel employed this analogy particularly forcefully. In his 1866 book *Generalle Morphologie der Organismen,* and elsewhere, Haeckel popularized the concept of developmental recapitulation. He observed that vertebrate embryos pass through a common series of stages that correspond to the evolutionary succession of the various groups. Early embryos of land animals resemble fish, with gill slits that later disappear. For a time, developing human fetuses have a tail, just as did our primate and earlier mammalian ancestors. Ontology—that is, development—seems to recapitulate the evolutionary history. This argument of recapitulation leads to a progressive view of evolution, because it assumes a linear sequence of vertebrate stages: fish → amphibians → reptiles → birds → mammals → humans. Although Haeckel called himself a Darwinian, Bowler considers him a pseudo-Darwinian, because he did not adopt one of Darwin's most fundamental propositions: that evolution is branching.

Haeckel's formulation is a predicate syllogism, a deductive argument based on the analogy of predicates rather than subjects. Gregory Bateson, an anthropologist and philosopher and the grandson of William Bateson, gives a simple example: Men are alive; grass is alive; therefore, men are grass.[11] Strictly logically, predicate syllogisms are nonsense. But Bateson noted that predicate syllogisms are the basis of much poetry. We thrive in tough places, at times we may feel as though we are forcing our way through cracks toward light and space.

This is not the style of reasoning we associate with science. Experiments must follow classical logic. But Robert Scott Root-Bernstein has shown that discovery, the moment of insight that precedes the experiments that test an idea, often occurs by metaphoric leaps. Science is distinguished from art by having a means—experiment—by which to test the value of its metaphors.[12] Haeckel reasoned that evolution is change; development is change; therefore, development is like evolution. His error was reversing cause and effect: he believed the developmental process encoded an evolutionary program, which it expressed in each embryo. In fact, evolution uses development. It sculpts developmental programs—shifting timing or rates of change, moving starting or ending points—to mold organisms to a changing environment.[13]

Bowler argues that Mendelism enabled biologists to frame evolution without the analogy with growth. Mendelism's "unit characters" provided an alternative conception of heredity that was particulate and nondevelopmental. Although at first many saw Mendelism and Darwinism as antagonistic, Mendelism divorced evolution from development and laid the foundation for the reintegration of evolution and heredity.[14]

Mendelians and old-fashioned zoologists and botanists shared a tendency to atomize nature. D'Arcy Thompson, in *On Growth and Form*, his classic book of 1917, expressed with characteristic grace the opposite, integrated view of nature. He noted that both the Mendelian and the morphologist tend to break the organism into parts, and he described the problems inherent in doing so.

With the "characters" of Mendelian genetics there is no fault to be found; tall and short, rough and smooth, plain or coloured are opposite tendencies or contrasting qualities, in plain logical contradistinction. But when the morphologist compares one animal with another, point by point or character by character, these are too often the mere outcome of artificial dissection and analysis. Rather is the living body one integral and indivisible whole, in which we cannot find, when we come to look for it, any strict dividing line even between the head and the body, the muscle and the tendon, the sinew and the bone. Characters which we have differentiated insist on integrating themselves again; and aspects of the organism are seen to be conjoined which only our mental analysis had put asunder.[15]

Thompson developed a powerful morphological tool for re-establishing the integration of animal forms. By tracing biological structures—skulls, limbs, fruits—on a grid, he demonstrated the unity of disparate forms. One form may be converted into another, quite different, by imagining the grid to be imposed on a sheet of rubber. Stretching or deforming the sheet transforms a basic vertebrate arm into a bat's wing, a bird's wing, a whale's flipper. Thompson argued that evolution often works this way. The stretching occurs by means of relative changes in timing and rate of growth, lengthening one bone or shortening another, or altering the developmental curve of a whorl of leaves or scales.

Thompson's argument is based on pattern. His coordinate grids are a means by which to compare patterns among different biological forms. Thompson understood that grasping the underlying pattern enabled the scientist to perceive the integration of the natural world; it was only the human mind that had differentiated that world. McClintock would have agreed emphatically with Thompson that "characters which we have differentiated insist on integrating themselves again."

Barbara McClintock's scientific career, too, is a story of integrating again that which had been differentiated. Her example is illuminating because of her deviations from the mainstream history of twentieth-century biology.

During the modern evolutionary synthesis of the 1930s, when heredity and evolution were integrated, McClintock was unconcerned with evolution. The modern evolutionary synthesis took place outside the domain of the cell. Sewall Wright, Ronald A. Fisher, J. B. S. Haldane, and the other architects of the synthesis accepted the string-of-pearls model of the gene. They held the gene constant within the individual in order to look at fluctuating allele frequencies among individuals.

The research program of the German geneticists in the 1930s was an attempt to integrate heredity with development. Theirs was a question of gene action: how does the activity of the genes modulate to produce the changing form and different structures of the organism? In this case, the domain was the cell, and the string-of-pearls gene was a limitation. Mendelism gave these early developmental geneticists no means to modulate gene action; genes were always "on." The developmental geneticists looked outside the nucleus to the cytoplasm for modulators of gene action. Although for a time in the late 1940s this line of research seemed poised to integrate biology, the grand vision of the cytoplasm or plasmagenes as the modulator of gene action has not been sustained.

McClintock's exposure to this German school, when she was in Berlin and continuing via her enduring friendship with Richard Goldschmidt, primed her to think in terms of gene action, to see pattern as the result of an underlying control. Her cytological background engendered in her an interest in the landscape of the chromosomes and prepared her to think of nongenic chromosomal elements. The combination resulted in a scientist unwilling to accept the conventional theory of the gene as a unit of mutation.

Yet when developmental genetics came to America, McClintock was not part of it. In this country, it split into cytoplasmic inheritance studies, exemplified by the work of Tracy Sonneborn, and biochemical genetics, à la Beadle and Tatum. McClintock was aware of this work and friends with the workers, but she did not fit into either camp. Her nucleus-centered view excluded her from the cytoplasmic inheritance group. She always sought chromosomal sources of regulation of gene action. And her lack of interest in biochemistry kept her from being integrated into the stream that led from the one gene–one enzyme model through the operon.

As long as the action of the cytoplasm could be described only in terms of mysterious influences and principles, cytoplasmic inheritance studies were open to charges of mysticism. But with the development of molecular biology—the double helix, messenger and transfer RNA, the genetic code—heredity began to be understood in terms of information. The genes were no longer particles that contained traits. They became information-storage devices. The early-twentieth-century debate over the relative importance of the nucleus and cytoplasm dissolved as molecules were discovered that shuttled between the two compartments, carrying hereditary information.

The modern developmental synthesis, then, came in the 1960s and 1970s. Operons were found in higher organisms. Ways were found to identify gene activity within the cell. Genes and proteins were identified that were crucial to developmental stages. Embryology went molecular, and became developmental biology.[16] The nucleus and cytoplasm were seen no longer as competing compartments but as integrated, interacting domains.

Ironically, just at this time McClintock had her personal evolutionary synthesis. The races-of-maize study expanded her vision from development to evolution. When she integrated this work with her ongoing studies of *Suppressor-mutator*, the result was a resurrection of the analogy with growth. She saw development as the expression of an unfolding of a pattern, and evolution as a shift from one pattern to another. In a sense, she was back to Darwin and D'Arcy Thompson, although she had arrived there, not by the path of natural history, anatomy, or even embryology, but rather by that of cytology and genetics.

McClintock's was not a nineteenth-century style. Unlike Haeckel, in applying the analogy with growth McClintock was not led to accept either the inheritance of acquired characters or a progressive view of evolution. Her understanding of genetics precluded the former; her avoidance of the latter is more complex. McClintock made the analogy with growth by means of pattern. She saw macroevolution as a shift from one developmental pattern to another. Thus, her conception of evolution remained of the branching sort. Nothing in evolution was inexorable; species were not ordained to execute a predetermined program. In her conception, the programs were responses to evolutionary forces (that is, natural selection). To McClintock, evolution was not the unfolding of a program; developmental programs were units of evolution.

Myth Redux

According to the myth described in Chapter 1, McClintock is a marginal, even deviant figure in the history of genetics. In that myth, her marginality is framed socially. Transposition is imagined to be the idea no one could understand; both transposition and McClintock herself are said to have been rejected by the genetics community. Gender, it is argued, gave McClintock an intuitive, empathic, feminine scientific style that enabled her to arrive at the concept of transposition and that colored other scientists' reception of her work.

Close examination of McClintock's research and writing reveals that she was far more deviant intellectually than the myth allows. By the mid-1940s she intended her work to overturn established dogmas. Further, reframing her work in the broader view of the reintegration of the problems of growth and form reveals the originality of her work—of both her correct ideas and her incorrect ones. The fact that she was not ignored by geneticists, that, contrary to the myth, she continued to win recognition, honors, and awards through the 1950s and 1960s, underlines her true marginality. McClintock was not marginalized because she was a woman. She received immense respect from geneticists in spite of her gender. She marginalized herself, through her deeply subversive project of undermining the autonomy of genetics, of integrating it with the rest of biology.

At the beginning of the twenty-first century, scientific opinion is curiously divided on McClintock. Some, such as Joshua Lederberg and Norton Zinder, continue to view her as a member—unorthodox and outstanding, to be sure—of the old school of classical cytogenetics. These scientists tend to think of McClintock as someone who embraced some of the concepts of molecular biology without adopting the language or tools. She was wrong about development, she was wrong about controlling elements. Her ideas were speculative; she failed to adopt the new techniques that could have guided her to a more correct interpretation.

Others—the University of Chicago's James Shapiro is the most articulate exemplar—see her as a visionary, a bridge to a new era of biological thought in which the integration of nature that Thompson described can at last be addressed empirically. They emphasize the grand sweep of her biological vision, her understanding of the integrated nature of nature. In that perspective, McClintock resembles the oracles of ancient Greece and Rome, mystical beings who spoke in allegory and metaphor, the truth of which depended on the seeker's finding the appropriate interpretation.

Late in her career, McClintock relied ever more heavily on photographs of corn kernels to communicate, as though there was little more she could add to the phenomena.[17] She came to believe that the connections among her insights were sufficient confirmation; she felt less and less need to test her ideas by experiment. Some of her last writings are more like works of art than science. They are open to interpretation, their truth emerging not from data and logic, but from their ability to change the way other scientists think. Jeffrey Strathern, a yeast geneticist, wrote in 1992, "In some senses, it does not matter whether I understood Barbara completely or correctly, it only matters that in trying to understand her, my thoughts traversed new ground."[18]

The Unity of Nature

Yet McClintock's most mystical-sounding ideas stemmed from observation and skepticism, not occult visitations. The ever-expanding variations produced by her pattern generator, the controlling elements in her corn, challenged her assumptions about biological stability. The more complex her research material became, and the more surprises it revealed to her, the more convinced she became that she actually knew little about nature. On theoretical grounds, she refused to deny anything's existence, because she understood that in biology disproof is impossible. No experiment can distinguish between, say, the nonexistence of ESP and the failure of the observer to detect it.

McClintock's apparent mysticism, I believe, was rooted ultimately in her sense that science could provide no final answers. Heredity, development, and other phenomena considered mysterious at the beginning of the century were "still just as mysterious as they were originally," she told Will Provine and Paul Sisco in 1980.[19] "Oh, yes, you can explain things with physics and chemistry, but you can't explain physics and chemistry. So all you're doing is seeing things in a different light." Science, she maintained, can never reach logical bedrock. "You can get all kinds of relationships. You can't get fundamentals," she said. "You never can find the cause of the cause." The "relationships" were merely a tool for integrating natural phenomena, for synthesizing a coherent world view. McClintock wanted to grasp the entire universe as a whole. "The point is we forget we're all one thing. It's all one. And we're just breaking it up into little things."[20]

This perception of the universe as "all one," that we, meaning scientists, are just breaking it up into little things, is exactly equivalent to Thompson's statement about the inherent integration of nature. It is very close to Darwin's metaphor of the entangled bank. McClintock belongs in that small group of visionary biological thinkers who perceived the unity of nature. No doubt what drew her to Buddhism and Taoism later in life was the realization that those religious philosophies describe this unity.

Another quality of those Eastern philosophies is that the practitioner seeks detachment from the material world. Mundane attachments are like shackles that restrain one, preventing the attainment of enlightenment. In Chapter 2 I ventured that an incapacity for intimacy was the figure under the carpet of McClintock's private myth of freedom. From letters she wrote in the 1940s to stories she told in the 1980s, McClintock seems to have viewed herself with similar detachment. It is not strange that she did not marry; what is strange is that she told two Evelyns, Witkin and Fox Keller, that she could not conceive of marrying. It is strange that she never, in the memory of anyone with whom I have spoken, had an intimate relationship. This incapacity for attachment and commitment, I believe, underlay her repeated insistence that her colleagues thought she was crazy, her resignations from Missouri and Cold Spring Harbor, her lack of formal students. If she were indeed detached from colleagues, friends, family, and even herself, it would have been an ideal preparation for her ultimate vision of the unity of nature.

Barbara McClintock might seem a sad figure, if not for her freedom and opportunities, the richness of her mental life, and the loyalty of her friends. Evidence from her research notes, her letters, and her friends suggests we can take her at her word when, during her Nobel Prize press conference at Cold Spring Harbor in 1983, she said to a journalist: "I've had such a good time, I can't imagine having a better one. No, no, that's true. I've had a very, very satisfactory and interesting life."[21]

APPENDIX:
A MOLECULAR EPILOGUE

The labilizing forces of recombination and transposition are just barely contained, giving a dynamic genetic system of ever increasing complexity that verges on the chaotic.

—NINA FEDOROFF

Thus far, I have avoided molecular explanations of controlling elements and their behavior. Every effort has been made to restrict the description of controlling elements to the terms and concepts McClintock herself employed. The reasons are two. First, molecular explanations can be misleading. To understand McClintock's approach to science and nature, it is essential to understand her findings on her own terms. Only by restricting our vision to the perspective McClintock had at the time can we reconstruct what controlling elements actually meant to her.

Second, applying modern understanding to history, though in some cases edifying, can lead to ahistorical conclusions. Classical and molecular genetics operate on different logical levels, and it is both unfair to McClintock and poor historical practice to apply standards from one level to science that operated on another. We may say, for example, that, had she applied molecular methods, she might have found such-and-such result, or even that her refusal to adopt molecular methods prevented her from seeing certain things. But it would be illegitimate to argue that *Spm* does not occur in different states, as McClintock described, but rather that it occurs in different lengths of nucleotides, which are treated differently by the machinery of the cell. Although the latter explanation is true, it does not refute the correctness of McClintock's observations, because it operates on a different logical level than the former. In fact, such a claim *confirms* McClintock's data. The phenomena themselves are the common denominator between the genetic and molecular explanations.

The risk, however, in remaining faithful to McClintock's observations and language is that her system will appear hopelessly complex. Surveying her work, with the piling of variations one upon another, with exceptions

271

found for every rule, with patterns overlaying patterns, one senses the horizon of chaos approaching rapidly. One could hardly blame a few geneticists or molecular biologists if they wondered whether McClintock was imagining some of these behaviors.

The firm establishment of a new interpretation of McClintock and her science, then, requires a brief survey of some of the recent molecular work on the action of controlling elements and their role in development and evolution. What follows is a synopsis of some of the findings of the last fifteen years or so, presented as an attempt to resolve the tension over the correctness of McClintock's increasingly bizarre results and complex interpretations. It is a fairly technical account; the language of molecular biology is specialized, and it is important to emphasize the precision with which some of McClintock's genetic phenomena are now understood. Amid the molecular language, however, the story of controlling elements simplifies in a satisfying way. The punch line is this: McClintock's elements are real, her mutants are real, their behavior is complex but comprehensible.

Using plants derived from some of McClintock's maize stocks, Nina Fedoroff, working at the Carnegie Institution of Washington's Department of Embryology in Baltimore, isolated the *Ac* element in 1983. Subsequent work has revealed it to be a segment of DNA about 4,600 nucleotides long, flanked by inverted repeats. The transposon encodes a messenger RNA of about 3,500 nucleotides. This message codes for a protein 807 amino acids long, an enzyme now called a transposase. The transposase can act upon itself, catalyzing its own removal from the chromosome and its subsequent reinsertion. It does so by a mechanism similar to that used by insertion sequence elements ("conservative" rather than "replicative" transposition). *Ac* contains multiple copies of the sequence AAACGG. The transposase *Ac* encodes recognizes this sequence and binds to it, catalyzing *Ac* transposition. As was suspected in the early 1950s, Brink and Nilan's *Modulator* of pericarp *(Mp)* is in fact identical to *Ac*.[1]

McClintock recognized by the early 1950s that *Ac* and *Ds* were similar and probably related. It is now known that *Ds* is simply a mutated *Ac*. As McClintock found, the *Ac* element can both transpose itself and catalyze the transposition of another mobile element—*Ds*. A *Ds* is simply any *Ac* element that has lost its ability to catalyze transposition but retains the ability to transpose when acted upon by an intact *Ac*. The *Ac* element is a discrete genetic element of consistent structure. *Ds* elements, however, vary widely in structure. The initial *Ds* element, which McClintock detected by its ability to break chromosomes, has been analyzed in detail and turns out to be peculiar. It is actually a double *Ds:* one *Ds* inserted, in reverse orientation, within a second. Transposition of double *Ds* leads to chromosome breaks and rearrangements. Single *Ds* elements do not cause chromosome breaks. Biologists have yet to unravel why transposition of double *Ds* involves both chromatids.[2]

The snowballing complexity of *Suppressor-mutator* has a molecular basis. Analogous to *Ds*, the "operator," or "element residing at" the affected locus, is a mutated form of the complete controlling element. Fedoroff refers to this form as transposition-defective *Spm (dSpm)*. A *dSpm* element can respond to *Spm* but cannot itself transpose. Like *Ds* elements, *dSpm* elements are heterogeneous. The *Spm* element is 8,300 nucleotides long, nearly twice the length of *Ac*. It is flanked by inverted repeats, 13 basepairs long, and internal to this region at either end lies a second region of repeating sequence. Peter Peterson's *Enhancer* element is essentially identical to *Spm*.[3]

Suppression and mutation arise from the action of *Spm* upon *dSpm*. When *dSpm* inserts into a gene, an intact *Spm* suppresses the expression of the gene by acting on the *dSpm*. Sometimes *Spm* causes *dSpm* to transpose out of the gene, leading to mutations that allow full or partial expression of *A1*. Nina Fedoroff explained to me her reconstruction of the original inception of *Spm* in McClintock's *a1-m1* mutant: "The initial mutation was caused by an insertion of a transposition-defective *Spm* [*dSpm*] into the *A1* locus in such a way that it did not completely abolish expression of the gene (today we know this is because it gets spliced out of the [mRNA] transcript). When a fully functional *Spm* was present in the same genome, it 'suppressed' expression of the mutant *A1* gene *(a1-m1)*." But, she continued, under these circumstances, "the defective element could also transpose out of the gene, giving rise to fully pigmented sectors on the kernels—and occasionally fully revertant kernels," as well as partial revertants "that expressed the gene at different levels. This is because the insertion is in a coding region of the gene and because excision is not precise."[4] *Spm*, in other words, often takes a bit of the surrounding gene with it when it transposes, affecting *A1* expression and leading to the gradations in pigment McClintock observed.

The *Spm* element generates defective versions of itself even more than does *Ac*. Several of McClintock's later mutable alleles have been examined and shown to be curious permutations of the basic *Spm* system. The weakened form of *Spm* is simply *Spm* with a 1,600-base-pair deletion that reduces transposition frequency but does not eliminate suppressor or mutator activity. McClintock was at first puzzled by her second *A1* mutable, *a1-m2*, because it showed the opposite phenotype from *a1-m1*. Genetically, the *Spm* element mapped very close to the *a1* locus; McClintock could only speculate that proximity of the element to the gene it affected could influence its behavior. The *a1-m2* mutant probably resulted from insertion of a *dSpm* in the control region of the *a1* gene. In the absence of intact *Spm*, the gene is suppressed, because its control region is disrupted. In the presence of *Spm*, transpositions of *dSpm* can occur, restoring gene function.[5]

The mutable allele *a2-m1* interested McClintock for two primary reasons. First, unlike the inceptions of *Spm* at the *A1* locus, *a2-m1* showed dosage effects: more *Spm* elements increased mutation frequency and delayed the onset of mutations. Second, it showed two classes of states. Class 2 states gave gene expression in the absence of *Spm* but were fully suppressed in the presence of *Spm*. This results from insertion of a *dSpm* just upstream from the *A2* gene. The DNA motifs that allow suppression by *Spm* are all missing, though the inverted repeats are intact. Thus, the element can suppress but cannot transpose. In the absence of *Spm*, this *dSpm* appears to be treated as an intron; it is transcribed into messenger RNA along with the rest of the *a2* gene but then is spliced out of the message. As a result, in the absence of *Spm*, *a2-m1* behaves like normal *A2* and produces full color.[6]

"Using this [class 2] state, McClintock was now able to clearly see the alterations in activity phase of the transacting *Spm*—this is where the dosage effect comes in," Fedoroff explained. "She had identified an *Spm* that cycles between active and inactive [phases] during development of the kernel, so this *Spm* combined with the non-excising *a2-m[1]* allele showed colored sectors ([suppression] off) within colorless areas (on) and in turn the colored sectors had colorless sectors within them (on again). Two *Spm* elements in the same nucleus went on and off independently. Since colorless resulted whenever one element was off and colored only when both were off, the total colored area of the kernel was reduced."[7]

The changes of phase, between active and inactive forms, are explained by modifications to the DNA. Cytosine residues, the Cs in the DNA alphabet, have a site that can accept a methyl group (CH_3). Methylation blocks enzymes such as polymerases from binding to the DNA at that site. Two regions of *Spm* rich in cytosine residues influence the expression of *Spm*. When these regions are heavily methylated, transcription of the *Spm* genes cannot occur. Active *Spm* has little or no methylation. Inactive, or cryptic, *Spm* is heavily methylated. Methylation is reversible, but it can also be quite stable; a given state of methylation can be passed on through cellular generations and even from plant to plant. Thus, the changes in phase are changes in the state of methylation of an *Spm* element, particularly the promoter and another region downstream of the promoter.

What alters the state of methylation? *Spm* itself. It thus acts as both a positive and a negative regulator. Active or inactive *Spm* may be passed down through the generations, but active may be converted to inactive or vice versa.[8]

Methylation could also explain other bizarre features of *Spm* behavior. McClintock hypothesized modifying elements, other genes that influenced *Spm*. Fedoroff explained, "These were probably *Spm* elements that were still methylated enough to stay off in the absence of an *Spm* that is fully active, but could be reactivated by an active one."[9]

Fedoroff has also described a third form, "programmable" *Spm*, which may be active or inactive. Programmable *Spm* has an intermediate degree of methylation and can show a variety of patterns of expression. These patterns are heritable yet mutable; programming patterns, through development and through plant generations, may represent changes in the extent of methylation. It seems likely that presetting and erasure, the most bizarre of all the phenomena McClintock described, will be explainable in terms of programming patterns acquired in one generation and either passed on or wiped out in a subsequent one.[10]

McClintock lived to see much of this explanation elucidated. Though large parts were yet hypothetical, the outline of the story sketched here was in place by 1992. That year, McClintock celebrated her ninetieth birthday. Cold Spring Harbor Laboratory fêted her with a quiet birthday party, at which she was presented with a *festschrift* volume, edited by Fedoroff and David Botstein. Articles in this volume by Saedler and Starlinger and by Fedoroff summarized the molecular findings on maize transposons to date, though McClintock had followed their papers all along and indeed collaborated with Fedoroff early in her study. McClintock died less than three months later, on 2 September.

NOTES

Abbreviations

Interviews with McClintock

CB   Interview with Charles Burnham, Ronald Phillips, and students, 1970, University of Wisconsin, transcript, from private collection of Ronald Phillips.

EFK9.24.78   Interview with Evelyn Fox Keller, 24 September 1978, transcript, deposited at the American Philosophical Society Library, Philadelphia, Pa.

EFK12.1.78   Interview with Evelyn Fox Keller, 1 December 1978, transcript, deposited at the American Philosophical Society Library, Philadelphia, Pa.

EFK1.13.79   Interview with Evelyn Fox Keller, 13 January 1979, transcript, deposited at the American Philosophical Society Library, Philadelphia, Pa.

EFK2.25.79   Interview with Evelyn Fox Keller, 25 February 1979, transcript, deposited at the American Philosophical Society Library, Philadelphia, Pa.

FHP   Interview with Franklin H. Portugal, 21 August 1980, Cold Spring Harbor, N.Y., transcript, deposited at the American Philosophical Society Library, Philadelphia, Pa.

HFJ   Interview with Horace Freeland Judson, June 1973, deposited at the American Philosophical Society Library, Philadelphia, Pa.

NPC   Press conference, 10 October 1983, Cold Spring Harbor Laboratory, transcript, deposited at the Cold Spring Harbor Laboratory Archives, Cold Spring Harbor, N.Y.

NS   Interview with Neville Symonds, undated but about 1980, in author's possession. Graciously provided, with the interviewer's permission, by Evelyn Witkin.

PS   Interview with William B. Provine and Paul Sisco, August 1980, Ithaca, N.Y., and Cold Spring Harbor, N.Y., transcript, deposited at Cornell University Archives, Ithaca, N.Y.

Archives

BC	George W. Beadle collection, California Institute of Technology Archives, Pasadena, Calif.
CSHL	Cold Spring Harbor Laboratory Archives, Cold Spring Harbor Laboratory, Cold Spring Harbor, N.Y.
EA	Edgar Altenburg collection, Lilly Library, Indiana University, Bloomington, Ind.
EC	Ernst Caspari collection, American Philosophical Society Library, Philadelphia, Pa.
GF	John Simon Guggenheim Foundation archives, New York, N.Y.
HJM	Hermann J. Muller collection, Lilly Library, Indiana University, Bloomington, Ind.
MBG	William L. Brown collection, Archives of the Missouri Botanical Garden, St. Louis, Mo.
MC	McClintock collection, American Philosophical Society Library, Philadelphia, Pa.
OEN	Oliver E. Nelson collection, Steenbock Library, University of Wisconsin, Madison, Wis.
RAB	Royal Alexander Brink collection, Steenbock Library, University of Wisconsin, Madison, Wis.
RC	Marcus Rhoades collection, Lilly Library, Indiana University, Bloomington, Ind.
RCC	Ralph Cleland collection, Lilly Library, Indiana University, Bloomington, Ind.
RP	Private collection of Dr. Ronald Phillips, University of Minnesota, St. Paul, Minn.
TMS	Tracy M. Sonneborn collection, Lilly Library, Indiana University, Bloomington, Ind.

Books

AFFTO	Evelyn Fox Keller, *A Feeling for the Organism: The Life and Work of Barbara McClintock* (New York: Freeman, 1983). Tenth Anniversary Edition published 1993.
DG	Nina Fedoroff and David Botstein, eds., *The Dynamic Genome: Barbara McClintock's Ideas in the Century of Genetics* (Cold Spring Harbor, N.Y.: Cold Spring Harbor Laboratory Press, 1992).
Peters	James A. Peters, ed., *Classic Papers in Genetics* (Englewood Cliffs, N.J.: Prentice-Hall, 1959).
SS	Curt Stern and Eva Sherwood, eds., *The Origin of Genetics: A Mendel Source Book* (San Francisco: Freeman, 1966).

1. Myth

Epigraph from "Hypothesis and imagination," in Peter Medawar, *The Art of the Soluble* (London: Methuen, 1967), p. 132.

1. Joseph Campbell, *Creative Mythology: The Masks of God* (New York: Penguin, 1968), pp. 4–5.
2. Roland Barthes, *Mythologies,* trans. by Annette Lavers (New York: Hill and Wang, 1972), pp. 111–126.
3. François Jacob, *The Statue Within: An Autobiography* (New York: Basic Books, 1988), pp. 18–19.
4. Leon Edel, "The figure under the carpet," in Marc Pachter, ed., *Telling Lives: The Biographer's Art* (Washington, D.C.: New Republic Books, 1979), pp. 16–34; Edel, *Writing Lives* (New York: Norton, 1984), pp. 28–29.
5. Edel, "The figure under the carpet," p. 30.
6. Gerald Geison, *The Private Science of Louis Pasteur* (Princeton, N.J.: Princeton University Press, 1995).
7. *Profiles of Women Past and Present,* vol. 2 (Thousand Oaks, Calif.: American Association of University Women, Thousand Oaks Branch, 1996), pp. 42–43.
8. Evelyn Fox Keller, "Photoinactivation and the Expression of Genetic Information in Bacteriophage-Lambda" (Ph.D. diss., Harvard University, 1963); Keller, "Women in science: An analysis of a social problem," *Harvard Magazine* 77 (October 1974): 14–19.
9. Evelyn Fox Keller, "McClintock's maize," *Science '81* 2, no. 6 (1981): 54.
10. *AFFTO.*
11. See Margaret Rossiter, *Women Scientists in America: Before Affirmative Action (1940–1972)* (Baltimore: Johns Hopkins University Press, 1995), chap. 16.
12. Carolyn Merchant, *The Death of Nature: Women, Ecology and the Scientific Revolution* (San Francisco: HarperCollins, 1980); Liliane Stehelin, "Sciences, women, and ideology," in Hilary Rose and Steven Rose, eds., *The Radicalisation of Science: Ideology of/in the Natural Sciences* (London: Macmillan, 1976), pp. 76–89; "Special issue: 'Women, science, and society,'" *Signs* 4 (1978); Brian Easlea, *Science and Sexual Oppression: Patriarchy's Confrontation with Woman and Nature* (London: Weidenfeld and Nicolson, 1981); Sandra Harding and Merrill B. Hintikka, eds., *Discovering Reality: Feminist Perspectives on Epistemology, Metaphysics, Methodology, and Philosophy of Science* (Dordrecht: D. Reidel, 1983); Jan Harding, ed., *Perspectives on Gender and Science* (London: Falmer Press, 1986); Sandra Harding, *The Science Question in Feminism* (Ithaca, N.Y.: Cornell University Press, 1986); Harding, "Feminism and theories of scientific knowledge," *Women: A Cultural Review* 1 (1990): 87–98; Harding, *Whose Science? Whose Knowledge?* (Ithaca, N.Y.: Cornell University Press, 1991); Cheris Kramarae and Dale Spender, eds., *The Knowledge Explosion: Generations of Feminist Scholarship*

(New York: Teachers College Press, 1992). The founding text in the feminist critique of science is usually taken to be Ruth Herschberger, *Adam's Rib* (New York: Pellegrini and Cudahy, 1948; reprint, Harper and Row, 1970). The history of the feminist critique is reviewed in Ruth Bleier, *Science and Gender: A Criticism of Biology and Its Theories on Women* (Elmsford, N.Y.: Pergamon, 1984), and, more recently, in Londa L. Schiebinger, *Has Feminism Changed Science?* (Cambridge, Mass.: Harvard University Press, 1999).

13. For a review and analysis of the influence of feminist approaches to medicine, primatology, archaeology, and biology, see Schiebinger, *Has Feminism Changed Science?* chaps. 6–8. Anne Fausto-Sterling, in *Myths of Gender: Biological Theories about Women and Men* (New York: Basic Books, 1985), examines biologically based explanations of gender differences from a feminist perspective.

14. See, for example, Emily Martin, "The egg and the sperm: How science has constructed a romance based on stereotypical male-female roles," *Signs* 16 (1991): 485–501 (reprinted in Evelyn Fox Keller and Helen Longino, eds., *Feminism and Science,* Oxford: Oxford University Press, 1996, pp. 103–117); Martin, *The Woman in the Body: A Cultural Analysis of Reproduction* (Boston: Beacon Press, 1992).

15. See, for example, Sarah Blaffer Hrdy, "Infanticide as a primate reproductive strategy," *American Scientist* 65 (1977): 40–49; Meredith F. Small, "Females without infants: Mating strategies in two species of captive macaques," *Folia Primatol* 40 (1983): 125–133; and Donna Haraway's very different *Primate Visions: Gender, Race, and Nature in the World of Modern Science* (New York: Routledge, 1989).

16. Ruth Hubbard, Mary Sue Henifin, and Barbara Fried, eds., *Biological Woman—The Convenient Myth* (Cambridge: Schenkman, 1982); Hubbard, "The emperor doesn't wear any clothes: The impact of feminism on biology," in Dale Spender, ed., *Men's Studies Modified* (New York: Pergamon, 1981), pp. 213–235; Mary Jacobus, Evelyn Fox Keller, and Sally Shuttleworth, eds., *Body/Politics: Women and the Discourses of Science* (New York: Routledge, 1990).

17. Carol Gilligan, *In a Different Voice: Psychological Theory and Women's Development* (Cambridge, Mass.: Harvard University Press, 1982); Mary Field Belenky, Blythe McVicker Clinchy, Nancy Rule Goldberger, and Jill Mattuck Tarule, *Women's Ways of Knowing: The Development of Self, Voice, and Mind* (New York: Basic Books, 1986). See also Ruth Schwartz Cowan's discussion of the mythologization of McClintock in Cowan, "Women and science: Contested terrain" (review of Sandra Harding, *Whose Science? Whose Knowledge?* and Linda Jean Shepherd, *Lifting the Veil*), *Social Studies of Science* 25 (1995): 363.

18. Evelyn Fox Keller, "Baconian science: A hermaphroditic birth," *Philosophy Forum* 11 (1980): 299–308; Keller, "Baconian science: The arts of mastery and obedience," in *Reflections on Gender and Science* (New Haven, Conn.:

Yale University Press, 1985), pp. 33–42. For a different interpretation of Bacon's rhetoric, see Peter Pesic, "Wrestling with Proteus: Francis Bacon and the 'torture' of nature," *Isis* 90 (1999): 81–94.

19. *AFFTO,* pp. 198–199.

20. *AFFTO,* Ibid., p. xii.

21. Evelyn Fox Keller, "Feminist critique of science: A forward or backward move?" *Fundamenta Scientiae* 1 (1980): 341–349; Keller, "The gender/science system, or, Is sex to gender as nature is to science?" *Hypatia* 2 (1987): 43; Keller, "How gender matters, or, Why it's so hard for us to count past two," in Gill Kirkup and Laurie Smith Keller, eds., *Inventing Women: Science, Technology, and Gender* (Cambridge: Polity Press/Open University, 1992), p. 50.

22. Keller, "How gender matters," pp. 188–195. See also the debate Evelleen Richards and John Schuster carried on with Keller in the pages of *Zygon* 19 (1989): 697–729.

23. Ruth Bleier, *Science and Gender: A Criticism of Biology and Its Theories on Women* (Elmsford, N.Y.: Pergamon, 1984), pp. 204–206; Joan Dash, *The Triumph of Discovery: Women Scientists Who Won the Nobel Prize* (Englewood Cliffs, N.J.: Julian Messner, 1991), p. 92; Mary Frank Fox, "Gender, environmental milieu, and productivity in science," in Harriet Zuckerman, Jonathan Cole, and John Bruer, eds., *The Outer Circle: Women in the Scientific Community* (New York: Norton, 1991), pp. 203–204; Harding, *The Science Question in Feminism*, p. 122; Hilary Rose, *Love, Power, and Knowledge: Towards a Feminist Transformation of the Sciences* (Bloomington: Indiana University Press, 1994), p. 157; Linda Jean Shepherd, *Lifting the Veil: The Feminine Face of Science* (Boston: Shambhala, 1993).

24. Dash, *Triumph of Discovery,* p. 92; Fox, "Gender, environmental milieu, and productivity in science," p. 203; Shepherd, *Lifting the Veil,* pp. 70, 88; Deborah Felder, "Barbara McClintock," in *The 100 Most Influential Women of All Time* (New York: Citadel Press, 1996), pp. 186–187.

25. Allan Spradling, "McClintock myths: Review of *The Dynamic Genome: Barbara McClintock's Ideas in the Century of Genetics,*" *Science* 259 (1993): 1208; James D. Watson, Nancy Hopkins, Jeffrey Roberts, Joan Steitz, and Alan Weiner, *Molecular Biology of the Gene,* 4th ed. (Menlo Park, Calif.: Benjamin Cummings, 1987), p. 22. See also Keller's take on scientists in "The gender/science system," p. 41.

26. Harding, *The Science Question in Feminism*, p. 31.

27. Helen E. Longino, "Subjects, power, and knowledge: Description and prescription in feminist philosophies of science," in Evelyn Fox Keller and Helen E. Longino, eds., *Feminism and Science* (Oxford: Oxford University Press, 1996), p. 264.

28. Susan Allen, "Cutting loose," in Angela M. Pattatucci, ed., *Women in Science: Meeting Career Challenges* (London: Sage, 1998), pp. 149–150.

29. Rossiter, *Women Scientists in America.*

30. Ana Barahona, "Barbara McClintock and the transposition concept," presented to the International Society for the History, Philosophy, and Social Sciences of Biology, 1995, Leuven, Belgium; Ana Barahona and Francisco Ayala, "La importancia del contexto y el trabajo de Barbara McClintock," *Arbor* 152 (1995): 9–26; Nathaniel C. Comfort, "Two genes, no enzyme: A second look at Barbara McClintock and the 1951 Cold Spring Harbor Symposium," *Genetics* 140 (1995): 1161–1166; Comfort, "'The real point is control': The reception of Barbara McClintock's controlling elements," *Journal of the History of Biology* 32 (1999): 133–162; Carla Keirns, "Seeing patterns: Models, visual evidence, and pictorial communication in the work of Barbara McClintock," *Journal of the History of Biology* 32 (1999): 163–196.

31. Victor Cohn, "Long-neglected woman scientist awarded Nobel," *Washington Post,* 11 October 1983, p. A1; John Noble Wilford, "A brilliant loner in love with genetics," *New York Times,* 11 October 1983, p. C7; Claudia Wallis, "Honoring a modern Mendel," *Time* 122 (1983): 53–54; David Zinman, "A secular monastery for pioneers of science," *Newsday,* 11 October 1983, part 2, p. 4. See also Matt Clark and Kristine Mortensen, "Genetic pioneer wins a Nobel," *Newsweek,* 24 October 1983, p. 97; Matt Clark and Dan Shapiro, "The kernel of genetics," *Newsweek,* 30 November 1981, p. 74; Kathleen Teltsch, "Award-winning scientist prizes privacy," *New York Times,* 18 November 1981, p. B3.

32. Deborah Heiligman, *Barbara McClintock: Alone in Her Field* (New York: Freeman, 1994); Mary Kittredge, *Barbara McClintock* (New York: Chelsea House, 1991); Edith Hope Fine, *Barbara McClintock, Nobel Prize Geneticist* (Springfield, N.J.: Enslow, 1998).

33. Among these are McClintock's friend Susan Gensel Cooper, McClintock's colleagues Robert Pollack and James Shapiro, and historian Horace Freeland Judson.

34. Evelyn Fox Keller, "A world of difference," in *Reflections on Gender and Science* (New Haven, Conn.: Yale University Press, 1985), p. 159.

35. Schiebinger, *Has Feminism Changed Science?* pp. 1, 3–6.

36. Noretta Koertge, "Feminism: A mixed blessing to women in science," in Angela M. Pattatucci, ed., *Women in Science: Meeting Career Challenges* (London: Sage, 1998), p. 195; Stephen Jay Gould, "The triumph of a naturalist," *New York Review of Books,* 29 March 1984; Horace Freeland Judson, *The Search for Solutions* (New York: Holt, Rinehart, and Winston, 1980; paperback ed., Johns Hopkins University Press, 1987), pp. 7–8.

37. For a recent critical appraisal of this approach, see Jan Golinski, *Making Natural Knowledge: Constructivism and the History of Science* (Cambridge: Cambridge University Press, 1998).

38. For a guide to some of the technical problems inherent in studying recent science, see Thomas Soderqvist, ed., *The Historiography of Contemporary Science and Technology* (Amsterdam: Harwood, 1997); Richard Yeo and Michael Shortland, eds., *Telling Lives: Studies of Scientific Biography* (Cam-

bridge: Cambridge University Press, 1995); Joan Warnow-Blewett, "Documenting recent science progress and needs," *Osiris* 7 (1992): 267–298.

2. Freedom

Epigraph from EFK9.24.78.

1. Susan Gensel Cooper interview, 31 January 1997.
2. Oswald T. Avery, Colin M. MacLeod and Maclyn McCarty, "Studies on the chemical nature of the substance inducing transformation of Pneumococcal types," *Journal of Experimental Medicine* (1944): 137–158, reprinted in Peters, pp. 173–192. For discussion and analysis, see Horace Freeland Judson, *The Eighth Day of Creation* (New York: Simon and Schuster, 1979), pp. 35–41; Michel Morange, *A History of Molecular Biology* (Cambridge, Mass.: Harvard University Press, 1998), chap. 3; Maclyn McCarty, *The Transforming Principle: Discovering that Genes Are Made of DNA* (New York: Norton, 1985).
3. R. H. Andrewes, "Peyton Rous," *Biographical Memoirs of the Royal Society* (1971): 644; Ralph W. Moss, *The Cancer Industry: The Classic Exposé on the Cancer Establishment* (1989; New York: Paragon House, 1991), p. 274; Nathaniel Berlin, "Conquest of cancer," *Perspectives in Biology and Medicine* (1979): 509.
4. Nathaniel Comfort, "Rous's reception: Tumor viruses in the context of the germ theory," presented to the annual meeting of the History of Science Society, 26–29 October 1995, Minneapolis, Minn.
5. Gunther Stent, "Prematurity and uniqueness in scientific discovery," *Scientific American* (1972): 84–93.
6. Leon Edel, *Writing Lives* (New York: Norton, 1984), p. 31.
7. *AFFTO,* p. 28.
8. EFK9.24.78.
9. Affadavit dated 18 June 1943, Folder "Birth Certificate," Series II, MC.
10. EFK9.24.78.
11. Quoted in *AFFTO,* p. 20. In this and following notes, I cite Keller's book for published quotes and the interviews for material not published, either whole or in part. In an extensive comparison of Keller's interviews with the quotations published in *A Feeling for the Organism,* I have found no instance of misquotation, although one of the central arguments of this chapter is that some of those quotations admit of other interpretations.
12. *AFFTO,* p. 20.
13. Sharon Bertsch McGrayne, "Barbara McClintock," in *Nobel Women in Science* (New York: Birch Lane, 1993), p. 147.
14. EFK9.24.78.
15. All quotes in this paragraph are from ibid.
16. *AFFTO,* p. 24.

17. McGrayne, "Barbara McClintock," p. 148.

18. *AFFTO,* pp. 23–24.

19. McGrayne, "Barbara McClintock," pp. 147–148.

20. EFK9.24.78.

21. Ibid.

22. *AFFTO,* p. 22.

23. Ibid., p. 26.

24. Ibid., p. 27.

25. EFK9.24.78.

26. FHP; *AFFTO,* p. 28; McGrayne, "Barbara McClintock," p. 149.

27. FHP.

28. Ibid.

29. Virginia Valian, *Why So Slow? The Advancement of Women* (Cambridge, Mass.: MIT Press, 1998), pp. 20–21.

30. EFK9.24.78.

31. David Botstein interview, 23 December 1996.

32. *AFFTO,* p. 33.

33. EFK1.19.79.

34. EFK9.24.78.

35. Cornell University, Office of the Registrar, "Explanation of Grading Systems," no date, received from the university September 1998.

36. *AFFTO,* p. 31.

37. Her grade point average, calculated from McClintock's college transcript, was 3.22.

38. EFK9.24.78.

39. Ibid.

40. HFJ.

41. "Never thought": HFJ. "Waking up": EFK9.24.78.

42. EFK9.24.78.

43. EFK1.13.79.

44. EFK9.24.78.

45. Evelyn Witkin interview, 19 March 1996.

46. EFK9.24.78.

47. Waclaw Szybalski interview, 14 October 1996. Other interview subjects speculated on this topic off the record.

48. EFK9.24.78.

49. Susan Gensel Cooper interview, 31 January 1997.

50. Evelyn Witkin interview, 19 March 1996.

51. Susan Gensel Cooper interview, 31 January 1997.

52. Robert Pollack interview, 30 July 1998.

53. Ibid.

54. Susan Gensel Cooper interview, 31 January 1997.

55. M. Susan Lindee, "The conversation: History and history as it happens," in Thomas Soderqvist, ed., *The Historiography of Contemporary Science and Technology* (Amsterdam: Harwood, 1997), pp. 43–44.

56. Evelyn Fox Keller, "How gender matters, or, Why it's so hard for us to count past two," in Gill Kirkup and Laurie Smith Keller, eds., *Inventing Women: Science, Technology, and Gender* (Cambridge: Polity Press/Open University, 1992), p. 49; *AFFTO*, p. 145; Linda Jean Shepherd, *Lifting the Veil: The Feminine Face of Science* (Boston: Shambhala, 1993), p. 70; Mary Rosner, Plotting a middle ground: An account of a failed science story," in John T. Battalio, ed., *Essays in the Study of Scientific Discourse: Methods, Practice, and Pedagogy* (Stamford, Conn.: Ablex, 1998), p. 38.

57. Helena Pycior, "Marie Curie's 'anti-natural path': Time only for science and family," in Pnina Abir-Am and Dorinda Outram, eds., *Uneasy Careers and Intimate Lives: Women in Science, 1789–1979* (New Brunswick, N.J.: Rutgers University Press, 1987), pp. 191–215; Jonathan R. Cole, *Fair Science: Women in the Scientific Community* (New York: Columbia University Press, 1987), p. 194; Jonathan R. Cole and Harriet Zuckerman, "Marriage, motherhood, and research performance in science," in Zuckerman, Cole, and John T. Bruer, eds., *The Outer Circle: Women in the Scientific Community* (New York: Norton, 1991), p. 159.

58. EFK9.24.78; quoted in *AFFTO*, p. 70, and the title of its fourth chapter.

3. Integration

Epigraph from EFK2.25.79.

1. Peter J. Bowler, *The Non-Darwinian Revolution: Reinterpreting a Historical Myth* (Baltimore: Johns Hopkins University Press, 1988), pp. 7–13.

2. Ibid., pp. 117–118; Bowler, *The Mendelian Revolution: The Emergence of Hereditarian Concepts in Modern Science and Society* (Baltimore: Johns Hopkins University Press, 1989), chaps. 5, 6.

3. Gregor Mendel, "Versuche über Pflanzenhybriden," *Verhandlungen des naturforschenden Vereins* 4 (1866): 3–47, reprinted in SS, pp. 1–48.

4. Hugo Iltis, *Life of Mendel*, English ed. (London: George Allen and Unwin, 1932), p. 179.

5. *Neuigkeiten*, 8 February and 8 March 1865, in Robert Olby, *Origins of Mendelism*. 2d ed. (Chicago: University of Chicago Press, 1985), pp. 220–221; Viteslav Orel, *Gregor Mendel: The First Geneticist* (Oxford: Oxford University Press, 1996), p. 273.

6. Focke's citations of Mendel are reproduced in SS, pp. 103–106. Discussion of Focke and other early Mendel-citers may be found in Olby, *Origins of Mendelism*, pp. 216–234.

7. Curt Stern, "Foreword," in SS, pp. x–xi; Alfred H. Sturtevant, *A History of Genetics* (New York: Harper and Row, 1965), pp. 26–29.

8. Hugo de Vries, "The law of segregation of hybrids," *Das Spaltungsgesetz der Bastarde* (1900), in SS, pp. 107–117; Carl Correns, "G. Mendel's law concerning the behavior of progeny of varietal hybrids" (1900), in SS, pp. 119–132. On Correns's contributions, see Margaret Saha, "Carl Correns and an

alternative approach to genetics: The study of heredity in Germany between 1880 and 1930" (Ph.D. diss., Michigan State University, 1984).

9. William Bateson and E. R. Saunders, "Experimental studies in the physiology of heredity," *Report to the Evolution Commission of the Royal Society* 1 (1902): 160; Bateson to Adam Sedgwick, 18 April 1905, reprinted in *William Bateson, F. R. S., Naturalist: His Essays and Addresses, Together with a Short Account of His Life* (Cambridge: The University Press, 1928), p. 93.

10. Wilhelm Ludwig Johannsen, *Elemente der Exakten Erblichkeitslehre* (Jena: G. Fischer, 1909). On the genotype-phenotype distinction, see Garland Allen, "Naturalists and experimentalists: The genotype and the phenotype," in William Coleman and Camille Limoges, eds., *Studies in History of Biology*, vol. 3 (Baltimore: Johns Hopkins University Press, 1979), pp. 179–209.

11. Charles Darwin, *The Variation of Animals and Plants under Domestication*, 1st ed. (London: John Murray, 1868), chap. 27. See also Bowler, *The Mendelian Revolution*, pp. 57–64.

12. William B. Provine, *The Origins of Theoretical Population Genetics* (Chicago: University of Chicago Press, 1972), chaps. 2, 3; B. J. Norton, "The biometric defense of Darwinism," *Journal of the History of Biology* 6 (1973): 283–316; Bowler, *The Mendelian Revolution*, pp. 64–73.

13. Hugo de Vries, *The Mutation Theory: Experiments and Observations on the Origin of Species in the Vegetable Kingdom*, English trans. (Chicago: Open Court, 1901–1903); Bowler, *The Non-Darwinian Revolution*, pp. 121–122.

14. Edmund Beecher Wilson, *The Cell in Development and Inheritance*, 2d ed. (New York: MacMillan, 1900), p. 6.

15. Scott F. Gilbert "The embryological origins of the gene theory," *Journal of the History of Biology* 11 (1978): 307–351, *passim*, enucleated egg experiment on p. 314.; August Weismann, *The Germ Plasm: A Theory of Heredity*, trans. W. N. Parker and H. Rofeldt (London: Walter Scott, 1893); Walter Sutton, "The chromosomes in heredity," *Biological Bulletin* 4 (1903): 231–251, reprinted in Peters, pp. 27–41; Theodor Boveri, "On multipolar mitoses as a means for the analysis of the cell nucleus," *Verhandlungen der physikalische-medizinischen Gesellschaft zu Würzburg* 35 (1902): 67–90, reprinted in Bruce R. Voeller, ed., *The Chromosome Theory of Inheritance: Classic Papers in Development and Heredity* (New York: Appleton-Century-Crofts, 1968); C. E. McClung "The accessory chromosome—sex determinant?" *Biological Bulletin* 35 (1902): 43–84, in Voeller, *The Chromosome Theory of Inheritance;* Stephen G. Brush, "Nettie M. Stevens and the discovery of sex determination by chromosomes," *Isis* 69 (1978): 163–172; L. C. Dunn, *A Short History of Genetics: The Development of Some Main Lines of Thought, 1854–1939* (New York: McGraw-Hill, 1965), pp. 114–115.

16. Henry Harris, *The Cells of the Body: A History of Somatic Cell Genetics* (Cold Spring Harbor: Cold Spring Harbor Laboratory Press, 1995), pp. 16–18; W. Flemming, "Beiträge zur Kenntnis der Zelle und ihrer Lebenserscheinungen," *Archiv für Mikroskopische Anatomie* 16 (1879): 302–436; W.

Waldeyer, "Über Karyokinese und ihre Beziehung zu den Befruchtungsvorgängen," *Archiv für Mikroskopische Anatomie* 32 (1888): 1.

17. F. A. Janssens, "La théorie de la chiasmatypie, nouvelle interpretation des cinéses de maturation," *La Cellule* 25 (1909): 387–406.

18. Jane Maienschein, *Transforming Traditions in American Biology, 1880–1915* (Baltimore: Johns Hopkins University Press, 1991), chaps. 2, 8; Garland Allen, *Thomas Hunt Morgan: The Man and His Science* (Princeton, N.J.: Princeton University Press, 1978), chap. 2.

19. Allen, *Thomas Hunt Morgan,* chap. 4; Robert Kohler, *Lords of the Fly: Drosophila Genetics and the Experimental Life* (Chicago: University of Chicago Press, 1994), pp. 42–43.

20. Kohler, *Lords of the Fly,* p. 23.

21. Thomas Hunt Morgan "Sex-limited inheritance in Drosophila," *Science* 32 (1910): 120–122; Kohler, *Lords of the Fly,* pp. 38–45.

22. Allen, *Thomas Hunt Morgan,* pp. 132–136.

23. For a brief firsthand account of the work in the fly room, see Alfred H. Sturtevant, *A History of Genetics* (New York: Harper and Row, 1965), chaps. 7, 8.

24. Dunn, *A Short History of Genetics,* p. 142.

25. S. Blixt, "Why didn't Gregor Mendel find linkage?" *Nature* 256 (1975): 206. For a review of the Mendel controversy, see Jan Sapp, "The nine lives of Gregor Mendel," in H. E. Le Grand, ed., *Experimental Inquiries* (Dordrecht: Kluwer Academic Publishers, 1990), pp. 137–166.

26. Sturtevant, *A History of Genetics,* p. 47.

27. Kohler, *Lords of the Fly,* pp. 46–49.

28. Ibid., p. 173.

29. Thomas Hunt Morgan, "Chromosomes and heredity," *The American Naturalist* 44 (1910): 449; Morgan, "The theory of the gene," *The American Naturalist* 51 (1917): 514.

30. Mendel to Nägeli, 3 July 1870, in SS, p. 93.

31. Diane Paul and Barbara Kimmelman, "Mendel in America: Theory and practice, 1900–1919," in Ronald Rainger, Keith Benson, and Jane Maienschein, eds., *The American Development of Biology* (New Brunswick, N.J.: Rutgers University Press, 1988), pp. 296–301; Richard Lewontin, *Biology as Ideology* (New York: HarperPerennial, 1991), pp. 53–57.

32. Paul and Kimmelman, "Mendel in America," pp. 298–301; Peter A. Peterson and Angelo Bianchi, *Maize Genetics and Breeding in the Twentieth Century* (Singapore: World Scientific, 1999), pp. 15–20.

33. Edward M. East, "A Mendelian interpretation of variation that is apparently continuous," *The American Naturalist* 44 (1910): 82.

34. Peter J. Bowler, *The Eclipse of Darwinism: Anti-Darwinian Evolution Theories in the Decades around 1900* (Baltimore: Johns Hopkins University Press, 1983), pp. 198–199.

35. Rollins A. Emerson, "The inheritance of a recurring somatic variation in var-

iegated ears of maize," *The American Naturalist* 48 (1914): 87; Elof Axel Carlson, *The Gene: A Critical History* (Philadelphia: W. B. Saunders, 1966), p. 98.

36. Emerson, "The inheritance of a recurring somatic variation."

37. Rollins Emerson, "Genetical studies of pericarp in maize," *Genetics* 1 (1917): 1–35. A biographical sketch of Emerson may be found in Marcus M. Rhoades, "Rollins Adams Emerson, 1873–1947," *Biographical Memoirs of the National Academy of Sciences* 25 (1949): 313–325.

38. Emerson, "Genetical studies of pericarp in maize," p. 34.

39. Rollins Emerson, "The frequency of somatic mutation in variegated pericarp of maize," *Genetics* 14 (1929): 488–511.

40. Emerson, "Genetical studies of pericarp in maize," p. 32.

41. Milislav Demerec, "Eighteen years of research on the gene," *Cooperation in Research,* Carnegie Institution of Washington Publication 501 (1938): 295–314; quotation from Anderson's letter to Emerson is on p. 297. Carlson, *The Gene,* p. 100.

42. A brief biography of Demerec may be found in Bentley Glass, "Milislav Demerec, 1895–1966," *Biographical Memoirs of the National Academy of Sciences* 42 (1971): 1–25.

43. Milislav Demerec, "Mutable genes in *Drosophila virilis,*" *Proceedings of the International Congress of Plant Science* (Ithaca) 1 (1926): 943–946; Demerec, "Rate of instability of miniature-3 gamma gene of *Drosophila virilis* in the males in the homozygous and in the heterozygous females," *Proceedings of the National Academy of Sciences USA* 18 (1932): 656–658; Demerec, "What is a gene?" *Journal of Heredity* 64 (1933): 369–378; Demerec, "The gene and its role in ontogeny," *Cold Spring Harbor Symposium on Quantitative Biology* 2 (1934): 110–115; Demerec, "Unstable genes," *Botanical Review* (1935): 233–248.

44. Alfred H. Sturtevant, "The effects of unequal crossing over at the *Bar* locus in *Drosophila,*" *Genetics* 10 (1925): 117–147, reprinted in Peters, pp. 124–155; Carlson, *The Gene,* chap. 13; Dunn, *A Short History of Genetics,* pp. 162–163.

45. Carlson, *The Gene,* chap. 14; Sturtevant, *A History of Genetics,* chap. 14.

46. Thomas Hunt Morgan, "The failure of ether to produce mutations in *Drosophila,*" *The American Naturalist* 48 (1914): 705–711; Hermann Joseph Muller, "Artificial transmutation of the gene," *Science* 66 (1927): 84–87, reprinted in Peters, pp. 149–155; Lewis J. Stadler, "Mutations in barley induced by X rays and radium," *Science* 68 (1928): 186–187; Stadler, "On the genetic nature of induced mutations in plants," *Proceedings of the Sixth International Congress of Genetics* 1 (1932): 274–294.

47. Bowler, *Non-Darwinian Revolution,* pp. 113–125.

48. Jan Sapp, *Beyond the Gene: Cytoplasmic Inheritance and the Struggle for Authority in Genetics* (New York: Oxford University Press, 1987), pp. 26, 72–80; Jonathan Harwood, *Styles of Scientific Thought: The German Genetics*

Community, 1900–1933 (Chicago: University of Chicago Press, 1993), chaps. 2, 4; Garland E. Allen, "Opposition to the Mendelian-chromosome theory: The physiological and developmental genetics of Richard Gold-schmidt," *Journal of the History of Biology* 7 (1974): 53–61.

49. Sapp, *Beyond the Gene*, chap. 4, esp. pp. 97–98.
50. Harriet Creighton, "Recollections of Barbara McClintock's Cornell years," in *DG*, p. 15.
51. EFK9.24.78.
52. Marcus M. Rhoades, "The early years of maize genetics," in *DG*, p. 62.
53. Barbara McClintock, "A cytological and genetical study of triploid maize," *Genetics* 14 (1929): 180–222.
54. EFK9.24.78.
55. Barbara McClintock, "Chromosome morphology in *Zea mays*," *Science* 69 (1929): 629.
56. PS.
57. Laboratory notes for cytology courses, handout 4, p. 1; handout 5, by Charles Burnham, "Revised Sept. 1978," pp. 8, 18; and "Steps for staining using the McClintock method," RP.
58. EFK9.24.78.
59. PS.
60. PS.
61. PS; George Beadle, "Biochemical genetics: Some recollections," in J. S. Cairns, G. Stent, and J. D. Watson, eds., *Phage and the Origins of Molecular Biology*, expanded ed. (Cold Spring Harbor, N.Y.: Cold Spring Harbor Laboratory Press, 1992), p. 24.
62. George W. Beadle and Barbara McClintock, "A genic disturbance of meiosis in *Zea mays*," *Science* 68 (1928): 433.
63. On Rhoades's life and science, see D. Schwartz, "Marcus M. Rhoades (1903–1991)," *Genetics* 133 (1993): 1–3; N. Fedoroff, "Marcus Rhoades and transposition," *Genetics* 150 (1998): 957–961.
64. EFK9.24.78; PS.
65. Schwartz, "Marcus M. Rhoades," pp. 1–3.
66. FHP.
67. FHP.
68. EFK9.24.78.
69. PS. McClintock said the same thing in interviews with Portugal and Keller.
70. Marcus M. Rhoades and Barbara McClintock, "The cytogenetics of maize," *Botanical Review* 1 (1935): 296; Rhoades, "The early years of maize genetics," p. 62.
71. Theophilus Painter, "A new method for the study of chromosome rearrangements and plotting of chromosome maps," *Science* 78 (1933): 585–586.
72. Barbara McClintock, "A cytological demonstration of the location of an interchange between two non-homologous chromosomes of *Zea mays*," *Proceedings of the National Academy of Sciences USA* 16 (1930): 791–796.

73. PS; Charles Burnham, "Barbara McClintock: Reminiscences," in *DG,* p. 20; EFK12.1.78.

74. Barbara McClintock, "The order of the genes C, Sh, and Wx in *Zea mays* with reference to a cytologically known point in the chromosome," *Proceedings of the National Academy of Sciences USA* 17 (1931): 485–491; Harriet Creighton and Barbara McClintock, "A correlation of cytological and genetical crossing-over in *Zea mays,*" *Proceedings of the National Academy of Sciences USA* 17 (1931): 492–497, reprinted in Peters, pp. 155–160, and in facsimile in *DG,* pp. 7–12.

75. Creighton and McClintock, "A correlation of cytological and genetical crossing-over," in *DG,* p. 12.

76. *AFFTO,* pp. 58–59; Sharon Bertsch McGrayne, "Barbara McClintock," in *Nobel Women in Science* (New York: Birch Lane, 1993), pp. 154–156.

77. Peters, p. 156.

78. Margaret W. Rossiter, *Women Scientists in America: Struggles and Strategies to 1940* (Baltimore: Johns Hopkins University Press, 1982), p. 270.

79. Barbara McClintock, "Cytological observations of deficiencies involving known genes, translocations, and an inversion in *Zea mays,*" *Missouri Agricultural Experiment Station Research Bulletin* 163 (1931): 1–30.

80. Ibid., pp. 22–23.

81. Donald F. Jones, ed., *Proceedings of the Sixth International Congress of Genetics* (Menasha, Wis.: Brooklyn Botanic Garden, 1932), pp. 55, 63, 71, 146–149, 257 ("brilliant investigations"), 279–280, 283, 295.

82. PS; EFK12.1.78; NS. The stool detail is in the Symonds interview.

83. "Ring chromosomes," in Rigomar Rieger, Arnd Michaelis, and Melvin M. Green, *Glossary of Genetics and Cytogenetics* (Berlin: Springer-Verlag, 1976); EFK12.1.78.

84. EFK12.1.78; PS; NS.

85. EFK1.13.79; C. D. Darlington, *Recent Advances in Cytology,* 2d ed. (Philadelphia: Blakiston's, 1937), p. 20.

86. Barbara McClintock, "The relation of a particular chromosomal element to the development of the nucleoli in *Zea mays,*" *Zeitschrift für Zellforschung und Mikroskopiche Anatomie* 21 (1934): 321.

87. Letters of support: McClintock, application for Guggenheim fellowship, 15 October 1932, McClintock file, GF. Number of applicants: personal communication, G. Thomas Tanselle, vice president, Guggenheim Foundation, July 1999.

88. Richard Goldschmidt, *In and out of the Ivory Tower: The Autobiography of Richard B. Goldschmidt* (Seattle: University of Washington Press, 1960), pp. 293–294.

89. PS.

90. Richard Goldschmidt, "Spontaneous chromatin rearrangements and the theory of the gene," *Proceedings of the National Academy of Sciences USA* 23 (1937): 622.

91. See, for example, Allen, "Opposition to the Mendelian-chromosome theory," pp. 60–61.

92. PS.

93. "Nobody smiled": FHP, p. 21. Freiburg and Oehlkers: McClintock to Ralph Cleland, 8 May and 16 May 1934, RCC; FHP; PS. "Hitlerite business": FHP. On Oehlkers, see Harwood, *Styles of Scientific Thought*, pp. 79–80.

94. Margaret Saha, "Spemann seen through a lens," in Scott F. Gilbert, ed., *A Conceptual History of Modern Embryology* (New York: Plenum Press, 1991), pp. 104–105. "Metaphysical concept": Johannes Holtfreter, "Reminiscences on the life and work of Johannes Holtfreter," in Gilbert, ed., *A Conceptual History of Modern Embryology*, p. 117. Sapp, *Beyond the Gene*, p. 68.

95. EFK1.13.79.

96. McClintock to Edgar Anderson, 15 November 1938, William L. Brown collection, collection 1, record group 4/2/1/1, folder 51, MBG.

97. McClintock to Cleland, 8 May 1934, RCC.

98. Harriet Creighton interview, 22 October 1996.

99. Employment records, University of Missouri, Columbia, courtesy of University of Missouri archives.

100. Horace Freeland Judson, *The Eighth Day of Creation: Makers of the Revolution in Biology* (New York: Simon and Schuster, 1979), pp. 356, 460; James F. Crow interview, 13 October 1996.

101. McClintock to Rhoades, 16 April 1935, 12 December 1938, folder 19, box 12, RC.

102. Helen V. Crouse, "Recollections of a graduate student," in *DG*, p. 29.

103. McClintock to Rhoades, 29 June 1939. "Fingers crossed!": 1 November 1940. Other examples of tension with Guthrie are found in McClintock to Rhoades, 30 January 1941, 28 April 1941, 16 May 1941, all in folder 19, box 12, RC.

104. Bentley Glass interview, 10 November 1993.

105. McClintock to Rhoades, 28 April 1941, folder 19, box 12, RC.

106. *AFFTO*, p. 81.

107. Stadler to Rhoades, 18 March 1941, box 1, RC.

108. McClintock to Rhoades, 11 May 1941, folder 19, box 12, RC.

109. FHP.

110. McClintock to Rhoades, 11 May 1941; McClintock to Rhoades, 16 May 1941, folder 19, box 12, RC.

111. "Memorandum regarding Dr. Barbara McClintock," Demerec, 12 March 1942, McClintock file, CSHL; EFK9.24.78; Demerec to W. M. Gilbert (executive officer, Carnegie Institution of Washington), 6 June 1942, McClintock file, CSHL.

112. McClintock to Edgar Anderson, 12 April 1942, collection 1, record group 4/2/1/1, folder 51, MBG.

113. Harriet Creighton interview, 22 October 1996; FHP.

114. PS.

115. Quoted in Robert Scott Root-Bernstein, *Discovering* (Cambridge, Mass.: Harvard University Press, 1989), p. 250.

116. EFK1.13.79.

117. EFK2.25.79.

118. C. P. Snow, foreword to G. H. Hardy, *A Mathematician's Apology* (1967; Canto ed., London: Cambridge University Press, 1992), pp. 32–38; Paul Hoffman, *The Man Who Loved Only Numbers* (New York: Hyperion, 1998), p. 86.

119. Sylvia Nasar, interviewed by Robert Siegel, *All Things Considered*, 18 June 1997 (Washington, D.C.: National Public Radio).

120. James Gleick, *Genius: The Life and Science of Richard Feynman* (New York: Pantheon, 1992), pp. 315, 317.

121. EFK12.1.78.

122. PS.

123. EFK2.25.79.

4. Pattern

Epigraph from "The challenge of pattern," in Nina Fedoroff and David Botstein, eds., *The Dynamic Genome: Barbara McClintock's Ideas in the Century of Genetics* (Cold Spring Harbor, N.Y.: Cold Spring Harbor Laboratory Press, 1992), pp. 214–215.

1. For Emerson's original studies, see Chap. 3, notes 35–37. For a review of the *P* system, see E. G. Anderson and R. A. Emerson, "Pericarp studies in maize, I: The inheritance of pericarp colors," *Genetics* 8 (1923): 466–476.

2. Carla Keirns, "Seeing patterns: Models, visual evidence, and pictorial communication in the work of Barbara McClintock," *Journal of the History of Biology* 32 (1999): 163–196, esp. pp. 168–176.

3. Barbara McClintock, "The production of homozygous deficient tissues with mutant characteristics by means of the aberrant mitotic behavior of ring-shaped chromosomes," *Genetics* 23 (1938): 315–376.

4. Barbara McClintock, "The fusion of broken ends of sister half-chromatids following chromatid breakage at meiotic anaphases," *Missouri Agricultural Experiment Station Research Bulletin* 290 (1938): 1–48.

5. Barbara McClintock, "The behavior in successive nuclear divisions of a chromosome broken at meiosis," *Proceedings of the National Academy of Sciences USA* 25 (1939): 406, fig. 1.

6. McClintock to Marcus Rhoades, 29 January 1939, folder 19, box 12, RC.

7. Barbara McClintock, "The stability of broken ends of chromosomes in *Zea mays*," *Genetics* 26 (1941): 234–282, esp. figs. 8, 9, 11 (pp. 249–252, 260–261).

8. Ibid., pp. 276–278.

9. Ibid., pp. 243–244.

10. Barbara McClintock, "Maize genetics," *Carnegie Institute of Washington Year Book* 41 (1942): 181–186.
11. See Philip J. Pauly, "Summer resort and scientific discipline: Woods Hole and the structure of American biology, 1882–1925," in Ronald Rainger, Keith Benson, and Jane Maienschein, eds., *The American Development of Biology* (New Brunswick: Rutgers University Press, 1991), pp. 121–150; Jane Maienschein, *One Hundred Years of Exploring Life, 1888–1988: The Marine Biological Laboratory at Woods Hole* (Boston: Jones and Bartlett, 1989); Ralph W. Dexter, Keith R. Benson, Jane Maienschein, and Robert C. Terwilliger, "History of American marine biology and marine biology institutions," *American Zoologist* 28 (1988): 1–34; Keith R. Benson, "Laboratories on the New England shore: The 'somewhat different direction' of American marine biology," *New England Quarterly* 61 (1988): 55–78; James D. Ebert, Keith R. Benson, Jane Maienschein, John A. Moore, Scott F. Gilbert, and Sharon Kingsland, "The role of Johns Hopkins University in the development of experimental and quantitative biology in America," *American Zoologist* 27 (1987): 749–817; "The Naples Zoological Station and the Marine Biological Laboratory: One hundred years of biology," *Biological Bulletin* 168, suppl. (1985): 1–207.
12. On the history of science at Cold Spring Harbor, see Lee-Richard Hiltzik, "The Brooklyn Institute of Arts and Sciences' Biological Laboratory, 1890–1924: A history" (Ph.D. diss., dissertation abstracts no. AAI9328162, State University of New York, 1993); Elizabeth Watson, *Houses for Science* (Cold Spring Harbor, N.Y.: Cold Spring Harbor Laboratory Press, 1992); E. Carleton MacDowell, "Charles Benedict Davenport, 1886–1944: A study of conflicting influences," *Bios* 70 (1940): 3–50.
13. MacDowell, "Charles Benedict Davenport," pp. 17–24; Nathan Rheingold, "National science policy in a private foundation: The Carnegie Institution of Washington," in A. Oleson and J. Voss, eds., *The Organization of Knowledge in Modern America, 1860–1920* (Baltimore: Johns Hopkins University Press, 1979), pp. 313–314, 318; Robert Kohler, *Partners in Science: Foundations and Natural Scientists, 1900–1945* (Chicago: University of Chicago Press, 1991), pp. 16–17.
14. MacDowell, "Charles Benedict Davenport," p. 34.
15. For a description of these projects, see the Long Island Biological Association, *Annual Report of the Biological Laboratory* (Cold Spring Harbor: Long Island Biological Association, 1945), pp. 16–17, 19, 22–25.
16. McClintock herself recalled doing this (Barbara McClintock, comments at ninetieth birthday party, 13 June 1992, Cold Spring Harbor, N.Y., transcript, in author's possession).
17. Bentley Glass interview, November 1993. On American eugenics and the Eugenics Record Office, see Mark H. Haller, *Eugenics: Hereditarian Attitudes in American Thought* (New Brunswick, N.J.: Rutgers University Press, 1963); Kenneth Ludmerer, *Genetics and American Society* (Baltimore: Johns

Hopkins University Press, 1972); Garland E. Allen, "The Eugenics Office at Cold Spring Harbor, 1910–1940," *Osiris* 2 (1986): 225–264; Daniel J. Kevles, *In the Name of Eugenics: Genetics and the Uses of Human Heredity* (Berkeley: University of California Press, 1985); Jonathan Marks, "Historiography of eugenics," *American Journal of Human Genetics* 52 (1993): 650–652; Frederick Osborn, "History of the American Eugenics Society," interview, American Eugenics Society papers, American Philosophical Society Library, Philadelphia.

18. Marcus Rhoades, "The effect of varying gene dosage on aleurone color in maize," *Journal of Genetics* 33 (1936): 347–354; Rhoades, "Effect of the Dt gene on mutability of the a1 allele in maize," *Genetics* 23 (1938): 377–397.

19. *Annual Report of the Biological Laboratory,* 1945, pp. 41–42. On Max Delbrück and the phage group, see John Cairns, Gunther Stent, and James Watson, eds., *Phage and the Origins of Molecular Biology* (1966; exp. ed., Cold Spring Harbor, N.Y.: Cold Spring Harbor Laboratory Press, 1992); Max Delbrück, "A physicist's renewed look at biology: Twenty years later," *Science* 168 (1970): 1312–1315; Ernst Peter Fischer and Carol Lipson, *Thinking about Science: Max Delbrück and the Origins of Molecular Biology* (New York: Norton, 1988); Lily Kay, *The Molecular Vision of Life: Caltech, the Rockefeller Foundation, and the Rise of the New Biology* (New York: Oxford University Press, 1993); William Hayes, "Max Delbrück and the birth of molecular biology," *Social Research* 51 (1984): 641–673; Horace Freeland Judson, *The Eighth Day of Creation: Makers of the Revolution in Biology* (New York: Simon and Schuster, 1979), esp. pp. 50–170; Robert Olby, *Path to the Double Helix* (Seattle: University of Washington Press, 1974), pp. 231–239.

20. FHP.

21. Barbara McClintock, "Maize genetics," *Carnegie Institution of Washington Year Book* 42 (1943): 148–152.

22. Barbara McClintock, "Maize genetics," *Carnegie Institution of Washington Year Book* 43 (1944): 128.

23. Milislav Demerec, "Summary," *Carnegie Institution of Washington Year Book* 43 (1944): 104.

24. McClintock, "Maize genetics," 1944, p. 131.

25. NS.

26. Barbara McClintock, "Cytogenetic studies of maize and Neurospora," *Carnegie Institution of Washington Year Book* 44 (1945): 108.

27. Ibid.

28. Ibid., p. 109.

29. PS.

30. Barbara McClintock, "The significance of responses of the genome to challenge," *Science* 226 (1984): 793; "Plantings, Summer 1944," box 11, series 5, MC.

31. She made a career-long tally of the cultures she produced. The cultures I examine here fall in the range 3,500–4,800. A complete set of McClintock's culture cards, beginning with culture number 1, is in box 72ff, series 5, MC.

32. Culture card for culture B-87, box 72, series 5, MC.
33. NS.
34. Participant-observation with Dr. Robert Martienssen, Cold Spring Harbor, N.Y., 1996.
35. "Chapter 3—The initial appearance of a number of mutable loci . . . ," p. 14, box 14, series 5, MC.
36. Culture card B-87, box 72, series 5, MC.
37. McClintock to Rhoades, 1 November 1940, folder 19, box 12, RC; EFK2.25.79; Robert K. Mortimer, "Carl C. Lindegren: Iconoclastic father of Neurospora and yeast genetics," in Michael N. Hall and Patrick Linder, eds., *The Early Days of Yeast Genetics* (Cold Spring Harbor: Cold Spring Harbor Laboratory Press, 1993), p. 18.
38. EFK2.25.79.
39. McClintock to Beadle, 8 May 1945, folder 5.26, BC.
40. NS.
41. NS.
42. McClintock to Rhoades, 23 April 1945, folder 19, box 12, RC.
43. Ibid.
44. McClintock, "Cytogenetic studies of maize and Neurospora," p. 110.
45. NS.
46. McClintock, "Cytogenetic studies of maize and Neurospora," p. 110.
47. NS.
48. McClintock, "The significance of responses of the genome to challenge," p. 793.
49. NS; PS.
50. PS.
51. Barbara McClintock, "Maize genetics," *Carnegie Institution of Washington Year Book* 45 (1946): 182.
52. Ibid., p. 180.
53. See, for example, Staff meeting—October 28, 1946," folder "Chromosome los[s]—Unclassified"; "To be done, 1946–47"; "The interpretation of patterns of variegation," all in box 8, series 5, MC.
54. "Patterns—conditions controlling pattern," box 8, series 5, MC (undated but probably January 1948).
55. "Conclusions to mid-summer 1947," box 8, series 5, MC.
56. "Ds-Ac general," box 8, series 5, MC; "Odd notes—to Sept 1948," box 9, series 5, MC.
57. "Outlines—June 1948—essay," box 9, series 5, MC.
58. "3/20/84," box 38, series 5, MC.
59. Barbara McClintock, "The control of gene action in maize," *Brookhaven Symposia in Biology* 18 (1965): 162–184, figs. 2–6.
60. Keirns, "Seeing patterns," pp. 166, 172–174, 176–177.
61. Evelyn Witkin, e-mail to the author, 9 December 1997.
62. Witkin e-mail, 9 December 1997.
63. PS.

5. Control

Epigraph from PS.

1. McClintock to Beadle, 28 January 1951, folder 5.26, BC.
2. Evelyn Witkin interview, 19 March 1996.
3. McClintock to Beadle, 22 February 1953, folder 5.26, BC.
4. McClintock to Rhoades, 14 September 1942, folder 19, box 12, RC.
5. McClintock to Rhoades, 4 August 1949, folder 20, box 12, RC.
6. McClintock to Beadle, 31 January 1950, folder 5.26, BC; McClintock to Rhoades, 9 February 1950, folder 20, box 12, RC.
7. McClintock to Beadle, 26 October 1947, folder 5.26, BC.
8. L. F. Randolph to Marcus Rhoades, 13 January 1942, and Randolph to McClintock, 13 January 1942, box 1, RC.
9. McClintock, AAUW Achievement Award acceptance speech, 1947 convention proceedings, transcript, deposited with the convention papers at the AAUW Archives, Washington, D.C.
10. McClintock to Beadle, 30 April 1947, McClintock to Zilske, 23 May 1947, folder 5.26, BC.
11. Barbara McClintock, "Maize genetics," *Carnegie Institution of Washington Year Book* 43 (1944): 131.
12. Unmarked folder, subfolder "Origin of transposed Ds 4864A," box 22, series 5, MC.
13. Ibid.
14. All quotes in this paragraph are from McClintock to Beadle, 6 October 1946, folder 5.26, BC.
15. Notes dated 20 October 1946, folder "Chromosome Los[s] (unclassified)," box 8, series 5, MC.
16. Notes dated 24 October 1946, folder "Chromosome Los[s] (unclassified)," box 8, series 5, MC.
17. Ibid.
18. Notes dated 25 October 1946, folder "Chromosome Los[s] (unclassified)," box 8, series 5, MC.
19. Ibid.
20. All quotes in this paragraph from "Staff meeting, October 28, 1946," folder "Chromosome Los[s] (unclassified)," box 8, series 5, MC.
21. Ibid.
22. McClintock to Beadle, 6 October 1946 and 11 October 1946, folder 5.26, BC.
23. McClintock to Beadle, 6 October 1946; Beadle to McClintock, 26 December 1946, folder 5.26, BC.
24. McClintock to Gertrude Zilske, 1 April 1947, folder 5.26, BC.
25. McClintock to Beadle, 14 May 1947, folder 5.26, BC.
26. "To be done, 1947," box 8, series 5, MC.
27. "Memorandum to Marcus Rhoades, Jan. 1949," folder 32, box 12, RC and

folder "Rhoades, M. M.," series 1, MC; folder "c-m1," subfolder "4204 c-m1," box 15, series 5, MC.

28. McClintock to Beadle, 30 July 1947, folder 5.26, BC.

29. Notes headed "Pattern—9/28/47," folder "Pattern: Sept. 1947; Dec. 1947; Oct. 1947; Jan. 1948," box 9, series 5, MC.

30. File "To be done—winter 1946–47. Chromosome 9 short arm detachment 10/25/46," folder "Winter 1946–47," box 8, series 5, MC.

31. "Memorandum to Marcus Rhoades, Jan. 1949"; McClintock to Drew Schwartz, 15 July 1972, folder "Schwartz, D.," series 1, MC.

32. "Modifiers of Ds action," undated, but between 28 November 1947 and 20 January 1948, box 8, series 5, MC.

33. "The Ds locus. Part III. Transposition of the Ds locus," April 1949, folder "Ac locus #1," box "Ac-Chromosoma," series 3, MC, pp. 3–4.

34. "Ac dosage tests," box 16, series 5, MC.

35. Ibid.

36. "Modifiers of Ds action."

37. McClintock to Beadle, 17 December 1947, folder 5.26, BC.

38. Beadle to McClintock, 19 December 1947, folder 5.26, BC.

39. McClintock to Beadle, 27 January 1948, folder 5.26, BC.

40. "Modifiers of Ds action."

41. "To be done c-m1," folder "c-m1," subfolder "4204 c-m1," box 15, series 5, MC.

42. "Points to be cleared—(1/23/48)," folder "Ds locus—outlines to Spring 1948," box 9, series 5, MC.

43. "Modifiers of Ds action," 11 March 1948, box 8, series 5, MC.

44. "Transposed Ds—Case 4306. Outline of evidence and conclusion," undated but after summer 1948, untitled folder, subfolder "4306—Transposed Ds locus," box 15, series 5, MC.

45. "The Ds locus. Part III. Transposition of the Ds locus," April 1949, pp. 2–4.

46. Ibid.

47. NS.

48. Evelyn Witkin interview, 19 March 1996.

49. Ibid.

50. Folder "Odd notes—to Sept 1948," box 9, series 5, MC.

51. Untitled folder, 28 April 1948, box 15, series 5, MC.

52. McClintock to S. G. Stephens, 28 June 1948, folder 27, box 12, RC.

53. "The Ac locus," undated but about 1950, folder "Ac locus #2," box "Ac-Chromosoma," series 3, MC.

54. NS.

55. PS.

56. Ibid.

57. NS.

58. Jan Sapp, *Beyond the Gene: Cytoplasmic Inheritance and the Struggle for Authority in Genetics* (New York: Oxford University Press, 1987), p. 91; Judy

Johns Schloegel, "From anomaly to unification: Tracy Sonneborn and the species problem in protozoa, 1954–1957," *Journal of the History of Biology* 32 (1999): 122–124 and n. 12; Schloegel, "Biology as biography and biography of biology: Intimacy, subjectivity, and 'understanding' in the experimental work of Herbert Spencer Jennings, Tracy Sonneborn, and *Paramecium aurelia*," paper presented to the annual meeting of the History of Science Society, 1996, Atlanta, Ga.

59. Sapp, *Beyond the Gene,* pp. 98–104.
60. C. D. Darlington, "Heredity, development, and infection," *Nature* 154 (1944): 64–169.
61. C. D. Darlington, "Mendel and the determinants," in L. C. Dunn, ed., *Genetics in the Twentieth Century* (New York: MacMillan, 1951), p. 331.
62. McClintock's correspondence with Sonneborn is in TMS.
63. Darlington, "Mendel and the determinants," pp. 324–325.
64. Sapp, *Beyond the Gene,* pp. 204–205.
65. McClintock to Rhoades, 3 April 1949, folder 20, RC.
66. For example, see the informal elegy for Sager, posted on 3 April 1997 by Ursula Goodenough, a geneticist who greatly admired Sager, in the bio.net archives: http://bionet.hgmp.mrc.ac.uk/hypermail/chlamy/ chlamy.199704/0001.html.
67. Sapp, "Concepts of organization: The leverage of ciliate protozoa," in Scott Gilbert, ed., *A Conceptual History of Embryology* (Baltimore: Johns Hopkins University Press, 1994), pp. 229–258; Sapp, *Beyond the Gene;* Edward Yoxen, "Form and strategy in biology: Reflections on the career of C. D. Waddington," in T. J. Horder, J. A. Witkowski, and C. C. Wylie, eds., *A History of Embryology* (Cambridge: Cambridge University Press, 1986), pp. 309–330; Burian et al., "Boris Ephrussi and the synthesis of genetics and embryology"; Scott Gilbert, "Induction and the origins of developmental genetics," in Gilbert, ed., *A Conceptual History of Embryology,* pp. 181–203.
68. George Beadle and Edward Tatum, "Genetic control of biochemical reactions in *Neurospora,*" *Proceedings of the National Academy of Sciences USA* 27 (1941): 494.
69. See Sapp, *Beyond the Gene,* chap. 5.
70. Lily Kay, *The Molecular Vision of Life: Caltech, the Rockefeller Foundation, and the Rise of the New Biology* (New York: Oxford University Press, 1993), pp. 125–132.
71. Sapp, *Beyond the Gene,* p. 132.
72. McClintock to Rhoades, 2 September 1950, folder 21, box 12, RC.
73. Rollin Hotchkiss interview, 23 October 1996.
74. James D. Watson, "Growing up in the phage group," in John Cairns, Gunther Stent, and James Watson, *Phage and the Origins of Molecular Biology* (1996; Cold Spring Harbor, N.Y.: Cold Spring Harbor Laboratory Press, 1992), p. 241.
75. James D. Watson interview, 12 January 2000.

76. "Problems general," 13 October 1942, folder "Stadler's #9 duplicator," box 11, series 5, MC.
77. "Thoughts on various genetic and cytological phenomena. 1945 to 1950 or earlier," box 9, series 5, MC.
78. Ibid.
79. Ibid.
80. Ibid.
81. Ibid.

6. Complexity

Epigraph from McClintock's letter to George Beadle, 17 December 1947, folder 5.26, BC.

1. Robert Martienssen provided several of these observations.
2. See, for example, Robert Kohler, *Lords of the Fly: Drosophila Genetics and the Experimental Life* (Chicago: University of Chicago Press, 1994), esp. chaps. 3, 6, 7; Bonnie Tocher Clause, "The Wistar rat as a right choice: Establishing mammalian standards and the idea of a standardized mammal," *Journal of the History of Biology* 26 (1993): 329–349.
3. Jerry Kermicle interview, 15 October 1996.
4. Kohler, *Lords of the Fly,* pp. 46–49.
5. Richard Goldschmidt, "The gene," *Quarterly Review of Biology* 3 (1928): 307–324; Goldschmidt, "Spontaneous chromatin rearrangements and the theory of the gene," *Proceedings of the National Academy of Sciences USA* 23 (1937): 621–623; Goldschmidt, *Physiological Genetics* (New York: McGraw-Hill, 1938); Goldschmidt, "The theory of the gene," *Scientific Monthly* 46 (1938): 268–273; Goldschmidt, *The Material Basis of Evolution* (New Haven, Conn.: Yale University Press, 1940). See also Elof Axel Carlson's discussion in his book *The Gene: A Critical History* (Philadelphia: Saunders, 1966), pp. 124–125; Michael R. Dietrich, "On the mutability of genes and geneticists: The 'Americanization' of Richard Goldschmidt and Victor Jollos," *Perspectives on Science: Historical, Philosophical, Social* 4 (1996): 321–345; Scott F. Gilbert, "Cellular politics: Ernest Everett Just, Richard B. Goldschmidt, and the attempt to reconcile embryology and genetics," in Ronald Rainger, Keith Benson, and Jane Maienschein, eds., *The American Development of Biology* (Philadelphia: University of Pennsylvania Press, 1988), pp. 311–346; Garland E. Allen, "Opposition to the Mendelian-chromosome theory: The physiological and developmental genetics of Richard Goldschmidt," *Journal of the History of Biology* 7 (1974): 49–92; Jonathan Harwood, *Styles of Scientific Thought: The German Genetics Community, 1900–1933* (Chicago: University of Chicago Press, 1993), pp. 84–85.
6. "General 4/14/48," folder "Ds-Ac General," box 8, series 5, MC.
7. 8 August 1948, folder "Odd Notes to Sept. 48," box 9, series 5, MC.

8. Ibid.

9. "Ds-Ac general," box 8, series 5, MC.

10. Notes headed "6/22/47," folder "Summer 1947," box 8, series 5, MC.

11. Notes headed "Origin and behavior of c-m1 11/19/48," subfolder "cm-1 Summer 1948," folder "c-m1," box 15, series 5, MC.

12. "Odd Notes to Sept. 1948," box 9, series 5, MC.

13. "Annual Report of the Director of the Department of Genetics," *Carnegie Institution of Washington Year Book* 46 (1947): 126.

14. McClintock to Stephens, 28 June 1948, folder 27, box 12, RC.

15. "Outlines—June 1948—essay," p. 11, box 8, series 5, MC.

16. "Odd Notes to Sept. 1948."

17. Ibid.

18. McClintock to Beadle, 23 October 1948, folder 5.26, BC.

19. Muller to McClintock, 27 October 1948; McClintock to Muller, 12 November 1948, both in HJM.

20. Muller to McClintock, 27 October 1948.

21. Barbara McClintock, "Mutable loci in maize," *Carnegie Institution of Washington Year Book* 47 (1948): 154.

22. "11/14/48—Types of abnormal results of Ds behavior; types of breaks," folder "Summaries and outlines, 1948–49," box 9, series 5, MC; Barbara McClintock, "Mutable loci in maize," *Carnegie Institution of Washington Year Book* 48 (1949): 143.

23. "Origin and behavior of c-m1 11/19/48," subfolder "cm-1 Summer 1948," folder "c-m1," box 15, series 5, MC.

24. "Origin of c-m1—a transposed Ds locus (11/15/49)," subfolder "Outline for written report on cm-1," folder "c-m1," box 15, series 5, MC; "Memorandum to Marcus Rhoades, Jan. 1949," box 12, folder 32, RC, and series 1, MC.

25. Robert Olby, *Path to the Double Helix* (Seattle: University of Washington Press, 1974), p. 6.

26. "Memorandum to Marcus Rhoades," pp. 27–28.

27. Ibid., p. 26.

28. "Outline for written report on cm-1."

29. Sewall Wright, "Color inheritance in mammals," *Journal of Heredity* 8 (1917): 224–235.

30. "Memorandum to Marcus Rhoades," p. 51.

31. Ibid.

32. McClintock to Beadle, 23 October 1948, folder 5.26, BC.

33. McClintock refers to giving Rhoades the *Ds* report in May and to having previously given him the *Ac* report in a letter of 7 December 1949; see folder 20, box 12, RC.

34. "The Ds locus. Part III. Transposition of the Ds locus," typescript dated April 1949, folder "Ac locus #1," series 3, MC.

35. McClintock to Rhoades, 4 August 1949, folder 20, box 12, RC.

36. McClintock to Beadle, 31 January 1950, folder 5.26, BC.
37. The chapters may be found in series 3, MC, as well as folder 20, box 12, RC.
38. Jerry Kermicle interview, 15 October 1996.
39. McClintock to Rhoades, 27 November 1949, folder 19, box 12, RC.
40. McClintock to Rhoades, 7 December 1949, folder 19, box 12, RC.
41. http://www.columbia.edu/cu/biology/dept/special_lectures/jesup.htm.
42. McClintock to Rhoades, 7 December 1949.
43. Moore to McClintock, 12 January 1950, folder 20, box 12, RC.
44. McClintock to Rhoades, 7 December 1949.
45. McClintock to Rhoades, 9 February 1950, folder 20, box 12, RC.
46. McClintock to Rhoades, 4 March 1950, folder 20, box 12, RC.
47. Ibid.
48. Ibid.
49. Ibid.
50. George Beadle, "A gene for sticky chromosomes in *Zea mays*," *Zeitschrift fur Induktive Abstammungs und Vererbungslehre* 63 (1932): 195–217.
51. McClintock to Rhoades, 4 March 1950.
52. McClintock to Rhoades, "Sunday night—March 4, 1950," 5 March 1950, folder 20, box 12, RC.
53. McClintock to Rhoades, 9 March 1950, folder "Rhoades, M. M.," series 1, MC.
54. McClintock to Rhoades, 12 March 1950, folder 20, box 12, RC.
55. McClintock to Rhoades, 31 March 1950, folder 20, box 12, RC.
56. McClintock to Rhoades, 3 April 1950, folder 20, box 12, RC.
57. McClintock to Rhoades, 31 March 1950.
58. McClintock to Rhoades, 6 July 1950, folder 20, box 12, RC.
59. McClintock to Beadle, 28 January 1951, folder 5.26, BC.
60. McClintock to Rhoades, 12 March 1950.
61. Immanuel Velikovsky, *Worlds in Collision* (London: Gollancz, 1950).
62. See Henry H. Bauer, *Beyond Velikovsky: The History of a Public Controversy* (Urbana: University of Illinois Press, 1984); Alfred De Grazia, ed., *The Velikovsky Affair: The Warfare of Science and Scientism* (New Hyde Park, N.Y.: University Books, 1966); Donald Goldsmith, ed., *Scientists Confront Velikovsky* (Ithaca, N.Y.: Cornell University Press, 1977).
63. Nina Fedoroff interview, 3 February 1997; James Shapiro interview, 27–28 January 1997.
64. *AFFTO*, p. 203; Susan Gensel Cooper interview, 31 January 1997.
65. "List of Books owned by Dr. Barbara McClintock located in her office," McClintock file, CSHL.
66. *AFFTO*, pp. 202–203; see also the discussion in Chapter 1 and notes therein.
67. Norton Zinder interview, 12 March 1996.
68. Waclaw Szybalski interview, 14 October 1996.
69. Ibid.

70. James Shapiro interview, 27–28 January 1997.
71. Rollin Hotchkiss interview, 23 October 1996.
72. Evelyn Witkin interview, 19 March 1996.
73. Nina Fedoroff interview, 3 February 1997.
74. PS.
75. Oliver E. Nelson interview, 14 October 1996.
76. David Botstein interview, 23 December 1996.

7. Reception

Epigraph from interview, 13 October 1996.

1. McClintock to Rhoades, 4 March 1950, folder 21, box 12, RC.
2. Barbara McClintock, "The origin and behavior of mutable loci in maize," *Proceedings of the National Academy of Sciences USA* 36 (1950): 344–345.
3. Ibid., p. 347.
4. Ibid., pp. 347–348.
5. Ibid., p. 352.
6. Ibid., p. 353.
7. Ibid., p. 355.
8. Richard Goldschmidt, *In and Out of the Ivory Tower: The Autobiography of Richard B. Goldschmidt* (Seattle: University of Washington Press, 1960), pp. 300–303.
9. Richard Goldschmidt, *Physiological Genetics* (New York: McGraw-Hill, 1938), pp. v, 200; Scott Gilbert, "Cellular politics: Ernest Everett Just, Richard B. Goldschmidt, and the attempt to reconcile embryology and genetics," in Ronald Rainger, Keith Benson, and Jane Maienschein, eds., *The American Development of Biology* (New Brunswick, N.J.: Rutgers University Press, 1988), pp. 334–335.
10. Richard Goldschmidt, "Marginalia to McClintock's work on mutable loci in maize," *The American Naturalist* 84 (1950): 437.
11. Ibid., p. 446.
12. McClintock to Beadle, 28 January 1951, folder 5.26, BC.
13. Ibid.
14. Ibid.
15. Norton Zinder interview, 12 March 1996. See also Norton Zinder, "Forty years ago: The discovery of bacterial transduction," *Genetics* 132 (1992): 292.
16. Milislav Demerec, "Foreword," *Cold Spring Harbor Symposia on Quantitative Biology* 16 (1951): 5.
17. Norton Zinder interview, 12 March 1996.
18. Thomas Brock, *The Emergence of Bacterial Genetics* (Cold Spring Harbor, N.Y.: Cold Spring Harbor Laboratory Press, 1990), p. 78; Norman H. Horowitz, "The sixtieth anniversary of biochemical genetics," *Genetics* 143 (1996): 1–4.

19. Richard Goldschmidt, "Chromosomes and genes," *Cold Spring Harbor Symposia on Quantitative Biology* 16 (1951): 4, 7.
20. Several interview subjects recalled this as the length of McClintock's paper.
21. Barbara McClintock, "Chromosome organization and gene expression," *Cold Spring Harbor Symposia on Quantitative Biology* 16 (1951): p. 36.
22. Ibid.
23. Ibid., p. 46; Nathaniel C. Comfort, "Two genes, no enzyme: A second look at Barbara McClintock and the 1951 Cold Spring Harbor Symposium," *Genetics* 140 (1995): 1163.
24. McClintock, "Chromosome organization and gene expression," p. 17.
25. PS.
26. "Puzzlement, even hostility": Barbara McClintock, "Introduction," in John A. Moore, ed., *The Discovery and Characterization of Transposable Elements: The Collected Papers of Barbara McClintock,* (New York: Garland, 1987), p. x; "nobody understood": NS; "really knocked me": EFK1.13.79 and quoted in part in *AFFTO*, p. 140; "stony silence": *AFFTO*, p. 139; "dense with statistics and proofs": Sharon Bertsch McGrayne, "Barbara McClintock," in *Nobel Women in Science* (New York: Birch Lane, 1993), p. 168.
27. See, for example, McClintock to Rhoades, 3 April 1950; McClintock to Rhoades, 6 July 1950, both in folder 21, box 12, RC.
28. McClintock to Rhoades, 6 July 1950.
29. Waclaw Szybalski interview, 14 October 1996.
30. Joshua Lederberg interview, 12 March 1996.
31. Norton Zinder, "Forty years ago," p. 292. Lederberg did not recall the meeting specifically but did say he and others engaged McClintock in detailed discussion after her talk.
32. Horowitz, "The sixtieth anniversary of biochemical genetics," p. 4.
33. See, for example, *AFFTO*, p. 145; Joan Dash, *The Triumph of Discovery: Women Scientists Who Won the Nobel Prize* (Englewood Cliffs, N.J.: Julian Messner, 1991), p. 90.
34. Ernst Caspari, "Quantitative Biology," *Science* 114 (1951), editorial.
35. A. Buzzatti-Traverso, "The state of genetics," *Scientific American* (October 1951): 23, 25.
36. Norton Zinder interview, 12 March 1996.
37. Joshua Lederberg interview, 12 March 1996.
38. Waclaw Szybalski interview, 14 October 1996.
39. Demerec to Beadle, 2 June 1953, folder 5.26, BC.
40. McClintock to Beadle, 13 September 1951, folder 5.26, BC.
41. McClintock to Beadle, 4 October 1951, folder 5.26, BC.
42. R. A. Brink to Marcus Rhoades, 11 May 1945; Brink to Rhoades, 17 December 1946, both in RAB.
43. Robert A. Nilan interview, 11 May 1998.
44. Ibid.
45. R. A. Brink and R. A. Nilan, "Studies on variegated pericarp," *Maize Genetics Cooperation News Letter* 25 (1951): 42–47.

46. R. A. Brink and R. A. Nilan, "The relation between light variegated and medium variegated pericarp in maize," *Genetics* 37 (1952): 537.

47. Ibid., p. 538; Peter A. Peterson, "A mutable pale green locus in maize," *Genetics* 38 (1953): 682–683.

48. Jerry Kermicle interview, 15 October 1996.

49. James F. Crow interview, 13 October 1996.

50. Jerry Kermicle interview, 15 October 1996.

51. R. A. Brink, "Paramutation and chromosome organization," *Quarterly Review of Biology* 35 (1960): 132.

52. Ibid., p. 132.

53. Oliver Nelson interview, 14 October 1996.

54. Interviews with James F. Crow (13 October 1996) and Jerry Kermicle (15 October 1996).

55. Edgar Altenburg to McClintock, 21 January 1953, EA.

56. Such an article was never listed in the *Science Citation Index*, and a scan of the major genetics journals turned up no such article during the 1950s.

57. Edgar Altenburg to McClintock, 21 January 1953; McClintock to Altenburg, 22 January 1953; McClintock to Altenburg, 26 January 1953; Altenburg to McClintock, 26 January 1953; Altenburg to McClintock 29 January 1953, all in EA.

58. Barbara McClintock, "Induction of instability at selected loci in maize," *Genetics* 38 (1953): 579–599.

59. McClintock to Rhoades, 2 September 1950, folder 21, box 12, RC.

60. All quotes in this paragraph are from McClintock, "Induction of instability," pp. 594, 596.

61. Ibid., p. 597.

62. Peterson, "A mutable pale green locus in maize"; Peter C. Barclay and R. A. Brink, "The relation between Modulator and Activator in maize," *Proceedings of the National Academy of Sciences USA* 40 (1954): 1118–1126; Brink and Nilan, "Light variegated and medium variegated pericarp"; Cheng-Mei W. Fradkin and R. A. Brink, "Crossing over in maize in the presence of the transposable factors Activator and Modulator," *Genetics* 41 (1956): 901–906.

63. Keller (*AFFTO*, p. 140) wrote that McClintock received only two requests; four years later, McClintock herself ("Introduction," in Moore, ed., *The Discovery and Characterization of Transposable Elements*, p. x) claimed she received three. McGrayne ("Barbara McClintock," p. 168), wrote that "only three scientists outside her field" requested the article.

64. McClintock to Altenburg, 22 January 1953, EA.

65. McClintock to Beadle, 22 February 1953, folder 5.26, BC.

66. Ibid.

67. McClintock to Beadle, 7 March 1953, 22 March 1953, 31 March 1953, folder 5.26, BC.

68. "Anti-female sentiment," "anti-female bias": McClintock to Rhoades, 17

April 1953, folder 22, box 12, RC. "Women were shocked": EFK1.13.79. Research-associate position: McClintock to Beadle, 15 April 1953, folder 5.26, BC; McClintock to Rhoades, 17 April 1953.

69. Margaret W. Rossiter, *Women Scientists in America: Struggles and Strategies to 1940* (Baltimore: Johns Hopkins University Press, 1982), pp. 204–214.

70. "Splendid!": McClintock to Beadle, 15 April 1953, folder 5.26, BC. "Footloose," McClintock to Rhoades, 17 April 1953. Offers from Caltech and Illinois: Beadle to Bush, 25 May 1953, folder 5.26, BC.

71. Beadle to Bush, 25 May 1953; Bush to Beadle, 27 May 1953; McClintock to Beadle, 1 June 1953; Demerec to Beadle, 2 June 1953, all in folder 5.26, BC; McClintock to Rhoades, 17 April 1953.

72. Lyle Lanier to Marcus Rhoades, 20 July 1959, box 1, RC.

73. Z. A. Medvedev to I. M. Lerner, 3 September 1966; memo from Rhoades to Max Delbrück, Bentley Glass, and James F. Crow, probably 21 October 1966, folder 12, box 12, RC.

74. Delbrück to Rhoades, 21 October 1966; Glass to Rhoades, 9 November 1966, both in folder 12, box 12, RC.

75. Folder 12, box 12, RC.

76. *AFFTO*, p. 141; McGrayne, "Barbara McClintock," p. 168.

77. "Caltech lectures 1954," series 3, MC.

78. M. Gerald Neuffer, "1. Additional *Dt* loci. 2. A new pale aleurone gene," *Maize Genetics Cooperation News Letter* 27 (1953): 67–68; Neuffer, "1. Redesignation of *pa*" and notes following, *Maize Genetics Cooperation News Letter* 28 (1954): 63–65; Neuffer, "3. Similarity of M and Ac mutator systems" and notes following, *Maize Genetics Cooperation News Letter* 29 (1955): 59–61; Neuffer, "Dosage effect of multiple *Dt* loci on mutation of *a* in the maize endosperm," *Science* 121 (1955): 399.

79. M. Gerald Neuffer interview, 20 July 1998.

80. Barbara McClintock, "Intranuclear systems controlling gene action and mutation," *Brookhaven Symposia in Biology* 8 (1956): 64–65.

81. All quotes in this paragraph are from McClintock, "Intranuclear systems," pp. 60 ("regular inheritance patterns"), 64 ("Are controlling elements present?").

82. McClintock, "Intranuclear systems," p. 67.

83. Ibid., pp. 71–74.

84. McClintock to Ernst Caspari, 5 April 1956, folder "McClintock, Barbara," EC.

85. Barbara McClintock, "Controlling elements and the gene," *Cold Spring Harbor Symposia on Quantitative Biology* 21 (1956): 197–216.

86. Jerry Kermicle interview, 15 October 1996.

87. Alex C. Fabergé to Marcus Rhoades, 29 January 1953 and 23 February 1960, box 1, RC.

88. McClintock, "Intranuclear systems," p. 67.

89. Some examples from the maize literature employing or considering control-

ling elements include Edward H. Coe, "Spontaneous mutation of the aleurone color inhibitor in maize," *Genetics* 47 (1962):780, 782; Margaret H. Emmerling, "An analysis of intragenic and extragenic mutations of the plant color component of the Rr gene complex in Zea mays," *Cold Spring Harbor Symposia on Quantitative Biology* 23 (1958): 406–407; Alexander C. Fabergé, "Relation between chromatid-type and chromosome-type break-age-fusion-bridge cycles in maize endosperm," *Genetics* 43 (1958): 737–740, 746–748; John Laughnan, "Structural and functional bases for the action of the *A* alleles in maize," *The American Naturalist* 89 (1955): 94; Paul C. Mangelsdorf, "The mutagenic effect of hybridizing maize and teosinte," *Cold Spring Harbor Symposia on Quantitative Biology* 23 (1958): 419; Oliver E. Nelson, "Intracistron recombination in the Wx/wx region in maize," *Science* 130 (1959): 795; M. Gerald Nuffer, "Dosage effect of multiple Dt loci on mutation of alpha in the maize endosperm," *Science* 121 (1955): 400; Drew Schwartz, "Analysis of a highly mutable gene in maize: A molecular model for gene instability," *Genetics* 45 (1960): 1141, 1143, 1144, 1148; S. H. Yarnell, "Cytogenetics of the vegetable crops," *Botanical Review* 20 (1954): 290–291.

90. O. E. Nelson to McClintock, 20 March 1968, OEN.

91. A few examples will suffice. From the nonmaize plant literature: G. Ledyard Stebbins and A. Vaarama, "Artificial and natural hybrids in the Gramineae tribe Hordeae, VII: Hybrids and allopolyploids between *Elymus* glaucus and *Sitanion spp.*," *Genetics* 39 (1954): 392; Harold H. Smith and Seward A. Sand, "Genetic studies on somatic instability in cultures derived from hybrids between *Nicotiana langsdorfii* and *N. sanderae*," *Genetics* 42 (1957): 560–583; Werner Braun, "Studies on population changes in bacteria and their relation to some general biological problems," *The American Naturalist* 86 (1952): 369. From the Drosophila literature: L. Sandler and Yuichiro Hiraizumi, "Meiotic drive in natural populations of *Drosophila melanogaster*, II: Genetic variation at the *Segregation-distorter* locus," *Proceedings of the National Academy of Sciences USA* 45 (1959): 1413; William D. Kaplan, "The influence of minutes upon somatic crossing over in *Drosophila melanogaster*," *Genetics* 38 (1953): 646. From the bacterial literature: H. B. Newcombe, "Radiation-induced instabilities in *Streptomyces*," *Journal of General Microbiology* 9 (1953): 34–35; Henry P. Treffers, Viola Spinelli, and Nao O. Belser, "A factor (or mutator gene) influencing mutation rates in *Escherichia coli*," *Genetics* 40 (1954): 1064. From the mammalian and other literature: Barry Commoner, "Roles of deoxyribonucleic acid in inheritance," *Nature* 202 (1964): 964; Kenneth Paigen, "The genetic control of enzyme activity during differentiation," *Proceedings of the National Academy of Sciences USA* 47 (1961): 1648.

92. J. A. Serra, *Modern Genetics*, vol. 1. (New York: Academic Press, 1965), p. 454; Ruth Sager and Francis Ryan, *Cell Heredity* (New York: Wiley, 1961); Robert King, *Genetics* (New York: Oxford University Press, 1962),

pp. 187–188; Adrian Srb, Ray Owen, and Robert Edgar, *General Genetics,* 2d ed. (San Francisco: Freeman, 1965), p. 341.

93. Ruth Sager to Marcus Rhoades, 28 July 1960, box 1, RC.
94. L. J. Stadler, "The gene," *Science* 120 (1954): 811–819, see p. 818 (Peters).
95. Torbjorn Caspersson and Jack Schultz, "Cytochemical measurements in the study of the gene," in L. C. Dunn, ed., *Genetics in the Twentieth Century* (New York: MacMillan, 1951), p. 168; George Beadle, "Chemical genetics," in Dunn, *Genetics in the Twentieth Century,* p. 235.
96. G. Pontecorvo, *Trends in Genetic Analysis* (New York: Columbia University Press, 1958), p. 66.
97. Lecture notes are located in series 3, MC.
98. "Mutation *Ac Ds,*" folder 27, box 8, RC.
99. Drew Schwartz interview, 6 December 1996.
100. Melvin M. Green interview, 21 December 1996.
101. E.g., *AFFTO,* p. 139; McGrayne, "Barbara McClintock," p. 168.
102. "Too much evidence": McClintock to Rhoades, 31 March 1950, folder 21, box 12, RC; "so little published evidence": McClintock to Altenburg, 22 January 1953, EA.

8. Response

Epigraph from letter to Oliver Nelson, 28 March 1968, folder "Nelson, O. E.," series 1, MC.

1. *AFFTO,* pp. 181–182.
2. McClintock sketched the genealogy of *a2-m1* in "a2-m1 origins 1985 incomplete," folder "a2-m1 Cultures 1946–1983," and "Origin of a2-m1 Summer 1946," folder "a2-m1 1948–1955," subfolder "a2-m1 general to 1950," box 32, series 5, MC.
3. McClintock to Marcus Rhoades, 4 March 1950, and "Sunday night—3/4/50" folder 21, box 12, RC; McClintock, "Mutable loci in maize," *Carnegie Institution of Washington Year Book* 49 (1950): 167.
4. Barbara McClintock, "Mutations in maize and chromosomal aberrations in Neurospora," *Carnegie Institution of Washington Year Book* 53 (1954): 255–256; "al-m2 *Early* summary. Seminar—March 31, 1952," folder "al-m2 1950–52," box 28, series 5, MC; "a-ml Summer 1953 5628-9, 5719A-1 derivatives," folder "al-ml 1950–1953," box 27, series 5, MC.
5. "Comparisons between Spm and Ac (2/9/55)," folder "al-ml 1955," subfolder "al-ml Feb. 1955," box 27, series 5, MC.
6. Barbara McClintock, "Controlled mutation in maize," *Carnegie Institution of Washington Year Book* 54 (1955): 253.
7. Photographs in folder "al-ml," box 53, series 5, MC.
8. "Types of Spm 4/8/58," folder "Spm action—Summary statements 1957 a2-ml," box 32, series 5, MC.

9. Barbara McClintock, "Mutation in maize," *Carnegie Institution of Washington Year Book* 55 (1956): 328–329; McClintock, *Carnegie Institution of Washington Year Book* 56 (1957): 396–398.

10. "A2-m1 origins 1985 incomplete" and "Origin of a2-m1 Summer 1946."

11. Notes headed "4/22/55," and "4/20/55," folder "a2-m1 1948–1955," subfolder "a2-m1 general to 1950," box 32, series 5, MC.

12. McClintock, "Genetic and cytological studies of maize," 1957, p. 399.

13. McClintock, "The suppressor-mutator system of control of gene action in maize," *Carnegie Institution of Washington Year Book* 57 (1958): 418–419; McClintock, "Some parallels between gene control systems in maize and in bacteria," *The American Naturalist* 95 (1961): 268–269, 271–274.

14. "Spm action—Summary statements 1957 a2-m1," 5 March and 15 March 1958, folder "a2-m1 1957," box 32, series 5, MC. Quotes are from "3/21/58 stability of Spm elements," this folder.

15. "3/21/58 stability of Spm elements." See also McClintock, "The suppressor-mutator system of control," pp. 416–418.

16. McClintock, "Genetic and cytological studies of maize," *Carnegie Institution of Washington Year Book* 58 (1959): 453.

17. Barbara McClintock, "Further studies of the suppressor-mutator system of control of gene action in maize," *Carnegie Institution of Washington Year Book* 60 (1961): 452.

18. *Annual Report,* Cold Spring Harbor Laboratory (1985), p. 6.

19. McClintock to Oliver Nelson, 15 April 1965, folder "Nelson, O. E.," series 1, MC.

20. Barbara McClintock, "Aspects of gene regulation in maize," *Carnegie Institution of Washington Year Book* 63 (1964): 592–601; McClintock, "Components of action of the regulators Spm and Ac," *Carnegie Institution of Washington Year Book* 64 (1965): 528–534.

21. Oliver E. Nelson to McClintock, 25 October 1962, and McClintock to Nelson, 7 November 1962, folder "Nelson, O. E.," box 1, series 1, MC.

22. Barbara McClintock, "The control of gene action in maize," *Brookhaven Symposia in Biology* 18 (1965): 162–184.

23. Ibid., p. 183.

24. Thomas Brock, *The Emergence of Bacterial Genetics* (Cold Spring Harbor, N.Y.: Cold Spring Harbor Laboratory Press, 1990), p. 172.

25. André Lwoff, Louis Siminovitch, and Niels Kjeldgaard, "Induction de la production de bacteriophages chez une bactérie lysogeène," *Annales de l'Institut Pasteur* 79 (1950): 815–859; Brock, *The Emergence of Bacterial Genetics,* pp. 176–178; André Lwoff, "The prophage and I," in John Cairns, Gunther Stent, and James D. Watson, eds., *Phage and the Origins of Molecular Biology* (1966; exp. ed., Cold Spring Harbor, N.Y.: Cold Spring Harbor Laboratory Press, 1992), pp. 92–94.

26. Alfred Hershey and Martha Chase, "Independent functions of viral protein and nucleic acid in growth of bacteriophage," *Journal of General Physiology*

36 (1952): 39–56; Hershey, "The injection of DNA into cells by phage," in John Cairns, Gunther Stent, and James D. Watson, eds., *Phage and the Origins of Molecular Biology*, exp. ed. (Cold Spring Harbor, N.Y.: Cold Spring Harbor Laboratory Press, 1992 [1966]), pp. 100–108.

27. James Dewey Watson and Francis Henry Compton Crick, "The structure of DNA," *Cold Spring Harbor Symposia on Quantitative Biology* 18 (1953): 123.

28. E.g., James D. Watson, *The Double Helix* (New York: Signet, 1968), p. 205.

29. For analyses of the impact of the double helix, see Horace Freeland Judson, *The Eighth Day of Creation: Makers of the Revolution in Biology* (New York: Simon and Schuster, 1979), pp. 185–195; Michel Morange, *A History of Molecular Biology* (Cambridge, Mass.: Harvard University Press, 1998), chap. 12; Gunther Stent, *The Coming of the Golden Age: A View of the End of Progress* (Garden City, N.Y.: Natural History Press [Doubleday], 1969), esp. chaps. 3, 4.

30. Brock, *The Emergence of Bacterial Genetics*, p. 137.

31. Guido Pontecorvo, *Trends in Genetic Analysis* (New York: Columbia University Press, 1958), pp. 128–129.

32. Joshua Lederberg and Tetsuo Iino, "Phase variation in Salmonella," *Genetics* 41 (1957): 755.

33. Joshua Lederberg interview, 16 July 1999.

34. Brock, *The Emergence of Bacterial Genetics*, pp. 180–181.

35. Benno Müller-Hill, *The Lac Operon: A Short History of a Genetic Paradigm* (Berlin: de Gruyter, 1996), p. 11; Brock, *The Emergence of Bacterial Genetics*, p. 282.

36. François Jacob, *The Statue Within: An Autobiography* (New York: Basic Books, 1988), pp. 290–295; Judson, *The Eighth Day of Creation*, pp. 200–220; Brock, *The Emergence of Bacterial Genetics*, pp. 294–299.

37. Jacob, *The Statue Within*, p. 297.

38. François Jacob, "Genetic control of viral functions," *Harvey Lectures* (1959): 1–39. "New molecular patterns," p. 14; "control the activity," p. 24; "position effect," p. 25.

39. Ibid., p. 32.

40. François Jacob, D. Perrin, C. Sanchez, and Jacques Monod, "L'Opéron: Group de gènes à expression coordonnée par un opérateur," *Comptes rendus des Academie des Sciences* 250 (1960): 1727–1729, trans. Edward A. Adelberg as "The operon: A group of genes whose expression is coordinated by an operator," in Jonathan Beckwith and Thomas J. Silhavy, eds., *The Power of Bacterial Genetics: A Literature-based Course* (Cold Spring Harbor, N.Y.: Cold Spring Harbor Laboratory Press, 1992), pp. 330–332. François Jacob and Jacques Monod, "Genetic regulatory mechanisms in the synthesis of proteins," *Journal of Molecular Biology* 3 (1961): 318–356.

41. Jacob et al., "The operon," p. 1729 (p. 332 in Beckwith and Silhavy, *The Power of Bacterial Genetics*); Judson, *The Eighth Day of Creation*, p. 461.

42. PS.

43. McClintock, "Some parallels between gene control systems in maize and in bacteria," *The American Naturalist* 95 (1961): 265–277.

44. See also McClintock's report to the Carnegie Institution that year: "Further studies of the suppressor-mutator system of control."

45. Jacques Monod and François Jacob, "Teleonomic mechanisms in cellular metabolism, growth, and differentiation," *Cold Spring Harbor Symposium on Quantitative Biology* 26 (1961): 394–395.

46. François Jacob and Jacques Monod, "On the regulation of gene activity," *Cold Spring Harbor Symposium on Quantitative Biology* 26 (1961): 207–208.

47. Ibid.

48. McClintock, "Further studies of gene-control systems in maize," *Carnegie Institution of Washington Year Book* 62 (1963): 487; McClintock, "The control of gene action in maize," *Brookhaven Symposia in Biology* 18 (1965): 170.

49. McClintock to Nelson, 28 March 1968, folder "Nelson, O. E.," series 1, MC.

50. EFK1.13.79.

51. Kim Kleinman, "Edgar Anderson: An authority on what was NOT known about corn," paper delivered to the annual meeting of the History of Science Society, 22–25 October 1998, Kansas City, Mo., pp. 4–7.

52. On Beadle, Mangelsdorf, and the corn war, see Betty Fussell, *The Story of Corn* (New York: Knopf, 1994), pp. 76–86; Catherine Dold, "The corn war," *Discover*, December 1997.

53. Deborah Fitzgerald, "Exporting American agriculture: The Rockefeller Foundation in Mexico, 1943–1953, in Marcos Cueto, ed., *Missionaries of Science: The Rockefeller Foundation and Latin America* (Bloomington: Indiana University Press, 1994), p. 77.

54. "Third Report of the NAS-NRC Committee on Preservation of Indigenous Strains of Maize," 1958. Folder "National Research Council. Committee on Preservation of Indigenous Strains of Maize," series 2, MC.

55. Cleland to McClintock, 27 June 1957, folder "National Research Council," series 1, MC.

56. McClintock to Alexander Grobman, 2 November 1957, folder "Grobman, Alexander," series 1, MC; McClintock to G. D. Meid, Business manager, NAS, 21 January 1958, folder "National Research Council," series 1, MC.

57. Alexander Grobman to McClintock, 4 July 1957, folder "Grobman, Alexander," series 1, MC.

58. Grobman to McClintock, 27 January 1958, folder "Grobman, Alexander," series 1, MC.

59. FHP.

60. Cleland to McClintock, 21 July 1958, folder "National Research Council," series 1, MC.

61. M. Gerald Nuffer, "Dosage effect of multiple *Dt* loci on mutation of alpha in the maize endosperm," *Science* 121 (1955): 399; Barbara McClintock,

"Intranuclear systems controlling gene action and mutation," *Brookhaven Symposia in Biology* 8 (1955): 65.

62. McClintock to William Hatheway, 20 October 1958, folder "Rockefeller Foundation #1," series 1, MC.

63. McClintock to Roberts, 13 November 1958, folder "Rockefeller Foundation #1," series 1, MC.

64. EFK1.13.79.

65. EFK1.13.79.

66. McClintock, "Genetic and cytological studies of maize."

67. PS.

68. EFK1.13.79.

69. FHP.

70. FHP.

71. McClintock, "Genetic and cytological studies of maize," p. 456.

72. McClintock to Robinson, 18 February 1962, Robinson to McClintock, 22 February 1962, folder "North Carolina State College," series 1, MC.

73. Fitzgerald, "Exporting American agriculture," p. 73.

74. "Outline of a Program for the study of the chromosome constitutions of Maize and its Relatives in the Americas: Nov.-Dec. meeting in Mexico City," folder "Project Outline Mexico City Conference, November-Dec. 1964," box 44, series 5, MC.

75. McClintock to Kato, 1 August 1970, folder "Kato Yamakake, Takeo Angel #1," series 1, MC.

76. Beadle to McClintock, 22 January 1972; McClintock to Beadle, 14 February 1972; Beadle to McClintock, 26 February 1972, all in folder "Beadle, George," series 1, MC; Kato to McClintock, 9 September 1972, folder "Kato Yamakake, Takeo Angel #1," series 1, MC.

77. Kato to McClintock, 24 June 1974, 16 July 1975, folder "Kato Yamakake, Takeo Angel, #2," series 1, MC.

78. Evelyn Witkin interview, 19 March 1996.

79. Barbara McClintock, "Genetic systems regulating gene expression during development," *Developmental Biology Supplement* 1 (1967): 108.

80. Ibid., pp. 95–103.

81. "Why a geneticist became fascinated with some of our local plants," notes from public lecture, Cold Spring Harbor Laboratory, 18 May 1971, box "Int—Po," series 3, MC.

82. Barbara McClintock, "Development of the maize endosperms as revealed by clones," paper presented at "The Clonal Basis of Development," Thirty-sixth Symposium of the Society for Developmental Biology, 1978. "Integrations," p. 218; "races of maize," p. 236; "project the mechanisms," p. 236.

83. "Why a geneticist became fascinated with some of our local plants."

84. PS.

85. Richard Goldschmidt, *The Material Basis of Evolution* (New Haven, Conn.: Yale University Press, 1940). See also the discussion of macroevolution in

Masatoshi Nei, *Molecular Evolutionary Genetics* (New York: Columbia University Press, 1987), p. 405.

86. For a recent review, see S. J. Gould and N. Eldredge, "Punctuated equilibrium comes of age," *Nature* 366 (1993): 223–227.

87. PS.

88. Barbara McClintock, "Mechanisms that rapidly reorganize the genome," *Stadler Genetics Symposium* 10 (1978): 27.

89. Ibid., p. 29.

90. Barbara McClintock, "Chromosome constitutions of Mexican and Guatemalan races of maize," *Carnegie Institution of Washington Year Book* 59 (1960): 464.

91. McClintock, "Mechanisms," p. 45.

92. Ibid., p. 27; Barbara McClintock, T. Angel Kato Y., and Almiro Blumenschein, *Chromosome Constitution of Races of Maize: Its Significance in the Interpretation of Relationships between Races and Varieties in the Americas* (Chapingo, Mexico: Colegio de Postgraduados, Escuela National de Agricultura, 1981), p. 15.

93. PS.

94. Barbara McClintock, "The significance of responses of the genome to challenge," *Science* 226 (1984): 792–801. "Stress, and the genome's reaction to it," p. 800; "highly sensitive organ of the cell," p. 800, heat shock and SOS, p. 792.

9. Renaissance

First epigraph from interview, 16 July 1999.

1. CB.

2. Horace Freeland Judson, *The Eighth Day of Creation: Makers of the Revolution in Biology* (New York: Simon and Schuster, 1979), p. 336.

3. François Jacob, "Genetic control of viral functions," *Harvey Lectures* (1959): 14.

4. A. Campbell, "Episomes," *Advances in Genetics* 11 (1962): 101–145.

5. Ibid., p. 107.

6. Austin Lawrence Taylor, "Bacteriophage-induced mutation in Escherichia coli," *Proceedings of the National Academy of Sciences USA* 50 (1963): 1044.

7. A. Lawrence Taylor interview, 23 July 1999.

8. Ibid.

9. A. Lawrence Taylor interview, 23 July 1999; Ariane Toussaint, "A history of mu," in N. Symonds, A. Toussaint, P. v. d. Putte, and M. Howe, eds., *Phage Mu* (Cold Spring Harbor, N.Y.: Cold Spring Harbor Laboratory Press, 1987), p. 3.

10. Taylor, "Bacteriophage-induced mutation in Escherichia coli," pp. 1049–1050; A. Lawrence Taylor interview, 23 July 1999.

11. Toussaint, "A history of mu," pp. 2–5.
12. Cold Spring Harbor Laboratory, *Annual Report,* 1972, pp. 28, 48.
13. Toussaint, "A history of mu," pp. 4–5.
14. Toussaint, "A history of mu," p. 3; A. I. Bukhari, "Bacteriophage mu as a transposition element," *Annual Review of Genetics* 10 (1976): 403, 405; Martha M. Howe, "Phage mu: An overview," pp. 25–39, and David Sherratt, "Transposable elements: An overview," pp. 191–200, both in N. Symonds, A. Toussaint, P. v. d. Putte, and M. Howe, eds., *Phage Mu* (Cold Spring Harbor, N.Y.: Cold Spring Harbor Laboratory Press, 1987).
15. Melvin M. Green interview, 21 December 1996.
16. Ibid.
17. Green to McClintock, 30 July 1968, folder "Green, M. M." series 1, MC.
18. McClintock to Green, 13 February 1967, folder "Green, M. M.," series 1, MC.
19. Melvin M. Green interview, 21 December 1996.
20. See Judson, *The Eighth Day of Creation,* chaps. 5–8.
21. James Shapiro, "Kernels and colonies: The challenge of pattern," in *DG,* p. 213.
22. R. H. Epstein, A. Bolle, C. M. Steinberg, E. Kellenberger, E. Boy de la Tour, R. Chevalley, R. S. Edgar, M. Susman, G. Denhardt and A. Leilausis, "Physiological studies of conditional lethal mutations in bacteriophage T4D," *Cold Spring Harbor Symposium on Quantitative Biology* 16 (1963): 445; Sydney Brenner and Jonathan R. Beckwith, "*Ochre* mutants, a new class of suppressible nonsense mutants," *Journal of Molecular Biology* 13 (1965): 629–637.
23. Shapiro, "Kernels and colonies," pp. 213–214.
24. Quotation from J. A. Shapiro, "Mutations caused by the insertion of genetic material into the galactose operon of Escherichia coli," *Journal of Molecular Biology* 40 (1969): 93. Other results are described in S. L. Adhya and J. A. Shapiro, "The galactose operon of E. coli K-12, I: Structural and pleiotropic mutations of the operon," *Genetics* 62 (1969): 231–247; J. A. Shapiro and S. L. Adhya, "The galactose operon of E. coli K-12, II: A deletion analysis of operon structure and polarity," *Genetics* 62 (1969): 249–264.
25. James Shapiro interview, 14 July 1999; Heinz Saedler and Peter Starlinger, "Twenty-five years of transposable element research in Köln," in *DG,* pp. 243–244; Elke Jordan, Heinz Saedler, and Peter Starlinger, "Strong-polar mutations in the transferase gene of the galactose operon in E. coli," *Molecular and General Genetics* 100 (1967): 296–306; H. Saedler, J. Besemer, B. Kemper, B. Rosenwirth, and P. Starlinger, "Insertion mutations in the control region of the Gal operon of E. coli, I: Biological characterization of the mutations," *Molecular and General Genetics* 115 (1972): 258–265; P. Starlinger and H. Saedler, "Insertion mutations in microorganisms," *Biochimie* 54 (1972): 177.
26. James Shapiro interview, 14 July 1999.
27. Waclaw Szybalski interview, 13 October 1996.

28. M. H. Malamy, M. Fiandt, and W. Szybalski, "Electron microscopy of polar insertions in the lac operon of Escherichia coli," *Molecular and General Genetics* 119 (1972): 221.

29. Starlinger and Saedler, "Insertion mutations in microorganisms," pp. 183–184.

30. Reviewed in English in T. Watanabe, "Infective heredity of multiple drug resistance in bacteria," *Bacteriological Reviews* 27 (1963): 87–155.

31. P. A. Sharp, S. N. Cohen, and N. Davidson, "Electron microscope heteroduplex studies of sequence relations among plasmids of Escherichia coli, II: Structure of drug resistance (R) factors and F factors," *Journal of Molecular Biology* 75 (1973): 235–255; R. W. Hedges and A. E. Jacob, "Transposition of ampicillin resistance from RP4 to other replicons," *Molecular and General Genetics* 132 (1974): 31–40.

32. F. Heffron, C. Rubens, and S. Falkow, "Translocation of a plasmid DNA sequence which mediates ampicillin resistance: Molecular nature and specificity of insertion," *Proceedings of the National Academy of Sciences USA* 72 (1975): 3623–3627; Nancy Kleckner, Russell K. Chan, B. K. Tye, and David Botstein, "Mutagenesis by insertion of a drug-resistance element carrying an inverted repetition," *Journal of Molecular Biology* 97 (1975): 561; D. E. Berg, J. Davies, B. Allet, and J. D. Rochaix, "Transposition of R factor genes to bacteriophage lambda," *Proceedings of the National Academy of Sciences USA* 72 (1975): 3628–3632; M. M. Gottesman and J. L. Rosner, "Acquisition of a determinant for chloramphenicol resistance by coliphage lambda," *Proceedings of the National Academy of Sciences USA* 72 (1975): 5041–5045.

33. Fred Heffron, e-mail to the author, 7 September 1999; David Botstein interview, 1 August 1999.

34. David Botstein interview, 1 August 1999.

35. Y. Oshima and I. Takano, "Mating types in Saccharomyces: Their convertibility and homothallism," *Genetics* 67 (1971): 327–335; Y. Oshima, "Homothallism, mating-type switching, and the controlling element model in *Saccharomyces cerevisiae*," in Michael N. Hall and Patrick Linder, eds., *The Early Days of Yeast Genetics* (Cold Spring Harbor, N.Y.: Cold Spring Harbor Laboratory Press, 1993), pp. 298–301.

36. R. Egel, "Rearrangements at the mating type locus in fission yeast," *Molecular and General Genetics* 148 (1976): 158.

37. James B. Hicks, Jeffrey N. Strathern, and Ira Herskowitz, "The cassette model of mating-type interconversion," in Ahmad Bukhari, James A. Shapiro, and Sankhar Adhya, eds., *DNA Insertion Elements, Plasmids, and Episomes* (Cold Spring Harbor, N.Y.: Cold Spring Harbor Laboratory Press, 1977), pp. 457–462.

38. J. Hicks and I. Herskowitz, "Interconversion of yeast mating types, II. Restoration of mating ability to sterile mutants in homothallic and heterothallic strains," *Genetics* 85 (1977): 373, 387–388.

39. Ira Herskowitz, "Controlling elements, mutable alleles, and mating-type

interconversion," in *DG,* pp. 291–294; Amar J. S. Klar, "The role of McClintock's controlling element concept in the story of yeast mating-type switching," in *DG,* pp. 310–312.

40. Shapiro, "Kernels and colonies," p. 214.
41. James Shapiro interview, 14 July 1999.
42. Program of meeting, "DNA Insertions," 18–21 May 1976, Cold Spring Harbor Laboratory; courtesy David Stewart, Cold Spring Harbor Laboratory. James Shapiro interview, 14 July 1999.
43. David Botstein interview, 23 December 1996.
44. "DNA Insertions" meeting program; A. Campbell, D. E. Berg, D. Botstein, E. M. Lederberg, R. P. Novick, P. Starlinger, and W. Szybalski, "Nomenclature of transposable elements in prokaryotes," in Ahmad Bukhari, James A. Shapiro, and Sankhar Adhya, eds., *DNA Insertion Elements, Plasmids, and Episomes* (Cold Spring Harbor, N.Y.: Cold Spring Harbor Laboratory Press, 1977), p. 16, n. 1.
45. Campbell et al., "Nomenclature of transposable elements in prokaryotes," p. 16.
46. Tracy Sonneborn to Marcus Rhoades, 21 March 1967; press release dated 17 April 1967, box 12, folder 12, RC.
47. Ernst Caspari to McClintock, 27 June 1973, folder "McClintock, Barbara," EC.
48. McClintock to J. R. S. Fincham, 16 May 1973, folder "Fincham, JRS," series 1, MC.
49. McClintock to G. R. K. Sastry, 25 October 1971, folder "Sastry, GRK," series 1, MC.
50. J. R. Fincham and G. R. Sastry, "Controlling elements in maize," *Annual Review of Genetics* 8 (1974): 45–46.
51. Judson John van Wyk interview, 18 July 1999.
52. "Description of the contributions of Barbara McClintock," scientific evaluation attached to nomination of Barbara McClintock for 1976 Nobel Prize in Physiology or Medicine, provided by Judson John van Wyk.
53. Van Wyk, nomination of Barbara McClintock for 1976 Nobel Prize in Physiology or Medicine.
54. "Description of the contributions of Barbara McClintock."
55. Judson John van Wyk interview, 19 July 1999; Diter von Wettstein interview, 1 July 1999.
56. H. Greer and G. R. Fink, "Unstable transpositions of his4 in yeast," *Proceedings of the National Academy of Sciences USA* 76 (1979): 4006–4010; G. S. Roeder and G. R. Fink, "DNA rearrangements associated with a transposable element in yeast," *Cell* 21 (1980): 239–249.
57. Gerald Fink, "Transposable elements *(Ty)* in yeast," in *DG,* p. 283.
58. J. E. Majors, R. Swanstron, W. J. DeLorbe, G. S. Payne, S. H. Hughes, S. Ortiz, N. Quintrell, J. M. Bishop, and H. E. Varmus, "DNA intermediates in the replication of retroviruses are structurally (and perhaps functionally) re-

lated to transposable elements," *Cold Spring Harbor Symposium Quantitative Biology* 45 (1980): 731–732, 735.

59. See the sessions "Retroviruses as Insertion Elements" and "Rearrangements in Antibody Genes," *Cold Spring Harbor Symposia on Quantitative Biology* 45 (1980).

60. James D. Watson, "Foreword," *Cold Spring Harbor Symposia on Quantitative Biology* 45 (1980): xiii.

61. Barbara McClintock, "Modified gene expressions induced by transposable elements," in W. A. Scott, R. Werner, D. R. Joseph, and J. Schultz, eds., *Mobilization and Reassembly of Genetic Information* (New York: Academic Press, 1980), pp. 15–16.

62. Marcus Rhoades to MacArthur Foundation, 4 November 1981, folder 14, box 12, RC.

63. Lawrence K. Altman, "Lasker Awards go to geneticist and a lab head," *New York Times,* 19 November 1981, sec. B, p. 2; "Foundation picks geneticist as its first fellow laureate," *New York Times,* 18 November 1981, sec. B, p. 3.

64. "Unappreciated," "flipped my top," "going in the right direction," Altman, "Lasker Awards"; "recombinant-DNA technology" and "other important medical implications," Matt Clark and Dan Shapiro, "The kernel of genetics," *Newsweek,* 30 November 1981, p. 74; "gene splicing and human engineering," Kathleen Teltsch, "Award-winning scientist prizes privacy," *New York Times,* 18 November 1981, sec. B, p. 3.

65. Teltsch, "Award-winning scientist prizes privacy."

66. Herschel Roman to Marcus Rhoades, 2 September 1981, folder 16, box 12, RC.

67. Judson, *Eighth Day of Creation,* p. 201.

68. James D. Watson interview, 12 January 2000.

69. Ibid.

70. Herschel Roman to Marcus Rhoades, 30 November 1981, folder 16, box 12, RC.

71. Joshua Lederberg interview, 16 July 1999.

72. Joshua Lederberg interview, 12 March 1996.

73. Herschel Roman to Joshua Lederberg, 1 October 1981; Lederberg to Roman, 5 October 1981; Roman to Lederberg, 30 December 1981, Folder 16, box 12, RC.

74. Joshua Lederberg, undated nomination letter, folder 16, box 12, RC.

75. Joshua Lederberg interviews, 12 March 1996, 1 July 1999.

76. Joshua Lederberg, undated nomination letter; van Wyk, nomination of Barbara McClintock for 1976 Nobel Prize in Physiology or Medicine.

77. Diter von Wettstein interview, 1 July 1999.

78. "Nomination: Nobel Prize for the year 1982: Dr. Barbara McClintock," submitted 25 January 1982. Courtesy Dr. Heinz Saedler, translated from the German by Carl-Henry Geschwind.

79. Jean L. Marx, "Antibody research garners Nobel Prize," *Science* 238 (1987):

484–485; Peter Newmark, "Nobel Prize for Japanese immunologist," *Nature* 329 (1987): 570.

80. S. Tonegawa, H. Sakano, R. Maki, A. Traunecker, G. Heinrich, W. Roeder, and Y. Kurosawa, "Somatic reorganization of immunoglobulin genes during lymphocyte differentiation," *Cold Spring Harbor Symposia on Quantitative Biology* 45 (1980): 839–840; Marx, "Antibody research garners Nobel Prize."

81. *Nobel Foundation Directory, 1997–1998* (Stockholm: Nobel Foundation, 1998), p. 146.

82. David Dickson, "Bumps and falls on the road to Stockholm," *Science* 238 (1987): 263–264.

83. NPC.

84. NPC.

85. Joshua Lederberg interview, 16 July 1999.

86. James Shapiro interview, 14 July 1999.

87. Claudia Wallis, "Honoring a Modern Mendel," *Time Magazine* 122, no. 18 (1983): 53–54.

88. Swedish translation kindly provided by Dr. Bengt Olle Bengtsson, Department of Genetics, University of Lund, Sweden.

89. Robert Olby makes a related argument in Olby, "Mendel no Mendelian?" *History of Science* 17 (1979): 53–72.

90. Susan Quinn, *Marie Curie: A Life* (New York: Simon and Schuster, 1995), p. 189.

91. Abraham Pais, *Subtle Is the Lord: The Science and the Life of Albert Einstein* (Oxford: Oxford University Press, 1982), pp. 502–511. Telegram to Einstein is quoted on p. 503.

92. James T. Patterson, *The Dread Disease: Cancer and Modern American Culture* (Cambridge, Mass.: Harvard University Press, 1987), p. 186; Stephen P. Strickland, *Politics, Science, and Dread Disease: A Short History of United States Medical Research Policy* (Cambridge, Mass.: Harvard University Press, 1972); Samuel Epstein, *The Politics of Cancer* (New York: Anchor Books, 1979); Daniel Kevles, "Renato Dulbecco and the new animal virology: Medicine, methods, and molecules," *Journal of the History of Biology* 26 (1994): 409–442.

93. See, for example, John Tooze and Joseph Sambrook, *Tumor Viruses* (Cold Spring Harbor, N.Y.: Cold Spring Harbor Laboratory Press, 1974), p. 3.

10. Synthesis

Epigraph from PS.

1. Barbara McClintock, "The fusion of broken ends of sister half-chromatids following chromatid breakage at meiotic anaphases," *Missouri Agricultural Experiment Station Research Bulletin* 290 (1938): 1–48; McClintock, "The

stability of broken ends of chromosomes in *Zea mays*," *Genetics* 26 (1941): 234–282; H. J. Muller, "The remaking of chromosomes," *The Collecting Net (M.B.L.)* 13 (1938): 181–198.

2. J. D. Watson, "Origin of concatameric T7 DNA," *Nature New Biology* 239 (1972): 197–201; C. B. Harley, A. B. Futcher, and C. W. Greider, "Telomeres shorten during ageing of human fibroblasts," *Nature* 345 (1990): 458–460; L. Hayflick and P. S. Moorhead, "The serial cultivation of human diploid strains," *Experimental Cell Research* 25 (1961): 585–621; A. G. Bodnar, M. Ouellette, M. Frolkis, S. E. Holt, C. P. Chiu, G. B. Morin, C. B. Harley, J. W. Shay, S. Lichtsteiner, and W. E. Wright, "Extension of life-span by introduction of telomerase into normal human cells," *Science* 279 (1998): 349–352; C. W. Greider and E. H. Blackburn, "Identification of a specific telomere terminal transferase activity in *Tetrahymena* extracts," *Cell* 43 (1985): 405–413.

3. Mary-Lou Pardue, "Do some 'parasitic' DNA elements earn an honest living?" in *DG*, pp. 146–152; M. L. Pardue, O. N. Danilevskaya, K. L. Traverse, and K. Lowenhaupt, "Evolutionary links between telomeres and transposable elements," *Genetica* 100 (1997): 73–74.

4. The Cold Spring Harbor Laboratory, Washington University Genome Sequencing Center, and PE Biosystems Arabidopsis Sequencing Consortium, "The complete sequence of a heterochromatic island from a higher eukaryote," *Cell* 100 (2000): 377–386.

5. See, for example, J. D. Boeke and J. P. Stoye, "Retrotransposons, endogenous retroviruses, and the evolution of retroelements," in H. Varmus, S. Hughes, and J. Coffin, eds., *Retroviruses* (Cold Spring Harbor, N.Y.: Cold Spring Harbor Laboratory Press, 1997), pp. 343–435; J. D. Boeke and S. E. Devine, "Yeast transposons: Finding a nice quiet neighborhood," *Cell* 93 (1998): 1087–1089.

6. P. Dimitri and N. Junakovic, "Revising the selfish DNA hypothesis: New evidence on accumulation of transposable elements in heterochromatin," *Trends Genet* 15 (1999): 123–124; E. S. th Buckler, T. L. Phelps-Durr, C. S. Buckler, R. K. Dawe, J. F. Doebley, and T. P. Holtsford, "Meiotic drive of chromosomal knobs reshaped the maize genome," *Genetics* 153 (1999): 415–426.

7. Nicholas Wade, "Human or chimp? Fifty genes are the key," *New York Times*, October 20, 1998, sec. D, p. 1.

8. Wade, "Human or chimp?" E. Nickerson and D. L. Nelson, "Molecular definition of pericentric inversion breakpoints occurring during the evolution of humans and chimpanzees," *Genomics* 50 (1998): 368–372.

9. Charles Darwin, *On the Origin of Species* (London: John Murray, 1859), pp. 489–490.

10. Peter Bowler, *Non-Darwinian Revolution: Reinterpreting a Historical Myth* (Baltimore, Md.: Johns Hopkins University Press, 1988), pp. 13, 116–120. On Weismann as completing the Darwinian revolution, see Ernst Mayr, *One*

Long Argument: Charles Darwin and the Genesis of Modern Evolutionary Thought (Cambridge, Mass.: Harvard University Press, 1991), chap. 8.

11. Gregory Bateson, *Men Are Grass: Metaphor and the World of Mental Process* (West Stockbridge, Mass.: Lindisfarne Press, 1980).

12. Robert Scott Root-Bernstein, *Discovering* (Cambridge, Mass.: Harvard University Press, 1989). See, for example, pp. 142–159.

13. For a similar idea, see Bowler, *The Non-Darwinian Revolution*, p. 85.

14. Ibid., pp. 117–120.

15. D'Arcy Thompson, *On Growth and Form* (Cambridge: Canto, 1992), p. 275.

16. Evelyn Fox Keller, *Refiguring Life: Metaphors of Twentieth Century Biology* (New York: Columbia University Press, 1995), pp. 24–28.

17. See Carla Keirns's analysis of McClintock's representations of her work in Keirns, "Seeing patterns: Models, visual evidence, and pictorial communication in the work of Barbara McClintock," *Journal of the History of Biology* 32 (1999): 163–196.

18. Jeffrey Strathern, "Thinking about programmed genome rearrangements in a genome static state of mind," in *DG*, p. 302.

19. PS.

20. PS.

21. NPC.

Appendix

Epigraph from abstract for "Transposable elements as a molecular evolutionary force," *Annals of the New York Academy of Science* 870 (1999): 251.

1. N. V. Fedoroff, S. Wessler, and M. Shure, "Isolation of the transposable maize controlling elements *Ac* and *Ds*," *Cell* 35 (1983): 235–242; Fedoroff, "About maize transposable elements and development," *Cell* 56 (1989): 181–191.

2. H-P Döring, E. Tillman, and P. Starlinger, "DNA sequence of the maize transposable element *Dissociation*," *Nature* 307 (1984): 127–130; Heinz Saedler and Peter Starlinger, "Twenty-five years of transposable element research in Köln," in *DG*, p. 250.

3. A. Pereira, H. Cuypers, A. Gierl, Z. Schwarz-Sommer, and H. Saedler, "Molecular analysis of the *En/Spm* transposable element system of *Zea mays*," *EMBO Journal* 5 (1986): 831–841; P. Masson, R. Surosky, J. A. Kingsbury, and N. V. Fedoroff, "Genetic and molecular analysis of the *Spm*-dependent *a-m2* alleles of the maize *a* locus," *Genetics* 117 (1987): 117–137; N. V. Fedoroff, "The suppressor-mutator element and the evolutionary riddle of transposons," *Genes to Cells* 4 (1999): 11–19.

4. Nina Fedoroff, letter to author, 17 August 1999.

5. Masson et al., "Genetic and molecular analysis of the *Spm*-dependent *a-m2*

alleles"; Fedoroff, "About maize transposable elements and development," p. 184.

6. A. Menssen, S. Hohmann, W. Martin, P. S. Schnable, P. A. Peterson, H. Saedler, and A. Gierl, "The En/Spm transposable element of Zea mays contains splice sites at the termini generating a novel intron from a dSpm element in the A2 gene," *Embo Journal* 9 (1990): 3051–3057; Saedler and Starlinger, "Twenty-five years of transposable element research," p. 257.

7. Nina Fedoroff, letter to author, 17 August 1999.

8. Fedoroff, "The suppressor-mutator element," pp. 13–16.

9. Nina Fedoroff, letter to author, 17 August 1999.

10. Ibid.

Interviews

The following interviews I conducted between 1996 and 2000.

David Botstein, 23 December 1996, Belmont, Calif., tape recording. Deposited at the American Philosophical Society Library, Philadelphia, Pa.
———, 1 August 1999, telephone interview.
Susan Gensel Cooper, 31 January 1997, Cold Spring Harbor, N.Y., tape recording.
Harriet Creighton, 22 October 1996, Wellesley, Mass., tape recording. Deposited at the American Philosophical Society Library, Philadelphia, Pa.
James F. Crow, 13 October 1996, Madison, Wis., tape recording. Deposited at the American Philosophical Society Library, Philadelphia, Pa.
Nina Fedoroff, 3 February 1997, telephone interview.
Bentley Glass, 3 and 10 November 1993, Old Field, N.Y. In author's possession.
Melvin M. Green, 21 December 1996, Davis, Calif., tape recording. Deposited at the American Philosophical Society Library, Philadelphia, Pa.
Rollin Hotchkiss, 23 October 1996, Lenox, Mass., tape recording. Deposited at the American Philosophical Society Library, Philadelphia, Pa.
Jerry Kermicle, 15 October 1996, Madison, Wis., tape recording. Deposited at the American Philosophical Society Library, Philadelphia, Pa.
Joshua Lederberg, 12 March 1996, New York, N.Y., tape recording. Transcript deposited at the American Philosophical Society Library, Philadelphia, Pa.
———, 16 July 1999, New York, N.Y., tape recording.
Oliver E. Nelson, 14 October 1996, Madison, Wis., tape recording. Deposited at the American Philosophical Society Library, Philadelphia, Pa.
M. Gerald Neuffer, 20 July 1998, telephone interview.
Robert A. Nilan, 11 May 1998, telephone interview.
Robert Pollack, 30 July 1998, New York, N.Y., tape recording.
Drew Schwartz, 6 December 1996, telephone interview.
James Shapiro, 27–28 January 1997, Chicago, Ill., tape recording.
———, 14 July 1999, telephone interview.

Waclaw Szybalski, 13 October 1996, Madison, Wis., tape recording. Deposited at the American Philosophical Society Library, Philadelphia, Pa.

A. Lawrence Taylor, 23 July 1999, telephone interview.

James D. Watson, 12 January 2000, telephone interview.

Diter von Wettstein, 1 July 1999, telephone interview.

Evelyn Witkin, 19 March 1996, Princeton, N.J., tape recording. Deposited at the American Philosophical Society Library, Philadelphia, Pa.

Judson John van Wyk, 18 July 1999, telephone interview.

———, 19 July 1999, telephone interview.

Norton Zinder, 12 March 1996, New York, N.Y., tape recording. Transcript deposited at the American Philosophical Society Library, Philadelphia, Pa.

Index

Problem-solving: mystic powers, 7, 154, 267; joy of, 22, 55; integration, 32–33, 67–68, 90; math problems, 67–68, 85–86; Feynman method, 68, 90; brute-force approach, 84–85; *Activator* dosage problem, 109–111, 123; lumper versus splitter, 124; assumptions, 226. *See also* Pattern

Proceedings of the National Academy of Sciences, 55, 150, 158, 160, 182, 185, 229, 237, 254

Progress in Biophysics, 182

Promoter, 232

Prophage, 200, 205, 209, 228

Provine, William, 153, 207, 221, 224, 268

Public view, 1–16, 30, 251, 253. *See also* Myth, public

Punctuated equilibrium, 221

Purdue, 182

Quantitative biochemistry, 142

Queen Anne's lace, 217, 219

Races of maize, 188, 189, 209–217, 221, 224, 266

Radium, 47, 255

Randolph, Lowell (Fitz), 50–51, 53, 101, 220

Recombination, 202–203, 226–227, 244, 246, 255

Regulation, gene model, 189, 205–209, 228, 241–253, 258–260. *See also* Controlling elements; Operons

Regulator, 181, 205, 207, 228

Rejection, 53, 166–168, 189, 266

Relativity, 255–256

Renaissance, 225–257

Repressor, 205–208, 232

Repulsive force, 134–144, 181

Retroviruses, 243

Rhoades, Marcus: Randolph–McClintock compared, 50; Cornell, 51, 53–54; maize research, 51–55, 113, 118, 127; Caltech, 53; friendship, 53, 63–64; prove it, 55; Ames, Iowa, 62; Stadler letter, 64; meteorology letter, 65; *Dotted* gene research, 80–81, 89, 172; revertant mutations, 92; Sager thesis letter, 118; Ephrussi-Taylor letter, 120; Stephens mutable gene letter, 133–134; *Dissociator*

letters, 136–137, 140; chemical analyses letter, 142; Carnegie trip letter, 143; mutable genes, 143, 144, 145, 172, 183, 185; Jesup Lectures, 143–144; transposition letters, 143–151, 158–159, 166; illness letter, 144–145; roving appointment, 175; Nobel nominating committee, 176–177, 246–249; veneration of, 226; Kimber Award committee, 240; MacArthur Award comments, 245

Ribosomes, 62, 232

Rice University, 171

R locus, 130, 181–182, 190

RNA: messenger RNA, 232, 251, 265, 272–274; transfer RNA, 232, 265; exons, 251; introns, 251; splicing, 251

Roberts, Richard, 152, 251

Robinson, H. F. (Cotton), 217

Rochaix, Jean-David, 237

Rockefeller Foundation, 63, 81, 211, 216

Rockefeller Institute, 18, 153, 256

Rockefeller Institute for Medical Research, 257

Rockefeller University, 163

Roman, Herschel, 246, 247, 249

Root-Bernstein, Robert Scott, 263

Rosner, Judah L., 237

Rosner, Mary, 30

Rossiter, Margaret, 8

Rous, Peyton, 18, 256–257

Roux, Wilhelm, 38

Roving appointment, 175

RP4 plasmid, 236

Rubens, Craig, 236

Rule-breaker, 24–26, 63

Saedler, Heinz, 234–235, 244, 248–249, 255, 275

Sagan, Carl, 151

Sager, Ruth, 118, 142, 167, 179, 180, 182

St. Lawrence, Patricia, 167, 182

Salmonella, 202, 247

Saltational theory, 35, 221

Sand, Seaward, 180

Sapp, Jan, 120

Sastry, G. R. K., 241–242, 259

Sato, Gordon, 242

Sax, Karl, 58

Scarecrow, 99